Integrated Circuit
Fabrication Technology

Other McGraw-Hill Books in Integrated Circuits

Other Books of Interest

*For more information about other McGraw-Hill materials,
call 1-800-2-MCGRAW in the United States. In other
countries, call your nearest McGraw-Hill office.*

Integrated Circuit Fabrication Technology

David J. Elliott
Cymer Laser Technologies
Wayland, Massachusetts

Second Edition

McGraw-Hill Publishing Company
New York   St. Louis   San Francisco   Auckland   Bogotá
Caracas   Hamburg   Lisbon   London   Madrid   Mexico
Milan   Montreal   New Delhi   Oklahoma City
Paris   San Juan   São Paulo   Singapore
Sydney   Tokyo   Toronto

Library of Congress Cataloging-in-Publication Data

Elliott, David J.
  Integrated circuit fabrication technology.

  Includes bibliographies and index.
  1. Integrated circuits—Design and construction.
I. Title.
TK7874.E49   1989   621.381′73   88-32022
ISBN 0-07-019339-8

1234567890   DOC/DOC   89432109

ISBN 0-07-019339-8

*The editors for this book were Daniel A. Gonneau and Galen H. Fleck,
the designer was Naomi Auerbach, and the production supervisor was
Richard A. Ausburn. This book was set in Century Schoolbook. It was
composed by the McGraw-Hill Publishing Company Professional and
Reference Division composition unit.*

Printed and bound by R. R. Donnelley & Sons Company.

To my wife, Jane, my son, Benjamin, and my daughter, Holly

Contents

Preface

This book was written as a reference for engineers, managers, and technicians involved in the many aspects of manufacturing integrated circuits. It is also written for newcomers to the field and for students, both of whom require technical understanding of ICs. An attempt has been made to present not only the practical aspects of the technology but also key concepts in all the major areas of device fabrication. Each chapter covers a key part of the IC manufacturing process.

I have benefited from contributions by many companies and individuals in the industry; their work and its results are noted throughout the book. In particular, I want to acknowledge my indebtedness to Ben Bosserman, United States Air Force; Charles R. and Lucia H. Shipley, Edwin F. Lewis, and John Frankenthaler of the Shipley Company; Michael W. Powell, Nikon Precision Inc.; and Bernard P. Piwczyk, Image Micro Systems. The Frontispiece is by courtesy of J. J. Donelon, J. E. Hurst, M. M. Oprysko, and R. J. Von Gutfeld, IBM Corporation, T. J. Watson Research Center, Yorktown Heights, New York.

David J. Elliott

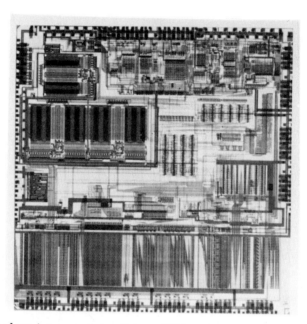

The above integrated circuit is an IBM u370 chip. The chip is a
32-bit microprocessor containing 200,000 transistors built in a 2-
micron silicon gate n-MOS process. The chip is 10 mm on a side
and is capable of implementing 102 instructions of an IBM
system 370 mainframe.

(Photo courtesy of J.J. Donelon, J.E. Hurst, M.M. Oprysko, and R.J.
vonGutfeld, IBM Corporation, T.J. Watson Research Center, Yorktown Heights,
New York.)

IC Fabrication:
An Overview

The materials, equipment, and processes used in integrated circuit (IC) fabrication are constantly being changed to meet increasing demands made by new chip designs. An analysis of the major areas of change in IC manufacturing processes will provide insight regarding how silicon device fabrication lines will be run in the future. In this section we will review these areas of major change and how they interact to produce new IC manufacturing approaches.

Wafer Diameter

The diameters of silicon wafers have been increasing dramatically in order to increase the number of chips processed in a single slice of silicon. This reduces the handling and the defects related to handling by reducing the number of wafers processed for a given number of chips. The economics derived by simply increasing wafer diameter are significant. In the short span of 10 years, wafer diameters have increased from 75-mm- (3-in-) diameter to 150-mm- (6-in-) diameter slices, providing a fourfold increase in area. The 200-mm- (8-in-) and 250-mm- (10-in-) diameter wafers are the next step in increasing production economy by increasing wafer area. Figure 1.1 shows wafer diameter versus time.

The IC processing implications of increasing wafer area are not trivial; they include solving major problems including new crystal-growing and -slicing strategies, better resist-coating techniques, and improved lens performance to enlarge the area printed in a single exposure step. Improved uniformity of lithography and pattern transfer (etching and doping) technologies is required to make the larger wafer

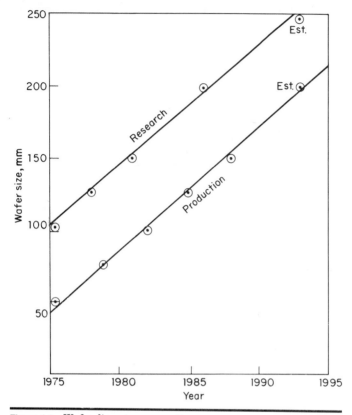

Figure 1.1   Wafer diameter versus time.

area truly more functional and more economical by yielding a proportional number of good, acceptable dies.

Minimum Feature Size in Production

Increased device complexity and density have continued to drive the minimum feature size lower, at an increasing rate. Whereas feature sizes of 1.5 μm on 100-mm wafers were once considered typical, a 200-mm-diameter wafer will now employ minimum feature sizes of 0.6 μm. Feature size reduction has the double benefit of providing more circuitry per unit area and increasing the speeds of devices. The incentive to reduce all feature sizes is further motivated by the potential of reducing chip area for an equivalent number of functions. This, in turn, increases yield potential due to a reduction of process-induced defects and increased yield in terms of more dies per wafer. Clearly, feature size reduction is the most highly leveraged parameter in improving the economics of IC production and reducing the cost per bit. Feature sizes of 0.2 μm will be used to produce gigabit-scale dynamic random access memory (DRAM) chips.

The implied costs of feature size reduction are more expensive lithography tools, because lithography inherits the task of producing even smaller structures. Pattern transfer equipment must be improved to translate these microstructures into functioning device structures in silicon or gallium arsenide. Finally, as with the cost to increase wafer diameter, the increased cost to produce smaller feature sizes must always be positively offset by the gain in overall performance and economy. Figure 1.2 shows average minimum feature size versus time.

Average Number of Mask Levels

The number of mask levels increases as the number of layers needed to manufacture increasingly complex chips increases. One example is interconnection layers, wherein single-level aluminum/aluminum alloy is giving way to two-, three-, and up to five-level metallization schemes. Each added metal layer adds another contact layer, and devices with up to fifteen levels (separate masking steps) are not uncommon. The average number of mask levels has increased steadily from

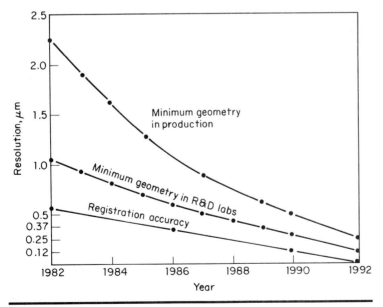

Figure 1.2 Minimum feature size versus time.

five to six to eight to ten and will most likely continue to increase as chips become more truly three-dimensional each year.

Added metal layers, incidentally, reduce the pressure for sizing metal lines too small and risking added rejects due to reaching too close to lithography limits. Features that are slightly wider and easier to control can be used when an additional masking layer is added. The cost is added processing and its attendant increases in defects and process cost. More mask levels also increase the problem of overlay registration per level and keep within the total overlay budget (all levels summed).

Registration and Overlay

Registration and overlay trends are shown in Fig. 1.3. The reduction in feature sizes, coupled with an increase in the number of mask levels, adds considerable burden to the area of layer-to-layer registration and the total overlay budget. The overlay, expressed as a 3-σ (sigma) value, is typically 0.25 μm for a corresponding feature size of 0.95 μm. The ratio of feature size versus registration increases as feature sizes decrease. Thus, registration for 0.5-μm features could be below 0.1 mm. The increase in resolution available from optical lithography has caused more attention to be focused on improved registration methods with an overall reduction in the overlay budget for a device. The overlay requirement for 16-Mbit memory chips is about 0.15 μm. Chips reaching the gigabyte density level will require overlay budgets of about 0.05 μm.

Alignment strategies using dark-field diffraction techniques are improving the capabilities in this critical area. Removing resist from alignment marks before alignment and exposure steps are taken also improves registration accuracy significantly. Improving the total stability of materials that impact the error budget is required. Thermally "quiet" environments at stable pressures (argon atoms) may be needed.

Particle Contamination

The number of particles landing on wafer surfaces in IC fabrication lines continues to be a major and increasingly critical problem, especially as viewed against the trend toward the larger chip area needed to accommodate the rising number of active devices per chip. Process operators and process equipment contribute the bulk of the particles that cause reject dies, and smaller geometries mean that smaller particles can qualify as "killer" defects. The 1-μm design rules having 0.5-μm processes then require ways to eliminate particles only ¼ μm in diameter. Defects are thus becoming associated with macro-molec-

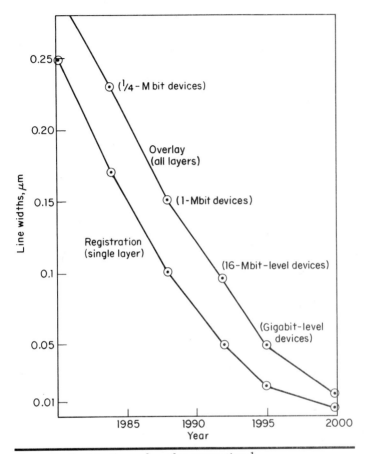

Figure 1.3   Registration and overlay versus time.[1]

ular-sized structures as lithography approaches 0.4-μm resolution levels. The Class 1 clean room must control particles down to 0.2 μm, as opposed to the 0.5-μm minimum for Class 10 clean rooms. This represents a hundredfold improvement, because the particles per cubic foot tolerated in Class 1 are only 1, whereas 10 particles are allowed in Class 10.

Chip Area

The increase in chip area has been as significant as the decreases in device geometries in calculating process economics. Reductions in area by overall lithography "shrinks" greatly reduce costs by improving yields. Chip area increased from 14K mils^2 (64K memory) to 36,000 mils^2 (256K memory) to 80,000 mils^2 (1 Mbit memory) in the short space of 6 years. It is estimated that a 16-Mbit memory will have a chip area of over 350,000 mils^2. Obviously, IC processes must remove both operators and particle-generating mechanical devices from the immediate wafer environment. This means increased use of robotics and modular automation in sealed cassettes. The effect of particles versus chip area is shown, along with clean-room class designation, in Fig. 1.4.

The importance of the z direction, or vertical section, of the chip is

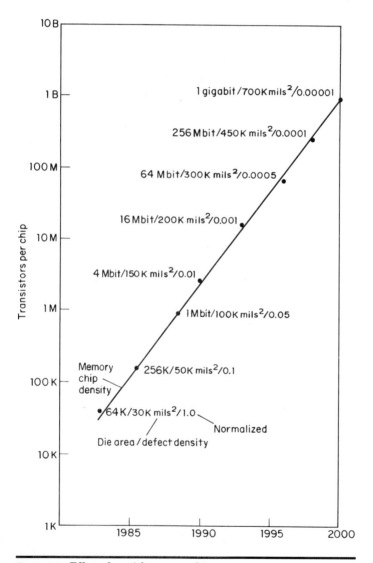

Figure 1.4 Effect of particles versus chip area.

becoming much greater. As more layers are added to the chip, trenches to interconnect and access the surface must be made deeper. This calls for higher aspect ratio structuring, more contrast and selectivity in imaging and etching, and increased use of planarization to offset high steps of topography. Vertical dimensions thus become as critical as horizontal dimensions.

Processing signals and effecting change state changes at increasing speeds call for inherently higher-speed or lower-resistance materials, including superconductors. Gallium arsenide, with an order of magnitude advantage over silicon in signal process speed capability, is a material certain to be selected for an increasingly important role in IC device fabrication. Similarly, refractory metal silicides and polycides are supplanting conventional aluminum metallization for reasons of high-speed performance capability. The parasitic resistance problems of aluminum at submicron dimensions preclude the wide use of aluminum in ultralarge-scale integration (ULSI). Titanium tungsten, for example, will be used in its place. In the sub-half-micron regime, silicides and self-aligned device technology will be required. The other major changes in materials needed in ULSI device fabrication are increased purity and bulk (x, y, z) uniformity.

The costs associated with producing higher-purity gases, filters, crystals, resists and developers, oxides, and etch chambers are well known. The need to improve purity and uniformity categorically in all these areas is increasing rapidly; it is driven by the lithography needs of the IC. In order to derive the cost advantage of higher density, rejects must be held to a level proportionate to the size of the device. Changes in operator-equipment interfaces are needed, along with more automated manufacturing. All materials entering the IC fabrication must be cleaner to meet Class 1 standards and to be compatible with the overall process.

Finally, layers are becoming thinner as the need for high-speed switching increases. This requires much more control in growing and depositing extremely thin (100- to 200-Å) gate layers consistently.

Pattern Transfer

The etching and doping aspects are undergoing changes to keep up with lithography changes. Etch bias, uniformity, selectivity, and aspect ratio are all impacted by the need for continued IC geometry reduction. Etch bias permits tighter sidewall angle control. More selectivity permits overetching to clean out residual materials that are harder to etch at nanometer-scale resolution and high aspect ratio. Higher aspect ratio is needed to conserve surface area, and yet planarity also is required to reduce feature size. This means adding steps for planarization and using surface-level lithography and processing to effect transfers deep in an underlying film. For example, imaging only the top 1 to 2000 Å of a layer with an etch mask can be followed by reactive ion etching (RIE) pattern transfer, with high aspect ratio, down through several microns of resist and oxide. This is where higher image contrast and etch selectivity are needed. Contrast, or gamma, values of 10 to 20 will be needed.

Metrology

Measurement accuracy and precision must have natural relationships with lithography such that accuracy is one-fourth to one-fifth of the resolution capability of lithography. Precision must in turn be smaller still and in proportion to resolution and accuracy. For example, a 10-in reticle with 0.5-μm dimensions requires 10-nm measurement accuracy. Nondestructive scanning electron microscopy (SEM) analysis and measurement at many stages of the process are needed to constantly monitor dimensions in wafers.

Equipment Reliability

Since it is imperative that operators become more isolated from wafer-processing equipment to reduce particulate contamination, equipment uptime, and mean time between failures/mean time between interrupts (MTBF/MTBI) be increased. As first one and then several steps become automated, increased uptime is required to the point at which 90 to 95% will be minimum levels. This may argue against linking or integrating too many individual pieces of equipment, but such integration becomes essential to keep defect levels low.

Introduction

The fabrication of ICs involves a large number of individual and generally complex interactive steps. In providing an overview of the IC fabrication process, we will outline each of these steps in approximately the order in which they occur. IC fabrication steps can be grouped into four general types of processes: (1) starting material processes, which produce the polished silicon wafer, (2) imaging processes, which replicate the IC pattern geometries on the various wafer surfaces, (3) deposition and growth processes, wherein various layers of semiconductive materials are applied to the wafer, and (4) etching-masking operations, in which selective removal or addition of the deposited or grown layers is effected. The result of these complex processes is the transformation of a thin cross section of a silicon ingot into many individual ICs, each containing up to several million individual circuit elements. Each complex IC is capable of orchestrating a variety of tasks that have an electromechanical basis. The microprocessor ICs, in which an entire computer "brain" is placed on a single chip, are good examples of how we are putting silicon to work. Figure 1.5 shows the major steps in IC fabrication.

Starting Material Processes

Silica purification

The starting material of all integrated circuits is silicon, a gray semiconductive material that, in crude form, is one of the most abun-

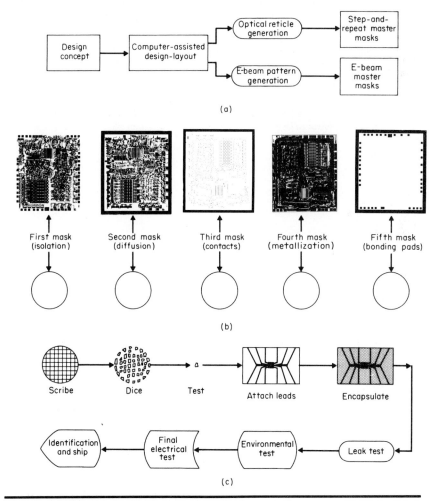

Figure 1.5 Major steps in IC fabrication. (*a*) Mask generation process. (*b*) Wafer fabrication. (*c*) Separation, testing, and packaging.

dant materials in the earth's crust. The properties of silicon are listed
in Table 1.1.

The first step in producing a silicon wafer is purifying raw silica,
which is mined as beach sand or taken in chunks from agatized rock
deposits. Small quantities are then heated in a high-temperature furnace. The result is the separation and removal of impurities, leaving
behind a chemically purified polycrystalline silicon material. This purified raw material is used to charge the "melt" or crucible of a machine from which a single crystal ingot is pulled (see Fig. 1.6).

Ingot growth

The ingot is grown by placing a single-crystal seed on the end of a
shaft and rotating it slowly through and away from molten silicon.
The melting and refreezing of the silicon at the seed interface allows
crystal formation to take place exactly following the crystal structure
of the seed. The conversion of molten polysilicon into monocrystalline
silicon cylinders continues as the seed is slowly withdrawn from the
melt, and the length of the silicon ingot is really dependent on a continued supply of molten silicon in the melt. The critical parameters in
single-crystal growth are rotation and withdrawal rate of the seed and
purity and temperature uniformity in the melt. Highly automated
computer-controlled crystal pullers are used to produce high-quality

TABLE 1.1   Properties of Silicon

Atomic number	14
Atomic weight	28.086
Atomic density	4.96×10^{22} atoms/cm^3
Density	2.328 g/cm^3
Dielectric constant	11.7 (± 0.2)
Energy gap	1.115 (± 0.008) eV
Temperature coefficient of energy gap	-2.3×10^{-4} eV/°C
Melting point	1417 (± 4) °C
Electron lattice mobility	1350 (± 100) cm^2(V•sec)
Hole lattice mobility	480 (± 15) cm^2(V•sec)
Refractive index	3,420
Thermal conductivity	1.57 W/(cm • °C)
Thermal expansion, linear	2.6 (± 0.3) $\times 10^{-6}$/°C
Lattice constant	5.4307 Å
Volume compressibility	0.98×10^{-12} cm^2/dyn
Photoemission work function	5.05 (± 0.2) eV
Hardness	7.0 MOH scale
Heat of fusion	1000 J/g
Heat of sublimation	18 (± 2) $\times 10^3$ J/g
Intrinsic carrier concentration (n_j)	1.5×10^{10} carriers/cm^3
Valence	4
Vapor pressure	2.8×10^{-4} mmHg (at m.p.)
Crystal structure	fcc, diamond

Figure 1.6 Crystal being pulled from a melt.

ingots. Even in the best process, however, crystal dislocations or imperfections occur in the growing process and may cause problems later
in imaging and etching processes. A process called "necking" is used
down near the seed to allow dislocations to grow out. Figure 1.7 illustrates the crystal-growing process in simple form.

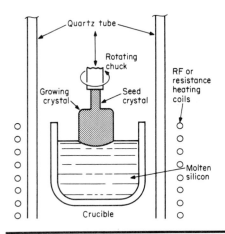

Figure 1.7 Single-crystal growth.

After crystal growth, the ingots are ground to produce a "flat" that parallels the growth axis. This flat will be used throughout the fabrication process as a reference for alignments during imaging and for electrical probe checks.

Wafer generation

After the flat is produced, the ingots are sliced into wafers 20 to 40 mils thick with diamond slicing saws. Sliced wafers are then lapped to remove the damage and irregularities caused by slicing. Several grades of lapping compound are used to produce successively smoother wafer surfaces. Mechanical lapping is performed in controlled lapping machines (Fig. 1.8). The final step in producing high-quality surfaces for imaging is chemical etching of the silicon. This produces a surface of optical quality by simultaneously removing debris from the last lapping operation and leveling the silicon surface, leaving a silicon "mirror." The sequence just described is shown in Fig. 1.9.

Figure 1.8   Mechanical wafer lapping.

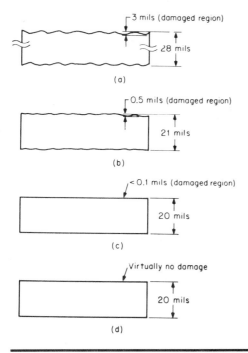

Figure 1.9 Wafer substrate preparation: (*a*) sawed slice; (*b*) lapped slice; (*c*) polished slice; and (*d*) etched slice.[1]

Imaging Processes

The basic sequence for imaging a wafer with either negative or positive resist is shown in Fig. 1.10.

Pretreatment

Wafers are chemically and mechanically cleaned to remove surface contaminants. They are then force-dried and given a short bake to remove residual surface moisture and thereby permit good resist adhesion. Careful control of wafer pretreatment is essential to maintaining high imaging quality. The step is critical for this reason: Any defects or contamination allowed at this stage of wafer processing can only be magnified by subsequent processing. The final stage of pretreatment usually involves the application of a resist adhesion promoter. This serves as an interfacial layer between the silicon dioxide and the photoresist. Wafers are coated immediately after priming.

Coating

Spin coating is the most widely used technique for applying photoresist to wafers. The resist is dispensed onto the surfaces of the wafers,

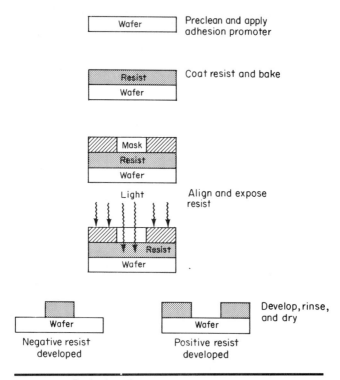

Figure 1.10 Resist imaging.

which are then accelerated on the coater to provide a thin uniform film across every wafer surface. Coating is always performed in a highly controlled clean-room environment with solvent exhaust and operators with clean-room clothing. Almost all wafers are coated in automatic cassette-loaded systems equipped with a computer program that predetermines the various coating parameters. The overall objective of this step is to provide a uniform and defect-free layer of a photo- or electron-sensitive masking material.

Softbaking

Softbaking is used to remove the solvents present in the spin-coated film of resist and is usually performed immediately downstream from the coater in a track-type wafer-handling system. Infrared, conduction, and microwave heating are used to drive the solvents from the resist coating and render it sensitive to exposure energy. Solvents left in the resist film from improper softbaking will cause poor imaging quality, usually from attack by the developer. This is one of the most common types of imaging problems. Properly baked wafers are loaded back into their cassettes and then transported, either automatically or manually, to the exposure station.

The solvent content can be measured as weight percent left in the film. For example, the weight percent of the residual casting solvent for Microposit 1470 is 3% at 80°C, 2% at 90°C, and 1.2% at 100°C. These data were generated for a 1-μm-thick resist film baked for 30 min in a convection oven. The goal of softbaking is to remove the solvent from the image formation ingredients or solids. The solids are comprised of the resin(s) and photosensitizer or inhibitor.

The reaction kinetics of the most commonly used positive resists is such that a relatively high ratio of differential solubility is created after a patterned exposure step. The first solubility parameter to test is that of the resin in solution without a sensitizer or inhibitor. This solution is spin-coated, and the developer dissolution rate is calculated after the standard softbake. The dissolution inhibitor (sensitizer) is then added to the sample, and a second dissolution test is run after spin-coating and softbaking. The final test to establish the developer solubility parameters of the resist system is the rate of exposed resist.

Typical novolak dissolution resists (AZ-1350J) have the parameters shown in Table 1.2. In highly softbaked (110°C) samples, the rate of developer dissolution decreases in both the exposed and unexposed areas and the actual differential solubility decreases only slightly. The solubility ratio also is changed by undersoftbaking, but it is seldom

TABLE 1.2   Positive Photoresist Component Solubility Parameters

Composition	
Photoactive compound	Speed and absorption
Base resin	Resistance, solubility, and flexibility
Solvent system	Dilution and coating properties
System Solubility Rates	
Resin	150–200 Å/sec
Resin and sensitizer (unexposed)	5 Å/sec
Resin and sensitizer (exposed)	1000–2500 Å/sec
Physical and Chemical Changes	
Color	Dark decay (no functional change)
Gas evolution	Diazo nitrogen release
Humidity	Greater than 30% RH at 21°C recommended
Storage	To maximize solvation and to minimize decomposition; 15–21°C recommended
Postbaking	Less than 135°C causes drying; greater than 135°C improves resistance. Typical thermal flow range, 125–140°C.

SOURCE: Shipley *Technical Manual.*

used because line geometry control is much more difficult with low softbakes. The amount of solvent left in underbaked samples greatly increases the attack rate of the developer on the unexposed areas. The same sample has little change in the rate of exposed resist dissolution, and so the ratio is lower.

Although softbaking is perhaps the most highly leveraged step for altering the differential solubility of a positive optical resist, there are alternatives. For example, postexposure chemical dips have been used to change the wetting angle and time of the developer. These surfactant, or surface-altering steps, change the initial rate of the developer reaction, and the initial rate happens to be greater and more critical than dissolution beneath the surface. Heat treatments after exposure (postexposure baking), which tend to rearrange the structure of the exposed resist chemistry, also are used to change differential solubility, specifically to remove standing-wave patterns. A 45-sec hotplate bake at 110°C is adequate for this purpose.

Exposure

The goal of exposure is to create a latent image of a desired pattern in the resist film. The role of the photomask is obviously paramount, since the mask is a passive tool that gives the resist its latent image. It must therefore be as clean and defect-free as the coated and softbaked wafer hopefully was. Wafers are either blanket exposed, step-and-repeat exposed, or scanned by a beam of exposing energy. In all cases, the result is a latent image closely matching the pattern of the photomask. Environmental cleanness is very important here as well, since contamination on either the wafer or the mask will be reproduced in the final developed image and cause a reject die or reject wafer, depending on the extent of the contamination. Again, properly exposed wafers are automatically reloaded into cassettes and moved to the developer station.

The wafer-handling equipment may provide for automatic wafer carriage to the developer, or the loaded cassettes may be carried manually and set on the front end of the developer. The important parameters to control are exposure energy uniformity and exposure time control.

Development

Batch and in-line wafer developing are widely used. In batch developing, the entire cassette(s) is (are) loaded into a chamber or tank, where the developer is sprayed or washed over the exposed resist coatings. In-line developing carries each wafer along a track, where it is placed under a stream or spray of the developer for a predetermined time. In both batch and in-line developing, rinsing and drying follow the developer step, and the parts are agitated, either by spinning (in-line) or cassette movement (batch), to provide uniform developing action. The concentration, temperature, and time of application of the developer must be closely controlled and monitored to ensure repeatable results. The development process is one in which the developer selectively attacks and removes either exposed regions (positive resist) or unexposed regions (negative resist), and leaves behind the image to serve as the mask for etching or, in some cases, metallization.

After developing, the wafers are rinsed and dried (unless plasma-developed) and sent for inspection prior to being postbaked. During inspection, defective or out-of-spec wafers are returned for reimaging and acceptable wafers are sent on. The overriding objective of development is to remove the resist that does not form the image without adversely affecting the resist that does form the image. In many processes, wafers are given a short plasma "descum," in which an oxygen plasma removes a thin layer of the resist image along with any residual films left in the developed areas.

Postbaking, or hardbaking

Postbaking, or hardbaking, is not always used in wafer imaging, especially when plasma etching is used on a simple undoped oxide. Even then however, it may be included to provide "insurance" that the resist is well bonded to the underlying oxide. It is accomplished in the same manner as softbaking, i.e., on track systems in which the heat is applied either above the track by a microwave or infrared source or beneath the wafer in a heated track. The heat removes any residual moisture from the developing operation and further bonds the resist to the wafer. Extensive postbaking additionally "hardens" the resist by making it chemically inert to the etchant fluid or gas. Although that makes resist removal more difficult, it may be required to achieve satisfactory etching results. Major factors to monitor and control in postbaking are bake time and temperature, because they will determine the nature of resist image thermal distortion. Wafers are sent directly to etching after postbaking, since delays may negate the effects of the bake. In fact, wafers that are delayed for more than 2 hr are typically rebaked before etching.

Etching

Wet- and dry-etching techniques are widely used, and as in developing, batch and in-line processing techniques are popular. The goal of etching is to precisely remove the semiconductor layer left exposed by the developing process. For this reason, *complete* removal of the developed resist is essential, and any residues left on the semiconductor layer may prevent or inhibit etching action, a phenomenon called "blocking." The etching control parameters are time, uniformity, temperature, and concentration of the etch species (liquid or gas). Numerous techniques, often called "black magic," are associated with the etching process. They include special spacing and positioning of wafers in the etchant, additives to buffer or otherwise alter the rate or direction of etching action, and special rotation or movement of wafers during etching that may enhance uniformity of etching. After etching, wafers are transported to the removal station.

Removal

Resist removal is typically done, in a batch mode, by immersing wafers in a heated resist-stripping solution or placing them in a batch plasma-stripping chamber, where an oxygen plasma removes the resist. The goal of removal is to leave behind a surface completely free of any resist material, since incomplete resist removal may cause defects in doping or metallization processes. Since residual resist films or particles are very hard to detect, relatively aggressive removal techniques are used, usually including corrosive resist strippers.

After removal, wafers are inspected for defects, and line-width measurements are taken. If the undercutting is excessive as a result of poor resist adhesion or overetching, wafers are etched to remove the defective layer, which subsequently must be regrown or redeposited and reimaged. Properly etched wafers are sent on for ion implantation or diffusion, after which a thin layer of oxide is generally deposited or grown and the entire cycle is repeated. Figure 1.11 shows an imaged wafer.

Figure 1.11 Imaged silicon wafer.

Deposition and Growth Processes

Oxidation

The oxidation of silicon is perhaps the most fundamental of all the processes used in IC fabrication. Silicon dioxide (SiO_2) is a very stable material that is almost universally used as the surface for resist-imaging operations. SiO_2 is grown in either an oxygen or steam atmosphere. Silicon interacts with the atmosphere in either of the following ways:

Oxygen atmosphere: $\qquad\qquad\qquad$ $Si + O_2 \rightarrow SiO_2$
Steam or water vapor atmosphere: $Si + 2H_2O \rightarrow SiO_2 + 2H_2$

The first reaction takes place when wafers are placed in an oxidation tube (also called "diffusion tube" or "oxide furnace") at temperatures of 950 to 1250°C. A typical oxidation tube (shown in Fig. 1.12) contains several sets of electronically powered heating coils surrounding the tube, which is either quartz, silicon carbide, or silicon. In O_2 gas oxidation, the wafers are placed in the tube in a quartz "boat" (or "elephant") and the gas flow is directed across the wafer surfaces to the opposite, or exhaust, end of the tube. The temperature in the furnace, or tube, is highly controlled, generally to $\pm$ ½°C, to ensure extremely uniform and predictable oxide layer growth. Preplotted charts are then used to determine the thickness of the layer produced.

In steam oxidation, oxygen or nitrogen carrier gas is directed into a "bubbler" that contains water. The gas is thereby saturated with wa-

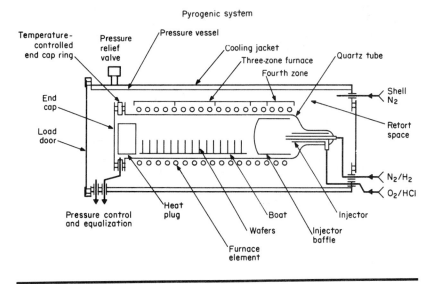

Figure 1.12   High-pressure oxidation. (*Courtesy of* Applied Materials.)

ter vapor and is then directed to the oxidation tube, or furnace. The water in the bubbler is kept at a constant temperature, below 100°C, to ensure a constant vapor pressure and thus uniform oxidation growth rates. Variations of water vapor oxidation are used, e.g., "flash" and "torch" oxidation. In flash oxidation, water is metered onto a heated surface, where it rapidly "flashes" into a vapor phase above the surface and is carried off with a gas to the furnace. In torch oxidation, the combustion of oxygen and hydrogen is used to produce water vapor.

Since rapid temperature changes in the wafer can produce considerable warpage, boats are always inserted and withdrawn slowly from oxide furnaces. The high temperature of silicon oxidation also causes redistribution of impurity ions. Consequently, several oxidation systems, in an attempt to avoid both warpage and redistribution, are operated at lower temperatures and increased pressure. High-pressure oxidation, operating at up to 25 atm, permits much better throughput as well, which reduces the time/temperature product. Typically, increasing the pressure by 1 atm permits a 25 to 35°C reduction in temperature. Comparison of this approach to the dry- and wet-oxygen processes is shown in Table 1.3. Pressure uniformity is critically necessary to prevent stress buildup in the quartz-lined tubes of the oxide furnace.

The long times normally associated with silicon oxidation are explained by the need for the O_2 and H_2O species to reach the silicon and react to form SiO_2. The actual atoms diffuse upward through the initially thin SiO_2 layer and meet the oxidizing species. As the oxide layer grows, the difficulty of this "chemical meeting" increases; hence, longer time is required for each incremental amount of oxide growth.

In all oxidation processes, extreme care must be taken in handling the wafers to prevent particulate or chemical (iron, sodium, potassium, etc.) contamination. For example, touching a wafer will inevitably result in sodium contamination from the salts on the skin and cause surface-leakage failure of devices. Clean dry wafers should always be handled with wafer tweezers and the special gloves designed for the handling of wafer boats.

TABLE 1.3   Comparison of Wet- and Dry-Oxygen Processes

Type of oxidation	Time required to grow a 1-μm-thick layer
Dry oxygen, 1200°C	2.5 hr
Wet oxygen, 1200°C	1.5 hr
High-pressure oxygen, 25 atm	15 min

The need for temperature, pressure, and other types of control in the oxidation process provides a natural application for IC technology. Microprocessor control systems are used to perform these functions, and as device parameters become more critical, the need for increased control becomes more acute.

The main use of SiO_2 layers is to provide a base for resist adhesion and imaging. The layers can also be used as an insulating dielectric.

Thermal oxidation of silicon.   One of the reasons silicon became the material of choice for solid-state transistor electronics was that its dioxide, SiO_2, has so many properties ideally suited to IC manufacturing. Silicon dioxide can be grown to a wide range of thicknesses for use as a dielectric, as a physical or structural barrier, as a thin ($\cong200$- to 400-Å) gate material, and as a pad mask. SiO_2 is used in thick layers (over 10,000 Å) as a field oxide and below 100 Å as a tunneling oxide. Oxides in the 1500- to 6500-Å range serve as masking oxides for ion implantation doping, diffusion masks, and as dielectric layers or passivation layers. Finally, the planarizing capability of SiO_2 permits structural complexity in the IC that would otherwise not be possible.

The thermal oxidation of silicon is the most common way to produce SiO_2, with spin-on glass, anodization, and chemical vapor deposition as alternative methods. Subjecting semiconductor silicon to oxidizing ambients, generally at elevated pressure and temperature, constitutes thermal oxidation. Examples of the applications of oxidation on an IC device are shown in Fig. 1.13.

The thin gate oxide grown for the metal oxide semiconductor (MOS) device illustrated is a critical application because the oxide layer is an

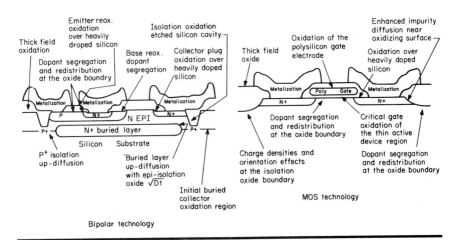

Figure 1.13   Applications of oxidation on an IC device.

electrically active part of the device. Gate and tunnel oxides need to be grown controllably to thicknesses below 700 Å, which requires a high degree of process control. Typically, low pressure (1 torr) and relatively low temperature (1000°C) can be used to grow thin oxides in a controllable manner. Thin dielectric layers can also be produced by thermal nitridation of silicon or silicon dioxide. This solves the high defect density associated with very thin SiO_2 layers, and nitride properties are well known for their durability. Increased control over thickness and reduced dopant diffusion are advantages of nitridation over oxidation.

Control of lateral diffusion is another critical parameter in thermal oxidation. Once again, varying pressure and temperature permits a wide range of end result possibilities. The movement of impurities is usually caused by high-temperature oxidation, and so reduced temperature combined with high pressure is used to reduce lateral diffusion of dopant. For example, temperature reduction from 1100 to 920°C and pressure increase from 0 to 25 atm will result in a 6 to 8 times reduction in lateral diffusion.

Lateral diffusion control is critical in conserving silicon area in the x dimension, especially as continued downward geometry scaling of IC devices prevails. Reduced oxidation operating temperature is advantageous in minimizing thermal distortion of the wafer. This helps maintain increasingly critical flatness parameters. Finally, the very nature of the oxide structure, including its physical and electrical properties, is changed when thermal growth parameters are radically varied. Process operators are advised to remap or recharacterize the oxide after making significant growth parameter changes.

The silicon crystal orientation is another variable affecting oxide growth parameters. Figure 1.14 shows silicon dioxide thickness as a function of the oxidation time, process temperature, and silicon crystal orientation. Note the higher growth rate over $\langle 111 \rangle$ oriented silicon.

Oxidation mechanism.   The oxidation of silicon results in the formation of a highly stable and useful material, silicon dioxide (SiO_2). The main reactions used to form SiO are shown above; see "Oxidation." The oxidation process begins at the surface of the silicon. As the oxide forms, it consumes silicon atoms first at the surface and then progressively deeper into the crystal. A rule of thumb in calculating final silicon consumed versus oxide thickness is that the amount of silicon consumed is 44% of the grown SiO_2 layer. For example, a 1-μm- (10,000-Å-) thick oxide layer will use 4400 Å of the silicon surface.

The basic mechanism of silicon oxidation is well understood and well characterized. The oxidizing material, being oxygen gas or steam, diffuses at a fixed rate calculated from the molecular weights and den-

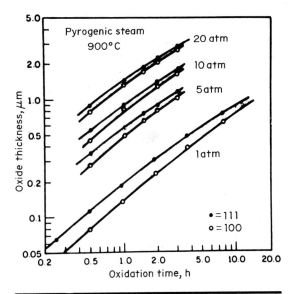

Figure 1.14 Oxide thickness versus oxidation time and crystal orientation.

sities of silicon and silicon dioxide. Linear parabolic models of the oxide growth parameters, with varying pressures and temperatures over a wide thickness range, exist. Silicon oxidation begins at room temperature, where a 10- to 20-Å-thick film will form. Increased temperature accelerates the reaction, and increased pressure also does. Oxides are typically grown thermally in the temperature range of 650 to 1400°C over pressures from 0.5 to over 30 atm.

Silicon dioxide properties. Crystalline silicon dioxide differs in structure from amorphous SiO_2 or fused silica and has a tetrahedral structure with a silicon atom surrounded by four oxygen atoms at a tetrahedral distance of 1.62 Å. These polyhedral structures are connected by oxygen ions called oxygen bridges. The amorphous state of silicon dioxide may have fewer bridging oxygen ions and therefore be less dense. The lower density of amorphous SiO_2 or fused silica (nonquartz) is more conducive to doping, because impurities penetrate the SiO_2 structure more easily.

One of the main properties of silicon dioxide is its effectiveness as a diffusion or ion implantation mask. Figure 1.15 shows the minimum oxide thickness required to mask against phosphorus and boron diffusions. Ion implantation masking requires different oxide thickness as a function of the incoming ion energy and mass.

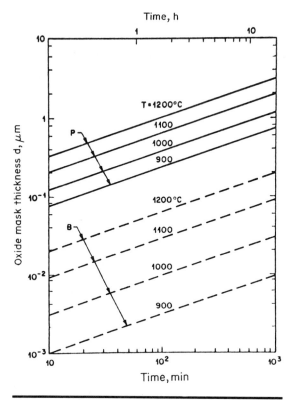

Figure 1.15 Minimum oxide thickness to mask against boron and phosphorus diffusions.

Oxide growth systems. Oxidation systems vary according to process requirements but, like all process equipment for a given step, have several common characteristics. They are very similar to the prevalent four-stack diffusion furnaces; in fact, many are converted diffusion systems. The main components of these furnaces include the growth chamber, a long, clear fused silica tube with an opening at each end for gas introduction or exhaust, as well as wafer loading or unloading. The process tube is surrounded by high-temperature ceramic heating elements connected to a control system, including thermocouples to sense chamber temperature. Wafers are loaded and unloaded automatically by low-friction arms or manually with quartz paddles and are stacked vertically in quartz boats. Automatic loading and unloading are required in many cases to eliminate particle generation caused by the friction of boats and paddles against the walls of the process tube.

A typical silicon oxidation system schematic is shown in Fig. 1.12. A key property required of such a system is precise control over temperature, pressure, and gas or steam concentration. Typical temperature control desired is in the 0.1 to 0.5°C range. Most systems can achieve 0.5°C control over the central part of the process tube, an area called the "flat zone." The ends of the process tube will vary outside this more uniform growth zone. Once the wafers are automatically or manually loaded into the process tube, the temperature drops below the set point. Rapid recovery to the desired temperature is important in maintaining good throughput. The loading and unloading rates also are important, because thermal shock caused by loading or unloading too quickly will reduce wafer flatness. Periodic checks of wafer flatness must be made before and after oxidation, along with the standard oxide thickness measurements that are made routinely. Particle counts on wafer surfaces before and after oxidation should also be made.

Common methods for checking oxide thickness include ellipsometry and optical interference fringe counting systems. They are nondestructive in nature and rely on the interference of light waves returning from two interfaces: the outside surface of the oxide (air/oxide interface) and the silicon/silicon dioxide interface. The light wavelength and intensity variations returning from these interfaces are predictable and are given by equations that fix the refractive index of the oxide at 1.46 at 5460 Å.

The ellipsometry method measures changes in the state of polarization of the light reflected from the sample. The main parameters used in systems that automatically compute the oxide thickness are refractive index, silicon optical constants, extinction coefficient, and the incident angle of the measuring energy.

These optical methods of oxide thickness measurement are accurate to $\sim +30$ Å over a thickness range of 300 to 12,000 Å. Table 1.4 is a color chart for thermally grown films observed perpendicularly under daylight fluorescent lighting. Note that some colors are repeated because they occur in a new order of reflection. This is a good quick reference chart for low-accuracy estimates of oxide thickness.

Epitaxy

"Epitaxy" is a deposition process wherein a single-crystal material is applied to a silicon wafer of the same crystal orientation. In order for the depositing layer of single-crystal silicon to atomically "fuse" with the wafer surface and, in effect, become an identical extension of that surface, two important conditions are required:

1. *Optimum surface.* The wafer surface must present to the depositing silicon a suitable number of nucleation sites where deposited

TABLE 1.4   Color Chart for Thermally Grown SiO$_2$ Films Observed Perpendicularly under Daylight Fluorescent Lighting[8]

Film thickness, μm	Color and comments	Film thickness, μm	Color and comments
0.05	Tan	0.63	Violet red
0.07	Brown	0.68	"Bluish" (Not blue but
0.10	Dark violet to red violet		borderline between vio-
0.12	Royal blue		let and blue green. It
0.15	Light blue to metallic blue		appears more like a mixture between violet
0.17	Metallic to very light yellow green		red and blue green and looks grayish)
0.20	Light gold or yellow—slightly metallic	0.72	Blue green to green (quite broad)
0.22	Gold with slight yellow orange	0.77	"Yellowish"
0.25	Orange to melon	0.80	Orange (rather broad for orange)
0.27	Red violet	0.82	Salmon
0.30	Blue to violet blue	0.85	Dull, light red violet
0.31	Blue	0.86	Violet
0.32	Blue to blue green	0.87	Blue violet
0.34	Light green	0.89	Blue
0.35	Green to yellow green	0.92	Blue green
0.36	Yellow green	0.95	Dull yellow green
0.37	Green yellow	0.97	Yellow to "yellowish"
0.39	Yellow	0.99	Orange
0.41	Light orange	1.00	Carnation pink
0.42	Carnation pink	1.02	Violet red
0.44	Violet red	1.05	Red violet
0.46	Red violet	1.06	Violet
0.47	Violet	1.07	Blue violet
0.48	Blue violet	1.10	Green
0.49	Blue	1.11	Yellow green
0.50	Blue green	1.12	Green
0.52	Green (broad)	1.18	Violet
0.54	Yellow green	1.19	Red violet
0.56	Green yellow	1.21	Violet red
0.57	Yellow to "yellowish" (not yellow but is in the position where yellow is to be expected. At times it appears to be light creamy gray or metallic)	1.24	Carnation pink to salmon
		1.25	Orange
		1.28	"Yellowish"
		1.32	Sky blue to green blue
		1.40	Orange
		1.45	Violet
0.58	Light orange or yellow to pink borderline	1.46	Blue violet
		1.50	Blue
0.60	Carnation pink	1.54	Dull yellow green

silicon atoms give up some energy and bond to the wafer surface. Nucleation sites are provided by bleeding in hydrogen chloride gas prior to epitaxial deposition and etching 0.2 to 1.0 μm of silicon from their surfaces, along with any crystal defects that would retard the process of crystal growth via deposition.

2. *Maximum lattice exposure.* Although there may be an optimum silicon surface, the depositing silicon atoms may have trouble "falling" into place on the existing crystal lattice structure. This difficulty can greatly extend the time needed for crystal growth. Consequently, wafers are tilted 3 to 8° off axis, thereby exposing a maximum number of crystal edges for the incoming silicon atoms.

Once these fundamental conditions are provided for, the epitaxy process can proceed. Prior to loading in the epitaxial reactor, wafers are solvent-cleaned and scrubbed in a typical brush-type scrubber to remove organic residues and particles that could interfere with the crystal growth process. They are then dried in filtered dry nitrogen. As with oxidation, all precautions must be taken to avoid any form of contamination on the wafer surface, especially particles that will cause crystal imperfections.

Epitaxial film growth. Epitaxy, meaning "arranged upon," is a process for the growth of thin film by deposition. Films grown by the epitaxial process are usually single-crystal ones in the thickness range of 1 to 20 μm, and deposition takes place on existing layers of single-crystal silicon. The ability to deposit precise silicon film thickness at very accurate doping levels is used in very large scale integration (VLSI) for increasing chip speed with only moderate increases in required current to power the chip. A typical application involving the deposition of a lightly doped oxide is used in MOS and bipolar processes, including gallium arsenide technology. Dopant concentration control is especially critical in high-speed devices, and although epitaxy is an added step, it is needed when the benefit of dopant control accuracy exceeds the cost of adding a process step.

Epitaxial film growth is carried out in the solid phase (SPE), liquid phase (LPE), and gas phase (GPE). All three are classified as chemical deposition processes, wherein the species to be deposited are vaporized and transported to the wafer and form a film and attendant by-products. A typical reaction would proceed as follows:

$$SiCl_4 + 2H_2 \rightarrow Si + 4HCl$$

This example uses silicon tetrachloride as the reactant in the hydrogen reduction reaction. Not shown are other gas and particle species formed during the reaction that may affect film quality. For example,

HCl and $SiCl_2$ are detected in many reactions and should be included to accurately portray the gas phase and surface phase reactions that are occurring simultaneously. In addition to silicon tetrachloride, silane (SiH_4), dichlorosilane (SiH_2Cl_1), and trichlorosilane ($SiHCl_3$) are used. The reaction for silane is

$$SiH_4 \text{ (gas)} \rightarrow Si \text{ (crystalline film)} + 2H_2 \text{ (gas)}$$

This reaction is popular because of the absence of hydrochloric acid and the relatively low deposition temperature. In reactions involving HCl, the acid acts as an etch-scavenger for metallic impurities such as copper and iron. The etching behavior or negative growth rate that occurs shows that the reaction can be reversible, with HCl etching the silicon instead of contributing to film growth. Figure 1.16 shows the growth rates of epitaxial silicon from various sources as functions of temperature.

Doping the epitaxial layers is accomplished by the addition of hydrides of commonly used dopant sources. Boron and phosphorus are introduced into the gas stream as diborane (B_2H_6) and phosphine (PH_3). Arsine (AsH_3) is used for arsenic doping, and the hydrides are diluted with hydrogen to relatively low (50- to 1000-ppm) concentrations. One of the benefits of epitaxial processing is the ability to provide highly controlled doping, which in turn controls device electrical resistivity and conductivity. Typical dopant concentration is 10^{12} to 10^{22} atoms/cm^3.

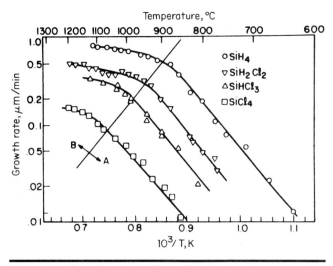

Figure 1.16   Growth rates of silicon films from various sources as functions of temperature.

Epitaxial process parameters. Epitaxial reactions take place in batch-type chemical vapor deposition (CVD) systems at temperatures in the 900 to 1300°C range. The growth rate of silicon varies according to temperature and the reactant gases, but typically it is 0.18 to 0.38 μm/min. Wafers are loaded into the reactor, which is purged first with argon or other rare gas. Hydrogen carrier gas is then bled in and heated to 1200°C to remove the surface oxide layer on the silicon, which is usually about 25 to 75 Å thick. Anhydrous HCl is then introduced to provide a mild etch of the silicon surface; then the HCl is purged and is followed by the silicon source reactant gases and appropriate dopants. This begins the epitaxial growth process. At the completion of the cycle, hydrogen buffer gas is once again used to purge the chamber. The entire cycle takes about an hour or less. Silane is widely used to produce extremely thin epitaxial layers at low temperatures (900°C). One of its disadvantages is its low-temperature decomposition and subsequent deposition of material on the walls of the epitaxial reactor. Frequent cleaning is then needed to prevent particulate contamination in the reactor. Silane is a good epitaxial gas because of low-temperature performance, however.

Since epitaxy is a crystalline material growth process, growth parameters in the epitaxial reactor need to be optimized to prevent misorientations in the "as-grown" film. Epitaxial films will actually shift as the crystalline lattice forms at an off-axis angle normal to the z axis. Wafers oriented at $\langle 111 \rangle$ are compensated by being supplied 3° off-oriented. The result of this phenomenon is called "pattern shift," because the underlying structures are displaced on the surface by an amount equal to the off-axis growth. Surface topography can change by lateral shifts (drift) or reduction of a geometry wherein a channel is either narrower or wider than the underlying, original topography before film deposition. In some cases the pattern is missing after film deposition, a phenomenon called "washout." Lowering the reactor pressure is one technique used to control pattern shift, distortion, and washout. Some reactant gases (e.g., silane) are also better than others in eliminating these epitaxial growth effects.

Epitaxial films are grown in both CVD and molecular beam epitaxy (MBE) reactors. CVD systems use either flat, single-wafer susceptors or hexagonally shaped, multiwafer susceptors that keep the wafers in a vertical position. The susceptor is usually graphite, carbide-coated material that is heated to process temperature with radio frequency (rf) inductive or infrared (IR) radiative heating elements. Thermocouples or IR pyrometers monitor and control process temperature; gas feed and mixing and exhaust rates likewise are controlled by a microprocessor. Gas entering the reactor is directed across the surface of the wafers uniformly to ensure a constant configuration to optimize wafer cleanliness, throughput, and uniform film deposition. A typical reactor schematic is shown in Fig. 1.17.

Molecular beam epitaxy is a technique used to grow silicon and other films under ultrahigh vacuum conditions (10^{-11} torr). MBE processing typically occurs at lower temperatures (600 to 900°C) than CVD processing (900 to 1200°C) and permits film layer growth control on an atomic level. Applications such as silicon-on-insulator (SOI) and silicon-on-sapphire are handled by MBE techniques. In these applications, the epitaxial reactor is used to grow an island of silicon on an insulating substrate (like sapphire), followed by subsequent interconnection. This technique reduces parasitic capacitance and improves radiation hardness. Single-crystal silicon films have also been grown on amorphous insulating layers.

MBE is so called because an electron beam is used to evaporate and deposit a film one molecular layer at a time. Dopant atoms also are introduced to the MBE chamber by evaporation or low-energy ion implantation. Dopants typically introduced with this technique are antimony (Sb) and gallium (Ga). The more common dopant sources such as boron and phosphorus, are implanted. The schematic drawing of a molecular beam epitaxial reactor is shown in Fig. 1.18. The use of multiple chambers facilitates substrate analysis, cleaning, film growth, and doping, all under ultrahigh vacuum in a single system.

Wafers are inspected for defects, after epitaxial growth, and electrical measurements are taken to check dopant profile, resistivity, and carrier lifetime. The deposited film thickness is measured destruc-

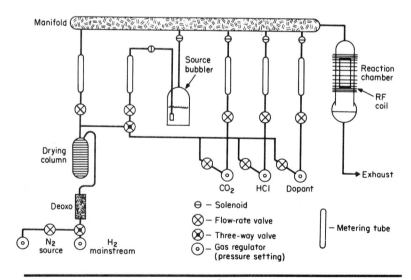

Figure 1.17 Epitaxial reactor schematic.[1]

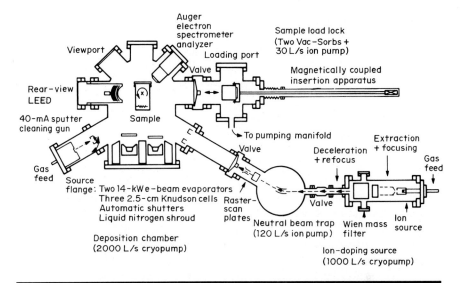

Figure 1.18 Schematic drawing of a molecular beam epitaxial growth chamber.

tively with cross-sectional SEM analysis or nondestructively by using Fourier transform infrared spectroscopy (FTIR). Infrared reflectance, based on a phase or interfering contrast of radiation reflected back from the top surface and substrate interface, also is used to measure film thickness. Surface inspection is made with various types of microscopes that highlight such effects as stacking faults and pits caused during deposition. Normarski phase contrast microscopy is useful for this application.

Epitaxy is used to provide both collector and base regions for transistors or in applications in which high-speed switching or high-current loads are required. Epitaxy is also commonly used in light-emitting diode (LED) fabrication.

Diffusion

"Diffusion," in a classical sense, is the uniform distribution of particles within a fixed volume of space according to a physical mechanism that begins with the same particles in a concentrated state in the same fixed volume. The integrated circuit analog of this physical behavior is the thermally induced distribution of impurity atoms throughout the silicon crystal lattice structure, thereby changing the electrical characteristics of the silicon. In IC processing, diffusion of several elements is common. Table 1.5 lists some of these elements and their maximum solubility temperatures in silicon. A series of doping, or diffusing, steps are used in the process of IC fabrication to gen-

TABLE 1.5   Some Elements Commonly Diffused in Silicon and Their Maximum Solubility Temperatures in Silicon

Element	Type	Maximum solubility temperature, °C
Antimony	n	1300
Arsenic	n	1150
Phosphorus	n	1150
Boron	p	1200
Gallium	p	1250
Aluminum	p	1150

erate a set of profiles and *pn* junctions that forms the basis for semiconductor devices. Diffusion as a process has two major steps: (1) predeposition and (2) diffusion per se.

The object of predeposition is to introduce a specific amount of dopant into the semiconductor surface. Predeposition processing begins by first cleaning the wafers to remove any contamination that might enter into the crystal structure with the dopant as an impurity and cause electrical problems later on. After cleaning, the wafers are loaded into a quartz boat and placed in the predeposition furnace. The basic type of equipment used is the same as that shown previously for wafer oxidation. The dopant is then introduced into the furnace in solid, liquid, or gaseous form.

In solid dopant predeposition, nitrogen is used as a carrier gas to transport a powder dopant that has been heated either in the quartz furnace or upstream in a small source furnace. Along with the nitrogen, some oxygen is often used to oxidize the dopant and thereby enhance the predeposition uniformity and other functional properties.

Gaseous predeposition and diffusion also are widely used. However, they pose a variety of potential problems, including toxicity, chemical stability, and insufficient concentration to achieve desired doping levels. Gaseous diffusion sources are easier to handle in the diffusion furnace than powder diffusion sources. As in solid source predeposition, the carrier gas is nitrogen. The source gas may be positioned upstream from the furnace, where it is bled into the furnace with the nitrogen.

When liquid source dopants are used, the carrier gas is bubbled through the liquid source chamber. Such a system is practically identical to the wet oxidation bubbling apparatus. As with solid and gaseous source dopants, oxygen gas is used along with the nitrogen carrier gas to oxidize the dopant before it reaches the wafers.

Solid, liquid, and gaseous doping sources are governed by the same laws of diffusion. As the heated dopant passes across the wafer surfaces, diffusion begins to take place in the predeposition process. Diffusion rates of the dopant are temperature-dependent. Increased dopant temperatures accelerate the movement of dopant atoms and permit the atoms to penetrate the surface of the wafer faster. Since each dopant has a different reaction and penetration rate that is a function of temperature, these relationships, or "diffusion coefficients," must be plotted for each type of dopant material.

As the dopants pass over the wafers in the predeposition furnace, the heated wafers become surface-saturated with the dopant. The point at which the crystal lattice of the wafer is saturated with dopant (maximum concentration) is described as the "solid solubility." Solid solubility is typically expressed in atoms per cubic centimeter and is plotted against temperature (700 to 1400°C range) for the commonly used dopants, which are boron, phosphorus, arsenic, gallium, aluminum, gold, and antimony. Thus, solid solubility data are used to see how hot wafers must be to achieve the desired predeposition level.

Surface dopant concentration, therefore, is also temperature-dependent, and the remaining parameter to complete predeposition is time. Short predeposition times leave a high dopant concentration near and at the surface, but dopant level falls off very rapidly below the surface. Increased predeposition times increase the dopant concentration below the surface until the profile becomes nearly flat.

Other types of predeposition sources include spin-on dopants; a solution containing the dopant is simply spin-coated onto the wafer. Low-pressure chemical vapor deposition of a doped oxide layer is another way to introduce dopant to wafers. Finally, the dopant source materials are used directly in diffusion furnaces, where they are spaced alternately with the wafers to be doped.

After predeposition, wafers are processed through the "drive-in" step, wherein the dopant applied and partly diffused in the predeposition step is driven to its final junction depth. The drive-in step takes place in a diffusion furnace with an ambient oxygen, or oxidizing, atmosphere, and no additional dopant is added beyond that applied in the predeposition. In the oxidizing environment, a silicon dioxide layer is grown over the newly diffused areas. The only areas not diffused are those previously protected by silicon dioxide, usually about 1 μm thick to prevent the dopant from penetrating in unwanted areas. The drive-in step uses temperature and time to determine the final dopant profile, along with the ambient gases in the diffusion furnace, which determine the thickness of the newly grown oxide layer.

After the diffusion operation, the wafers are checked for their sheet resistance and for the newly formed junction depth. Since multiple diffusions are made on individual wafers and the depth of the impurity ions is greater with subsequent diffusions, careful calculation of the results of these multiple interactive steps must be made. One technique to minimize diffusion depth changes during each step is to perform the higher-temperature diffusions first (impurities with the highest solubility parameter) and subsequent diffusions in declining solubility order.

Vacuum and sealed-tube diffusion techniques are used for special critical applications, and new diffusion furnaces for standard diffusions are incorporating microprocessor controls to achieve more precise doping of very thin layers.

Ion implantation

Ion implantation is used to place impurity, or doping, ions in semiconductor layers at various precisely controlled depths and with accurate control of dopant ion concentration. Its main advantage is the precision it allows in placing ions and the control it offers of ion concentration. It is an extremely reproducible process that achieves a high level of doping uniformity. Typical values are less than 1% uniformity across the wafer and less than 2% reproducibility from wafer to wafer and lot to lot.

The ion implanter (Fig. 1.19) works by providing an ion source in which collisions of electrons and neutral atoms result in a large quantity of various ions. The ions required for doping are selected out by an analyzing magnet and sent through an acceleration tube that goes directly onto the portion of the wafer where the doping is desired. The beam is usually scanned and/or the wafer is rotated to achieve uniformity, and thick layers of silicon dioxide or thick coatings of positive photoresist images are used as implant masks. A variation of this masking approach is to implant *through* a thin layer of silicon dioxide. The depth of the dopant is determined by the total energy present in the dopant ions, which is adjustable by varying the acceleration chamber voltage. The dosage, or number of dopant ions that reach the wa-

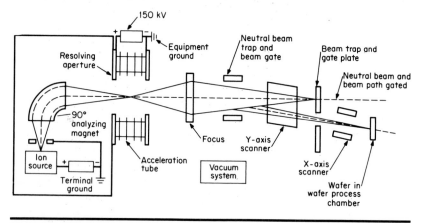

Figure 1.19 Ion implanter.[1]

fer, is determined by monitoring the number of ions as the ions pass through the detector.

Consequently, energy control affords a new capability not achievable with conventional tube diffusion techniques: precise control of junction depth. Dose control also offers new capabilities in terms of precise control of impurity, or dopant, concentration.

After being removed from the ion implanter, wafers are typically placed in a diffusion furnace to charge ions falling onto nonactive wafer regions. When the high-energy ion implantation beam enters the perfect crystal below the wafer surface, it knocks atoms out of the crystal lattice and renders the top few thousand angstroms amorphous. The physical change from crystalline to amorphous structure radically changes the electrical properties and essentially renders these areas useless.

Repair of this damage and reactivation of the dopant ions are accomplished by annealing the wafers in a furnace or with lasers. Furnace annealing is performed at temperatures below the melting point of the silicon. The heat allows the crystal to regrow along the original crystal planes, essentially putting the lattice back together again. Structurally, the crystal is almost the same as it was originally, except that dopant ions are "stuck" nicely between the silicon atoms.

Furnace annealing is being replaced by laser annealing for the following reasons:

1. Wafers can be contaminated in the furnace.

2. Wafers are warped more in a furnace than by lasers, because the entire wafer is heated in the furnace.

3. Bulk heating of the wafer in a furnace will cause lateral diffusion, which significantly increases the size of small geometries. A 1-μm geometry, for example, could be increased to 1.6 μm (line width) by this type of "heat diffusion."

4. Dopant activation is limited.

5. Process sequencing is limited (i.e., low-temperature steps must follow high-temperature steps).

Laser annealing provides less wafer warpage because the laser sweeps over only the areas implanted. One type of laser annealing utilizes a 4880-Å argon laser and heats the wafer to 1200 to 1300°C, below the melting temperature, to regrow the damaged silicon. This solid phase epitaxy results in a much higher quality lattice, much like the result from furnace annealing.

A liquid phase epitaxy also is used in laser annealing. In this process, infrared lasers heat the wafer surface to 1600 to 1700°C, and the silicon is regrown from the underlying layer interface up toward the surface, much like the growth of silicon crystal in a melt of molten silicon.

Originally, continuous wave lasers were used. They heated a significant amount of the area below the laser beam to about 1000°C. In such a process, crystal regrowth is solid-phase, much like furnace annealing. The more recent technique, called "laser cold processing," is accomplished by switching, or pulsing, a laser beam. This permits the wafer surface to melt and refreeze within microseconds and leaves the underlying wafer surface at ambient. The crystal growth is therefore liquid-phase, providing near-perfect growth. A key advantage of the rapid crystallization in laser cold processing is the "freezing" of ions within the lattice sites, something not possible with continuous laser or furnace annealing.

Laser annealing, whether solid or liquid phase epitaxy, is less costly and is claimed to produce better conductivity in the resulting silicon. Widespread use of the laser for annealing out implant damage has lead to laser use in such other applications as supplying heat for diffusing spin-on dopants. A relatively new application avoids the use of photolithography to define circuit paths. Wafers are placed in a chamber into which a dopant gas is bled; a laser then irradiates the wafer and causes the gas to break down in selected areas on the silicon surface. The laser also causes enough surface melting to allow the atoms of dopant gas to migrate into the silicon. Lasers are also used to melt and provide diffusion of dopant with solar cells.

Diffusion versus ion implantation. The objective of wafer doping is to place, within the crystal structure, known quantities of selected impurities that will predictably vary the circuit's electrical properties. Photolithography interfaces with doping by providing dimensional boundaries within which dopant ions must be placed. Photoetching and photoimaging of oxide and resist masks serve to provide the barriers for dopant ions.

The major change in doping technology is the shift from diffusion processes to high-energy ion implantation. Figure 1.20 is a schematic diagram of an ion implanter.

The diffusion approach, in which dopant ions are first predeposited and then thermally driven in by high-temperature processing, results in a greater consumption of silicon "real estate" in the critical horizontal direction. Ion implantation provides more vertical wall profiles by placing dopant ions into the wafer at steeper angles. The result is dopant profiles that tend to be shallow and deep compared to those produced by conventional diffusion processing, which is similar to the advantages of anisotropic dry etching over wet etching.

The basic strategy to increase IC chip density (more circuit func-

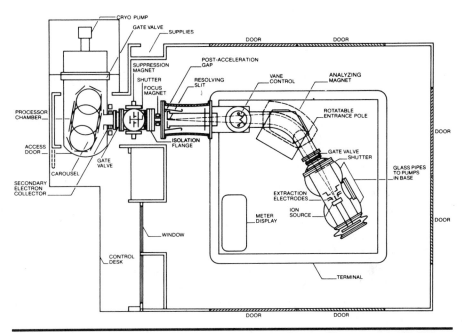

Figure 1.20 Schematic diagram of a high-current ion implanter. (*Courtesy of* Applied Materials.)

tions per unit area) is to reduce line width, add more layers of materials, and, essentially, move process technology in the direction of three-dimensional ICs.

High-energy ion implantation.   High-current, high-energy ion implantation causes damage to the crystal. Laser annealing, in which rapid heating and cooling of the defects caused by ion implant are possible, permits melting of the damaged crystal and resolidification into a natural, ordered structure. Ion implanters, with beam currents in the milliampere region and energy up to 600 keV, are widely used production tools for VLSI. The main device requirements to permit production use of these tools are as follows:

1. Generate wafer throughput for economic production.

2. Provide good doping uniformity for high yields.

3. Provide good dose control to meet electrical specs.

4. Maintain relatively low process temperatures to allow use of resists as implant masks and minimize unwanted thermal effects (diffusion, etc.).

5. Ensure low contamination to keep yields high.

6. Perform with thin and thick oxides.

Technology advances that have made all of the above requirements possible are high-temperature resists, laser annealing, and good wafer cooling during implantation.

Major requirements of ion implanters are maintaining dose uniformity (necessary to keep device yields high) and keeping the dose of dopant ions highly repeatable from wafer to wafer. Uniformity is helped by the use of secondary emission suppressors (positive and negative), and repeatability is controlled through the use of such devices as the Brookhaven dose of current integrators.

Doping uniformity is further improved by using large processor sizes, a feature that also helps keep throughput high. Scanning the ion beams in two dimensions increases uniformity and also spreads the power of the beam over a larger area. This keeps the process temperatures low. Doping uniformity on a single wafer will be in the range of 1 to 2%, and wafer-to-wafer uniformity is held to the same degree of accuracy. Run-to-run doping uniformity is generally in the range of 0.5 to 1.0%.

Typical elements implanted, along with beam current ranges, are shown in the following table:

Elements	Current ranges	Elements	Current ranges
Antimony	2–8 mA	Boron (doubly charged)	40 A
Arsenic (singly charged)	6–15 mA	Phosphorus (singly charged)	5–15 mA
Arsenic (doubly charged)	7–900 A	Phosphorus (doubly charged)	400 A
Boron (singly charged)	2–5 mA		

Implant application.   The ion doses used in MOS and bipolar IC processing range broadly. Some of the typical applications are measuring beam flux, in ions per square centimeter, versus kiloelectronvolts of energy. MOS manufacturers use ion implantation in the following areas:

Dielectric isolation

Source and drain doping

Complementary MOS (C-MOS) *p*-well doping

Self-aligning gates

C-MOS [bulk silicon, small-scale integration (SSI)]

Depletion, enhancement mode

Threshold voltage adjustment

Buried layers

High-dose polysilicon for low-resistance interconnect

One notable example in MOS is using the polysilicon gate as an implant mask, a critical step in conventional diffusion doping processes. The implanted source and drain allows a smaller gate and avoids common overetching problems in the diffusion approach.

The source-to-drain distance is a critical parameter in almost all high-density VLSI chips, since it determines gain and operating voltage parameters. It is much easier to control the source-to-drain distance in the ion implant example, since, in diffusion processing, three parameters must be closely watched at once: undercut of the gate oxide, gate length control, and lateral dopant diffusion.

Applications in bipolar technology for ion implantation are critical to new device design and fabrication. An example is the ability to use a photoresist as a direct mask for ion implant masking with resist instead of oxide as a mask.

High-current, high-dose ion implantation is also used for thick oxide applications which are followed by rare gas atmosphere annealing, processes that yield better than their diffusion counterparts. One of the major differences between ion implantation and diffusion is the relative process impact of each. In diffusion technology, a change in the dose profile results in a thermal process change that typically affects all subsequent thermal processes by altering dose profiles. Ion implantation avoids this thermal domino effect by allowing a single dose change *without* requiring process changes at other steps.

Applications for ion implantation are not limited to doping and include the following:

1. *Gettering.* Phosphorus and argon beams will getter metal impurities (medium- and especially high-current systems).

2. *Isolation.* Use of ion implant, oxygen, nitrogen, or noble metals replaces more complex reverse-biasing and thermal oxidation. Isolation "burial" is possible by using high-current systems.

3. *Dielectric formation.* Implantation of neon or argon to form amorphous dielectric layers.

4. *Electrical improvements.* Ion implant "damage" over implanted doped resistors improves temperature coefficient, linearity, and reproducibility.

All-ion implant process . Ion implantation as a tool for assuming all of the doping steps in IC fabrication is possible and useful. Several masks, including resists, silicon dioxide, silicon nitride, and aluminum, are available for ion implantation. Lower operating temperature is another major reason for ion implant to replace conventional thermal diffusion and thereby reduce wafer warpage and the risk of changing existing doping profiles in the crystal. The difficulties to overcome with ion implant include ensuring that all implanted species remain electrically active and all implant-related defects are annealed out. Overall, the high-energy (kiloelectronvolt range) bombardment of wafers with dopant ions, resulting in their implantation within the crystal lattice, offers more advantages than disadvantages compared to the diffusion process. One example of an all-ion implant process for a silicon-gate C-MOS/SOS device is shown in Fig. 1.21.

The basic implant steps used are

1. Phosphorus implant in silicon

2. Boron implant in silicon

3. High-dose boron implant in polysilicon gate

4. High-dose phosphorus implant in n-type source and drain areas

5. High-dose boron implant in p-type source and drain areas

Resist under ion implantation. Resists provide some interesting possible applications for use with ion implantation. One example, cited earlier, is their use as direct implant masks. To follow this example further, not as a wafer implant mask but as an optical mask, positive resist is used to make a photomask by ion implanting the resist image to change its optical density.

Simple patterning of a quartz or glass substrate is followed by ion implantation, which changes the optical transmission properties of the resist. The plot of optical density versus dose and transmittance versus wavelength completes the sequence for creating a simple high-resolution photomask without etching any metal or oxide layers. The resists used in the example are AZ-1350 and OSR.

The ion implantation of resist results in the formation of a carbonized layer of material ranging in thickness from 0.2 μm at 50 keV to 0.4 μm at 150 keV for AZ-111 when the implant is argon, arsenic, or phosphorus. The phosphorus creates about 20 to 30% more penetration than the arsenic, with argon in the middle.

Resist carbonization during ion implant is caused by ion species and ion energy. Therefore, the actual carbonized layer thickness is mainly a function of the physical scattering mechanisms in the resist layer. Future use of resists for implant masking will therefore encourage denser resist materials.

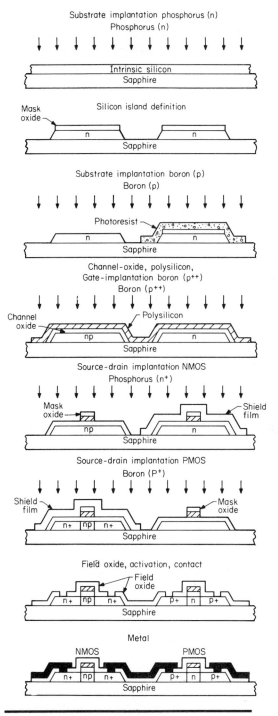

Figure 1.21   All-ion implant process.

Metallization

Aluminum "metallization" is used as the interconnection layer on ICs, making ohmic contact to the devices formed in the silicon and connecting these to the bonding pads on the chip's edge. Aluminum is used because it adheres well to both silicon and silicon dioxide, it can be easily vacuum-deposited (it has a low boiling point), and it has high conductivity. In addition to pure aluminum, alloys of aluminum are used to form IC interconnections for different performance-related reasons. For example, small amounts of copper are added to reduce the potential for electromigration effects, where current applied to the device induces mass transport of the metal. Small amounts of silicon also are added to aluminum metallization to reduce the formation of metal "spikes" that occur over the contact holes. In some cases, titanium is added for the same reason.

Aluminum metallization layers are vacuum-deposited onto wafers by one of the following methods:

1. Flash evaporation

2. Filament evaporation (tungsten, other)

3. Electron beam evaporation

4. Sputtering (planar and cylindrical)

5. Induction evaporation

Deposition of the films in VLSI device manufacturing.   Thin film deposition in VLSI is one of the major process areas required to manufacture an integrated circuit, the others being lithography, etching, and doping. Successful deposition calls for the following criteria: high uniformity of the deposit across all areas of the wafer containing active devices, uniform coverage of previously etched topography (step coverage), high purity of the material forming the deposit, good adhesion of the deposited film to the wafer or other substrate, sufficient thickness of the film to perform the required device function.

Economic considerations must be added to these basic physical requirements. They include good deposition rates for the process or method chosen, because throughput is a key parameter in calculating return on investment (ROI) for capital equipment. Operator safety also is necessary, and the deposition technique and attendent equipment must accommodate all required safety regulations. The cost of VLSI process floor space is high, so equipment must be constructed to have the smallest configurable footprint.

Evaporation is restricted from wider use by at least three drawbacks. One is the lack of control over deposition parameters (rate, grain structure, alloy, composition control). Second, the adhesion of the deposited atom is low compared to sputtering, primarily because of the difference in energy transfer kinetics. The high landing energy of sputtering provides higher surface bonding. Finally, the damage caused by x-rays in electron beam evaporators is not encountered in sputtering.

Evaporation is performed in one of three types of systems: resistance-heated sources, inductive-heated sources, and electron-beam-heated sources. Resistance-heated evaporation is severely limited to lower-melting-point metals (no refractory metals), filament contamination, and short runs due to small charges, thus, refractory metals and their silicides cannot be used. Inductive-heated sources are limited by equipment complexity and contamination of the deposit from the crucible. Electron beam evaporation, despite the problem (charges left in gate oxides by x-rays) of ionizing radiation and its effects on wafers, is widely used for evaporation.

High-energy electrons are streamed toward a target source at 3- to 30-keV energy levels. This high kinetic energy is converted to heat at the surface of the source, which causes the source to melt around the area of beam incidence. The area of the source outside the beam is considerably cooler, so a high-purity crucible is formed inside the source itself. One problem, however, is occasional dislodging of a molten drop of source material that lands as a mass on the wafer.

Evaporation of aluminum and chromium are by far the most common VLSI applications, the former for interconnect metallization, the latter for mask blank coating in mask manufacturing. Evaporation permits high deposition rates, up to 6000 Å/min, for example, for aluminum. The main benefit of evaporation, compared to sputtering, are very high purity of the deposit, little or no substrate damage, and relative simplicity of the equipment.

Deposition of thin films by evaporation. Thin films of several conductive materials are deposited by evaporation. The technique is based on heating the source material in a high-vacuum environment and thereby causing the source to evaporate. The result is that vaporized atoms and molecules are deposited onto the wafer.

Metals commonly evaporated for VLSI applications include platinum, niobium, aluminum, molybdenum, tungsten, titanium, tantalum, and chromium.

Sputtering. Sputtering is a physical (nonchemical) method for depositing thin films used in IC manufacturing. It is performed by ionizing inert gas particles in an electric field (producing a gas plasma or slow discharge) and then directing them toward the source or target, where the energy of the incoming particles dislodges or "sputters off" atoms of the target. The dislodged atoms then condense and are deposited

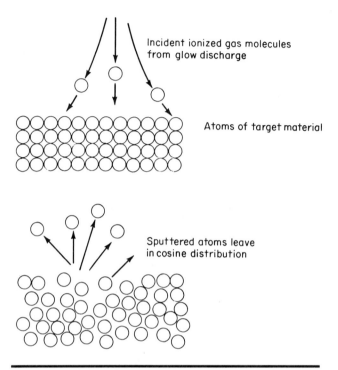

Incident ionized gas molecules
from glow discharge

Atoms of target material

Sputtered atoms leave
in cosine distribution

Figure 1.22 Physics of sputtering by momentum transfer.

onto the wafer. Sputtering, as a physical deposition method, is used to produce thin films of aluminum and aluminum/silicon alloys, refractory metal silicides and gold for interconnect metallization, titanium/tungsten, tungsten, molybdenum, and, in some applications, silicon dioxide and silicon.

Sputtering is widely used for VLSI film deposition. Compared to electron beam evaporation, sputtering is less likely to radiation-damage the wafer (no x-rays), and surface cleaning by sputtering is possible. Although there may be more particulates because a medium vacuum is used, sputtering typically offers a wider range of control over thin film properties than any other deposition technique. For example, sputtering temperature, pressure, power, and bias are all used to control physical properties like grain structure, film stress, adhesion, step coverage, uniformity, and alloy composition.

Mechanism of sputter deposition.   The basic physics of sputtering, outlined in Fig. 1.22, are explained by momentum transfer, wherein a target material (aluminum, for example) is struck by energetic incident ions. A wide range of energies in the incident ions will produce specific reactions on the target:

Energy range of incident ions	Reaction at target
2 eV	Physisorption
4–10 eV	Sputtering of surface
10 eV–5 keV	Sputtering range
	Monolayers
10–20 keV	Ions driven into target
	lattice structure

The energized atoms from the direct-current (dc) glow discharge accelerate toward the target, strike the surface, and dislodge atoms in the target by overcoming their binding energy. The required or minimum energy needed to initiate sputtering reactions roughly equals the heat of sublimation of the target. The heat of sublimation of silicon, for example, is 13.5 eV. Atoms in the target strike each other, are converted to the gas phase, and escape to condense and form a thin film on the wafer. The precise mechanisms at work in sputtering are still a matter of conjecture, but it is known that atoms leave the target in a cosine distribution.

The atoms sputtered from the target material leave with a velocity of approximately 3 to 7×10^5 centimeters per second (cm/sec). A large number of atoms undergo collisions before reaching the wafer, despite the fact that the distance to the wafer is only about 7 cm. These collisions reduce the velocity and energy (to less than 2 eV) of the sputtered atoms. The low-energy species will reach the wafer by diffusion as a result. Collisions also cause atoms from the target to be reflected back to the target, a problem that may be overcome by increasing the system pressure. Increased pressure is used primarily to increase the rate of deposition; it ensures that as many atoms as possible reach the wafer. It also keeps the deposition of atoms on chamber sidewalls to a minimum.

The sputtered atoms arrive at the wafer and begin to form a film, atom by atom. Atomic thickness (2- to 4-Å diameter approximately) and rates of deposition (150 to 7000 Å/min) determine film growth rates. Arriving along with the sputtered atoms are replicas including high-energy electrons, negative- and neutral-charge ions, photons from the glow discharge, x-rays, thermal electrons, and impurity gas ions. Therefore, gas and other contaminants causing unwanted film impurity and oxidation result. The sputtered atoms and other species are condensed onto the wafer by electrically attractive forces (negative charge). The atoms attach to the surface, nucleate, and, by chemisorption (including electron transfer) and physisorption (no electron transfer), blend together to form a film. The progressive growth of the film is ensured by the naturally decreasing free energy of any system of molecular species, based on the nucleation theory of Gibbs.[3] Thus, the nuclei become stable and grow, which results in the film. Some atoms, arriving before the process is underway, may reevaporate into the plasma and later return to become part of the film.

Sputtering methods.   Thin films of conductors and semiconductors for VLSI applications are typically sputtered by two methods: magnetron and radio frequency (rf). In magnetron sputtering, a magnetic field is used to concentrate otherwise stray electrons around the target, thereby providing current density in the 10 to 100 mA/cm^2 range. The result is very high deposition rates, since particles leaving the magnetic field toward the wafer create useful ionizing collisions instead of useless heat. The concept of using a magnetic field to change the direction and velocity of electrons is simple but highly useful for sputtering, and magnetron sputtering has become very popular as a result.

Variously shaped electrodes are used to optimize the number of ionizing events, including square, rectangular, and circular shapes. For an equivalent voltage, the planar surface, rectangular electrodes are favored. Maximizing deposited film uniformity and purity also are important, as is sputtering onto larger-diameter wafers. Balancing these often conflicting needs calls for flexibility in electrode design and use. Magnetron sputtering is used for applications that require its high deposition rates, such as aluminum metallization deposition.

Radio frequency sputtering is used more for the deposition of dielectric films, such as thick planarizing layers of silicon dioxide for pad masks, intermetal dielectric, and doping masks. It is characterized by an alternating electrical polarity of the electrodes (ac, not dc, voltage) in order to sustain a supply of electrons from insulator surfaces.

RF sputtering of dielectrics uses an insulative cathode, a condition that leads to a loss of potential between anode and cathode. In a single cycle of an rf waveform, enough electrons are gathered to sustain the needed anode/cathode potential. An rf frequency of 13.56 MHz at the electrodes is used to keep the insulator surfaces charged. RF-excited glow discharges are used for both additive (deposition) and subtractive (etching) processing. A typical planar magnetron sputtering system schematic is shown in Fig. 1.23.

Chemical vapor deposition

Chemical vapor deposition (CVD) is a gas reaction process in which any number of semiconductive layers can be formed by the heat-induced decomposition of selected gases. The gases most commonly used in IC fabrication and their respective reactions to produce compounds by vapor deposition are listed in Table 1.6. Silicon dioxide can be produced by chemical vapor deposition in both doped and undoped forms. Deposited oxides are not as dense as thermally grown oxides,

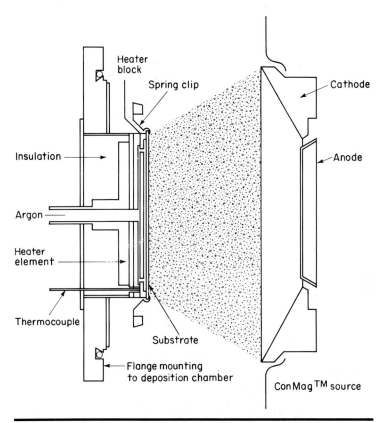

Figure 1.23 Planar magnetron sputtering schematic.

but they can be densified by heating at 1000°C for 25 to 40 min, depending on the thickness of the layer. Undoped deposited oxides are used in the same ways that thermally grown oxides would be used, e.g., as protective layers, predeposition layers, etching masks, or diffusion masks. When deposition is performed over a layer of aluminum metallization, the process temperatures are kept below 450°C to prevent microcracking or the driving of existing dopants further into the devices.

A variety of reactions are used to produce CVD silicon dioxide. They include the following:

$$SiH_4 + 4CO_2 \ (500–900°C) \rightarrow SiO_2 + 4CO + 2H_2O$$
$$SiH_4 + 2O_2 \ (200–500°C) \rightarrow SiO_2 + 2H_2O$$

These reactions use nitrogen as the carrier gas and occur at the temperatures shown in the equations.

TABLE 1.6   CVD Reactions for Major Films

Process reactants	Reaction conditions	Resultant deposit and evolved by-products
$SiH_4 + O_2$	$\xrightarrow[300 - 500°C]{N_2, Ar}$	$SiO_2 + 2H_2$
$Si(OC_2H_5)_4$	$\xrightarrow[740°C]{N_2, Ar}$	$SiO_2 + 2H_2O + 4C_2H_4$
$SiX_4 + 2H_2 + 2CO_2$	$\xrightarrow[1200°C]{H_2}$	$SiO_2 + 4HX + 2CO$
$3SiH_4 + 2N_2$	$\xrightarrow[250 - 500°C]{N_2, \text{Plasma}, 0.5 \text{ Torr}}$	$Si_xN_yH_z + xH_2$
$3SiH_4 + 4NH_3$	$\xrightarrow[750 - 1000°C]{NH_3, H_2}$	$Si_3N_4 + 24H_2$
$3SiCl_4 + 4NH_3$	$\xrightarrow[750 - 1100°C]{NH_3, H_2}$	$Si_3N_4 + 12HCl$
$2Al(OC_3H_7)_3$	$\xrightarrow[420°C]{N_2, O_2}$	$Al_2O_3 + H_2O + C_xH_y$
$Al_2Cl_6 + 3CO_2 + 3H_2$	$\xrightarrow[900 - 1200°C]{H_2}$	$Al_2O_3 + 6HCl + 3CO$
$(1 - 2x)SiH_4 + 2xPH_3 + 3[O_2]$	$\xrightarrow[300 - 500°C]{N_2, O_2}$	$(SiO_2)_{1-x}(P_2O_5)_x + [H_2]$
$(1 - x)SiH_4 + xB_2H_6 + 2[O_2]$	$\xrightarrow[300 - 500°C]{N_2, O_2}$	$(SiO_2)_{1-x}(B_2O_3)_x + [H_2]$

Higher-temperature reactions are used with hydrogen as the carrier gas. Thickening a field oxide is a typical application of CVD silicon dioxide.

Silicon nitride is very popular as a dielectric in ICs because of its density. CVD silicon nitride is commonly used as a passivation layer. Silicon nitride is deposited by the following reaction:

$$3SiH_4 + 4NH_3 \rightarrow Si_3N_4 + 12H_2$$

This reaction takes place with both nitrogen (approximately 650°C) and hydrogen (approximately 1000°C) carrier gases.

Polycrystalline silicon (poly) is deposited by the following reaction:

$$SiH_4 + \text{heat} \rightarrow Si + 2H_2$$

This occurs at approximately 900°C with a hydrogen carrier gas and at approximately 650°C with a nitrogen carrier gas. Various crystal structures can be achieved by varying the deposition rate, the temperature, or both. Various crystal structures are selected for specific electrical properties.

Reaction chambers. Several types of reaction chambers are used in all CVD processes, and they are similar to those used for epitaxial growth. In all cases, a susceptor, with wafers placed on its surface, is generally rotated or moved in the gas environment to provide improved uniformity. Aside from the standard bell jar type of chamber, in which gas is pumped up through the susceptor (vertical CVD), horizontal systems, in which the gas is passed from one end of the chamber containing the wafers to the other, are used. Barrel systems with wafers placed on a rotating cylinder are similar to barrel epitaxial systems. Finally, a vertical system, with gas flowing downward and contained by a surrounding blanket of nitrogen gas, is used in conjunction with a conveyorized wafer holder.

Heat sources. The heat sources that provide the gas decomposition are classified as either cold- or hot-wall. The cold-wall system uses either ultraviolet (uv) or rf energy to heat the susceptor, which in turn heats the wafers by conduction. The term "cold-wall" stems from the fact that the chamber walls are at a relatively lower temperature than they are in a hot-wall system. In hot-wall systems, thermal resistance heating is used and deposition rates on the chamber walls actually exceed those on the wafers. Obviously, the reverse is true of cold-wall systems. The chamber walls in the hot-wall system are heated by energy conducted and radiated from the susceptor. Cold-wall systems are advantageous for their greater throughput, which results from their higher gas flow velocity and more rapid heating and cooling.

CVD gases. The many gases used in CVD are often hazardous and must be respected. Except for inert argon and usually inert nitrogen, most CVD gases are toxic, flammable, pyrophoric, or corrosive. Many are combinations of two or more gases, most being flammable. The CVD gases are tabulated here.

CVD gas	Hazard*	CVD gas	Hazard*
Silane	T,F,P	Ammonia	T,C
Dichlorosilane	T,F,C	Hydrogen	F
Phosphine	HT,F	Nitrous oxide	
Diborane	HT,F	Nitrogen	
Arsine	HT,F	Argon	
Hydrogen chloride	T,C		

*T = toxic; HT = highly toxic; F = flammable; C = corrosive.

The reactant gases are energized by various means, including thermal, laser, ultraviolet light, and plasma. The reactions produced are complex chemical types; they include simple pyrolysis, hydrolysis, chemical transport oxidation, and combinations of these. A main objective of CVD is to produce heterogeneous reactions and avoid homogeneous reactions in which gas phase nucleation results in particle formation in the chamber and subsequent contamination. Variation of CVD process parameters allows for suppression of homogeneous reactions.

CVD chemistry.   The information listed in Table 1.6 shows additional process reactants, reaction temperature and gas "carriers," and the end product of the CVD reaction. The reaction temperatures vary with the wafer restrictions. Aluminum, for example, cannot tolerate CVD reactions much above 480°C. The other metallization systems, including polysilicon and refractory metal silicides, allow CVD processes in which oxide passivation layers are overcoated up to 900°C. The increased use of low-temperature reactors reflects the need for better control and tighter device specifications.

CVD glass types.   The materials listed in Table 1.6 are the building blocks for a range of CVD glass types widely used in the industry. They are used to synthesize the basic types of glasses used in devices including

1. Binary silicates

   - Phosphosilicate glass (PSG)
   - Borosilicate glass (BSG)
   - Arsenosilicate glass (ASG)

2. Nitrides

   - Silicon nitride (Si_3N_4)
   - Nitride polymers

3. Oxynitrides

   - Silicon dioxide ($Si_xN_yH_z$)
   - Al_2O_3, TiO_2, Z + O_2, HFO_2, Ta_2O_5, Nb_2O_5, GeO_2, Fe_2O_3

4. Ternary silicates

   - Germanium borosilicate glass (GBSG),
   - AlPSG, PBSG, AlBSG, LBSG, ZBSG.

   Many of the metal oxides and nitrides are used in market-limited custom device applications, but all of them are prepared by chemical vapor deposition.

CVD reactors: Hot-wall type.   Several reactor types (Fig. 1.24) are used to produce the variety of glasses mentioned previously. Four main types include hot-wall and reduced-pressure reactors used to deposit polysilicon, SiO_2, and Si_3N_4. Essentially this is a heated quartz tube with a three-zone heater that produces temperatures in the 300 to 1000°C range. Gas is pumped into one end and out the other. The substrates are placed perpendicular to the gas flow in a vertical quartz support holder, and up to 200 wafers are processed per run. Hot-wall reactions permit good throughput, large wafer diameters, and good film uniformity. The drawbacks are the flammable and toxic gases and the relatively low deposition rates that are due to the reduced pressures.

CVD reactors: Continuous.   Continuous-type CVD reactors are used primarily to deposit silicon dioxide when wafers are transported by conveyor through the reaction chamber in which wafers are convection-heated. Gas curtains are used to contain the reactant gases, and the results are high throughput and good uniformity on large-diameter wafers. The drawback of the continuous atmospheric pressure CVD reactor is contamination from the rapid gas flow rates needed to feed the system and the frequent cleaning operations.

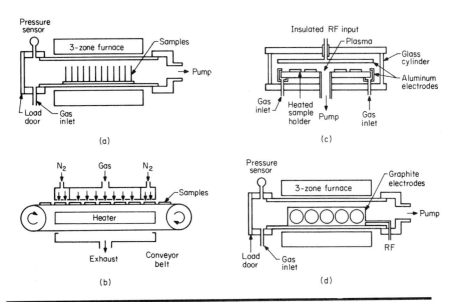

Figure 1.24 Schematic diagrams of CVD reactors. (*a*) Hot-wall, reduced-pressure reactor. (*b*) Continuous, atmospheric-pressure reactor. (*c*) Parallel-plate. (*d*) Hot-wall.

CVD reactors: Parallel-plate.   Parallel-plate CVD reactors are generally of the plasma-assisted type with top and bottom electrodes. The wafers rest on the bottom, grounded electrode (an aluminum plate), and rf voltage applied to the top aluminum plate electrode generates a flow discharge.

The reactant gas flows radially through the gas discharge from entrance ports typically at the edges of the reactor. In some cases, counterflow gas patterns are used. Heating is done with lamps or resistance heaters to 100 to 500°C, which is suitable for CVD of silicon nitride and SiO_2. The relatively low deposition temperature is highly useful for advanced devices in which dopant profile control is especially critical. The problems with parallel-plate CVD reactors are the same as with parallel-plate plasma etchers: contamination buildup on the chamber and wafer surfaces and low throughput because wafers are loaded individually.

CVD reactors: Hot-wall plasma deposition.   Hot-wall plasma deposition reactors offer high throughput at low deposition temperatures, but they suffer from contamination buildup and limited throughput from individual loading. The hot-wall plasma deposition reactor uses a heated quartz tube with vertically placed wafers parallel to the gas flow. The electrodes are aluminum or graphite slabs alternating with physical spaces in the reaction chamber. The power generates a glow discharge.

The use of low-pressure chemical vapor deposition (LPCVD) has become successful because of its wafer-packing density feature. This density is achieved by providing a maximum mass transfer rate. The transfer rate is optimized by an increase in the reactant concentration in the gas stream. LPCVD is widely used for deposition of SiO_2, phosphorus-doped silicon, glass, and Si_3W_4. Plasma-enhanced CVD (PECVD), on the other hand, is more widely used for application of the glassy $Si_xN_yH_z$ layers.

Despite a wide variance in deposition temperature, the CVD films are remarkably similar in density. As expected, etch rate increases with decreased deposition temperature, and vice versa. A good example of one of these films, shown as etched contacts in phosphorosilicate glass (PSG), reveals the smooth tapering from reflow in Fig. 1.25. The smooth tapering permits uniform and continuous coverage of subsequent metallization.

CVD film applications.   The variety of CVD reactions include a range of materials from basic insulators to semiconducting layers, conducting layers, and superconducting films. These materials need to meet a va-

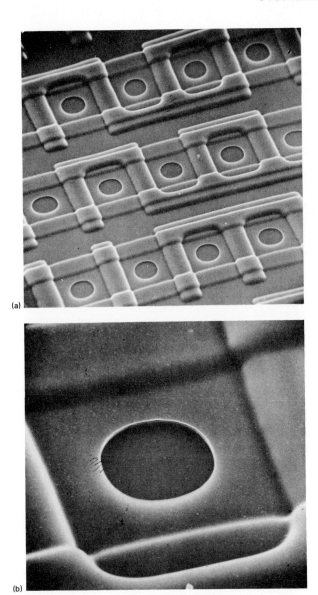

Figure 1.25 PSG contacts at (*a*) 1k × and (*b*) 5k × showing smooth reflowed surface. (*Courtesy of Shipley Company.*)

riety of primary requirements in order to function on advanced IC devices with high reliability. The prerequisites include the following:

Relatively low deposition temperatures to minimize doping profile impact

Uniform layer coverage of multistep topography to ensure film continuity over the lifetime of the device

Film thickness control to maintain uniform electrical parameters

Relatively low stress to prevent cracking or film discontinuity and deformation

Good film density to guarantee dielectric insulation and integrity

Ultrahigh purity that ensures consistent electrical performance within the layer

Structural uniformity

High dielectric insulation (low leakage, high breakdown voltage)

Process control of the film's chemical composition

High substrate adhesion to prevent undercutting, lifting, or separation effects

The categories of CVD films according to applications and their respective nomenclatures are as follows.

Insulators and dielectrics	SiO_2, Si_3N_4, Al_2O_3, Ta_2O_5, TiO_2, Fe_2O_3; PSG, BSG, AsSG, AlSG, LABSG; $MgAl_2O_4$.
Semiconductors	Si, Ge, GaAs, GaP, GaN, AlN, GaSb, AlP, InAs, AlAs; $GaAs_{1-x}P_x$, $In_{1-x}Ga_xAs$, $In_{1-x}Ga_xP$; ZnS, ZnSe, CdS, CdTe; SnO_2, SnO_2:Sb, In_2O_3:Sn, V_2O_3.
Conductors	Al, Ni, Pb, Au, Pt, Ti, W, Mo, Cr; WSi_2, $MoSi_2$, doped polySi.
Superconductors	Nb_3Sn, NbN, Nb_4N_5.
Magnetics	Ga:YIG, GdlG.

The most widely used films on ICs applied by CVD are silicon dioxide (SiO_2), silicon nitride (Si_3N_4), and polysilicon (Si). The major applications for these films are

1. *Silicon dioxide.* Dielectric insulation between multilevel metallization, diffusion and ion implantation masking, field oxide layers on MOS devices, and capping layers for doped oxides to reduce dopant loss. Doped oxide layers with up to 15% by weight of boron, phosphorus, arsenic, and antimony are used as solid-state diffusion sources. The phosphorus-doped silicate glass (PSG) is also used as surface passivation layers, gettering layer metal impurities, and dielectric layers for reflow on steps between multilevel interconnection.

2. *Silicon nitride.* Surface passivation; gate dielectric for MOS ICs; oxidation mask for isoplanar and locally oxidized silicon (LOCOS) de-

vices, based on the relatively slow oxidation rate of Si_3N_4 at high temperatures; and antireflective coatings in solar devices and electro-optical structures.

3. *Polysilicon.* Polysilicon (polycrystalline) is used without doping. When doped with oxygen, it is used as a conductive layer for multi-level interconnect, a field electrode for high breakdown voltage chips, and as silicon gates on MOS devices.

Molecular beam epitaxy

Molecular beam epitaxy is an ultrahigh vacuum evaporation process used to grow extremely thin films with highly controlled dopant profiles. The environment for MBE must be ultraclean to prevent unwanted defects in film surfaces and ensure elimination of any dopant contamination. The basic technique involves generating an atomic or molecular "beam" of the elements to be used in the film. In an ultrahigh vacuum environment, heated crucibles are used to thermally evaporate the material. As the evaporated material leaves the crucible, it impinges onto the heated substrate to make an epitaxial film. The actual beam shape is created by placing an aperture between the crucible and the substrate to "focus" the stream of material. A series of crucibles may be used, and control of the deposited film structure, composition, and thickness is regulated by varying, among other things, the crucible temperatures.

Substrates in the growth chamber should be placed face down to prevent the picking up of flakes or other contamination from the vacuum system walls. The substrate should also be rotated during film growth to increase uniformity. One important aspect of MBE is the ability to monitor and characterize the film surface and growth environment during the process. The need for control in epitaxy and all IC substrate and film material increases along with the critical specifications for the IC. Advanced IC processes will no doubt make good use of molecular beam epitaxy. See Fig. 1.18.

Photochemical vapor deposition

The use of ultraviolet light as the energy source to decompose reactant gases is a proven CVD process. Ultraviolet photons excite the gases and thereby increase their chemical reactivity. The reaction temperatures are low, and the wafers are heated slightly (50 to 200°C) to promote adhesion. The uv photons do not reach gas ionization energy, and thus no charged particles which could degrade the device are created. A system for production photochemical CVD is shown in Fig. 1.26.

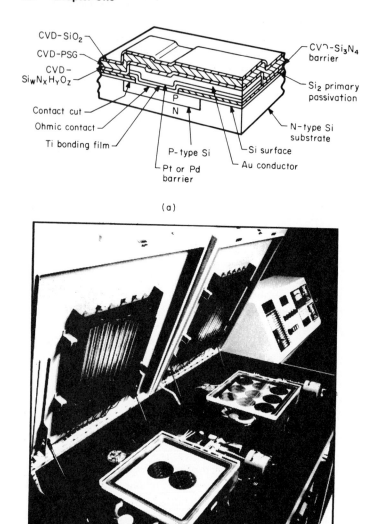

CVD-SiO₂
CVD-PSG
CVD-Si_WN_XH_YO_Z
Contact cut
Ohmic contact
Ti bonding film
P-type Si
Pt or Pd barrier
P
N
CV⌐-Si₃N₄ barrier
Si₂ primary passivation
N-type Si substrate
Si surface
Au conductor

(a)

(b)

Figure 1.26 Various layers of CVD glasses used in an IC and photochemical deposition reactor. (*Courtesy of Tylan Corporation.*)

The low-temperature photochemical vapor deposition (LTPCVD) reactor produces excellent step coverage with low defect density, reduces thermally induced mechanical stress, and avoids undesirable charged-particle damage. Applications for this technology are finding their way into optoelectronic devices made from silicon, indium antimonide, mercury cadmium telluride, and gallium aluminum arsenide. The

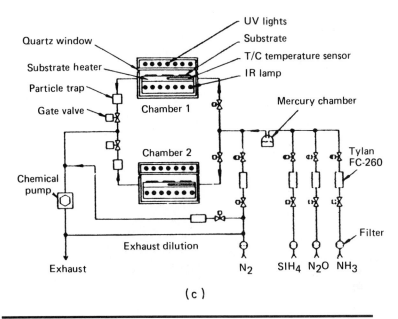

Figure 1.26   (*Continued*)

conformal nature of LTPCVD films is amply demonstrated in Fig. 1.27.

A variation of CVD technology is low-pressure CVD. The lower-pressure processes offer several advantages over conventional CVD systems, the most important being better uniformity of the deposited materials on etched steps. The major reaction mechanisms are as follows:

Silicon dioxide: $SiH_2Cl_2 + 2N_2O \rightarrow SiO_2 + 2N_2 + 2HCl$

Silicon nitride: $3SiH_2Cl_2 + 10NH_3 \rightarrow Si_3N_4 + 6NH_4Cl + 6H_2$

Polysilicon: $SiH_4 \rightarrow Si \ (poly) + 2H_2$

Figure 1.28 shows the LPCVD chamber scheme.

One area in the application of dielectrics is plasma deposition (PD); it involves the use of gas reactants that are converted into very reactive chemical species with the help of an rf glow discharge and result in vapor deposition of the desired material. Thus the plasma deposition process is a variation of the CVD process in which a gas plasma replaces the heat-induced decomposition reactions.

The most significant difference lies in the lower temperature of the PD process (200 to 400°C), which provides better wafer throughput and less physical distortion of the wafer. The plasma reactions in PD

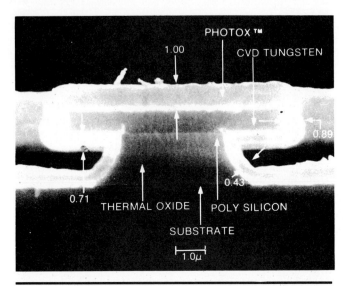

Figure 1.27 SEM cross section of photochemical vapor deposition of silicon dioxide film conforming to an existing structure. (*Courtesy of Tylan Corporation.*)

processes occur at approximately 5 torr of pressure. Even more significant to IC fabrication is the PD advantage of providing conformal coatings of the deposited materials over the previously etched steps, which is made possible by the low operating pressure.

In all CVD deposition processes, whether CVD, LPCVD, or PD, the biggest problems are keeping contamination out of the reaction cham-

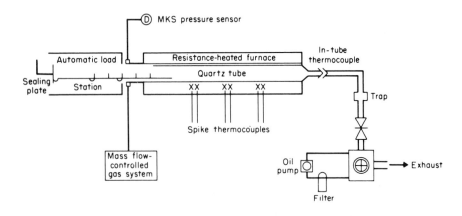

Figure 1.28 Low-pressure chemical vapor deposition chamber scheme.[3]

ber, providing good deposition rates, keeping temperatures low, and operating consistently.

Etching and Masking Processes

Undoped silicon dioxide etching

The most common etching application in IC fabrication is etching undoped SiO_2. The first etching step in making ICs is removal of select areas of the initial undoped oxide from the wafer. This first step involves imaging the oxide with a resist material to protect the areas on which the oxide will not be etched. Since most resists bond readily to undoped silicon dioxide and the etchant is not corrosive to the resists commonly used, the etching application is relatively easy to perform and simple to control. Figure 1.29 illustrates the silicon dioxide etching sequence.

The thickness of undoped silicon dioxide generally etched is about 1 μm, and with buffered hydrofluoric acid, etch time is about 10 min (1000 Å/min etch rate). The usual resist thickness is about 1 μm, and both positive and negative resists can be used. Positive resists are advantageous for this application because thicker coatings can be used

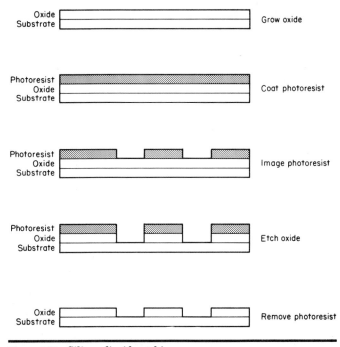

Figure 1.29 Silicon dioxide etching sequence.

than with negative resists, which provides superior pinhole protection. This is especially important because the first oxide cut sets the stage for future etch operations. Defects at this stage often become more serious to the device parameters as each subsequent etching and doping step is performed.

Doped silicon dioxide etching

Etching doped silicon dioxide is more difficult than etching undoped oxide mainly because resist adhesion is more difficult to obtain and etch rates are greater, making etch control more critical. Phosphorus-doped silicon dioxide is the most common application in this category; openings are made in the doped layer to provide for aluminum ohmic contacts. After etching the phosphorus-doped contact holes, aluminum is evaporated over the wafer to make electrical contact with the silicon.

Negative and positive resists are used to mask or image doped oxides, and the etchant species used is the same as that used for undoped silicon dioxide. The use of resist adhesion promoters is recommended for improved resist adhesion, particularly when positive resists are used.

The main functions of silicon dioxide throughout the IC process are as masks for the diffusion of dopants and as dielectric insulation layers.

Polysilicon etching

Polycrystalline silicon etching is relatively more difficult than doped oxide etching because the surfaces do not bond resists well (fewer sites for chemical bonding) and the etchant is much more corrosive to resists than are doped and undoped oxides.

Polycrystalline silicon is used mainly as a dielectric gate material. The use of plasma processes is more common for polysilicon etching because the resists will withstand plasma environments more easily than they will liquid etchants. Since the gate of the device controls the flow of electrons and is vital to proper device operation, gate dimensions are critical. Etching is therefore a dimensionally critical operation, and extreme control is required. Aside from plasma etching, in which a positive resist is preferred for better geometry control and resolution, wet etching is used with the resist as a mask. A common technique used to avoid this difficulty is to deposit a layer of silicon dioxide over the polysilicon, image and etch the SiO_2 (a simple operation), and then use the etched SiO_2 pattern as the etch mask for the polysilicon. The polysilicon etchant does not etch the SiO_2. Figure 1.30 illustrates this approach.

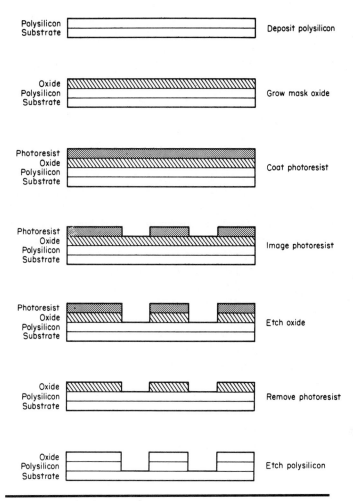

Figure 1.30 Polysilicon masking.

Silicon nitride etching

The etching requirements of silicon nitride are similar to those of polysilicon. As with polysilicon, there are three approaches to etching nitride layers: (1) direct imaging with resist and wet etching, (2) direct imaging with resist and dry etching, and (3) indirect imaging by depositing an oxide layer over the nitride, imaging and etching the oxide, removing the resist, and then etching the nitride with the SiO_2 as an etch mask.

Silicon nitride (Si_3N_4) is used primarily as a dielectric layer in IC fabrication. A common application is to use it as the final passivation layer of the IC, applying it over the aluminum layer. In this application, openings in the deposited nitride must be made to provide areas for bonding pad contact. Thus the resolution requirements are not as critical as they are with polysilicon gates. The direct imaging with resist is possible if resist hardbaking or deep uv curing is used, as well as oxygen plasma resist removal. Plasma etching and resist removal are the preferred approach for more advanced devices because they make the process simpler and easier to control.

Aluminum etching

All IC fabrication processes contain an aluminum or silicide metallization etching step, since metallization serves as the interconnection layer for all the devices on the chip. Aluminum etching is a relatively well understood and controllable etching step, yet it has difficulties different from those previously mentioned for nitride and polysilicon etching.

Aluminum can be imaged with positive and negative resists, but positive resists are greatly preferred because they are less sensitive to light reflection and have fewer negative side effects in image dimensional control. Negative resists often cause "metal bridging," in which light reflections cause the resist to cross-link over areas where etching should take place. This cross-linking causes a resist layer and a subsequent metal bridge (short) between the aluminum lines after etching. Positive resists do not suffer from this reflection problem and are therefore well suited to this application. Figure 1.31 shows an aluminum etching sequence.

Aluminum etching is accomplished with both wet and dry etchants, and both approaches are capable of meeting IC fabrication requirements. Wet aluminum etching is more mature and better understood as a process, but plasma aluminum etching is also a production-proven technology.

One of the key objectives of aluminum etching is to provide uniformly etched conductor lines across a variety of previously etched steps or wafer topography. This requires close control of etching parameters to avoid excessive undercutting and etched line variation.

Metal lift-off

Another method for creating aluminum patterns is photoresist metallization masking, or metal "lift-off," a technique seldom employed but useful for extremely fine geometries that are very closely spaced. Metal lift-off has been used for a variety of special applications in which undercutting is simply not tolerable and straight metal

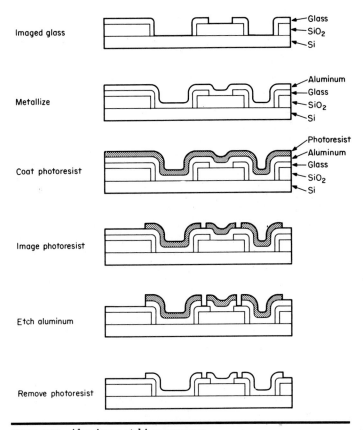

Figure 1.31 Aluminum etching sequence.

sidewalls, perpendicular to the oxide surface, are required. Positive photoresist is almost always used for lift-off techniques because a thick coating can be applied without resolution loss. Thick resist coatings, up to 2 μm, will easily contain the amount (masks/reticles) of aluminum or, in some cases, chromium to be deposited. This process begins by cleaning the wafer or glass surface to provide sufficient resist adhesion for the metallization but not enough to prevent the resist from being easily removed in the lift-off step. After cleaning, the parts are placed in an evaporator and the entire structure is coated with metal.

After metallization, the wafer or glass substrate is immersed in warm acetone or a solvent that will swell and remove the resist and break the metal film cleanly at the bottom corners of the substrate where resist and metal meet. Figure 1.32 shows a lift-off sequence.

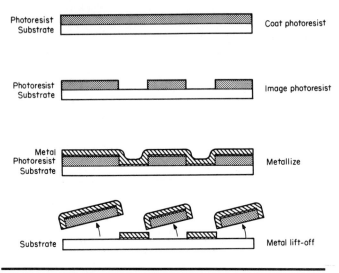

Figure 1.32   Lift-off sequence.

The disadvantages of this process include both the difficulty of obtaining the same metal pattern consistently and the complexity of automating the process.

Recent work on lift-off processes has served to improve the reliability and control problems associated with this technique. A variety of processes are used to obtain negative sidewall profiles in resist coatings, which in turn permit metal discontinuity in evaporation. Lift-off is a specialized technique that has limited but occasional use in production. Multilayer techniques have largely replaced lift-off processing.

Polyimide etching

The use of polyimide layers is important because polyimide provides excellent dielectric insulation not only between metal layers (aluminum metallization) but also as a final passivation coating over aluminum metallization.

Polyimide layers are etched in aqueous solvent solutions or are reactive-ion-etched by using very thick coatings of resist as a sacrificial mask layer. "Sacrificial" etching involves the application of a resist coating (usually always positive) that is thicker than the layer to be etched. After the resist is imaged, the wafer is placed in the reactive ion etch chamber and both polyimide and resist are removed or etched at the same approximate rate. Since the resist is thicker than the underlying polyimide, the polyimide layer is etched through first and the etching is terminated, leaving a thin layer of resist that was *not* sacrificed in the etching to protect the polyimide layer needed for insulation or protection of the IC.

This type of process will probably evolve into a direct imaging of polyimide, where a photosensitive polyimide will eliminate the need for sacrificial etching.

Another application of polyimide is in "planarizing," or using the polyimide to fill in between existing topography and leave a planar surface behind. This simplifies the imaging process. Often the polyimide can be left on as a dielectric as it is in pad masking, shown in Fig. 1.33.

Resist implant masking

Positive optical resists are widely used for masking in low and medium ion implantation dosages. This direct use of a photoimagable layer eliminates the extra steps of depositing, imaging, and etching a layer of silicon dioxide to serve as the ion implant mask. In the high-dose ion implantation applications, this extra operation is necessary because the high energy from the ion beam will completely distort the resist images and cross-link, or react, them into practically insoluble species. This application will evolve in a manner similar to that of the polyimide masking application, in which a technological break-through in the imaging area will shorten and simplify the process. In ion implantation masking, the need is to provide a resist imaging material that will withstand the energy and heat of higher-dose implants and still allow complete removal after implantation.

The key factor in using resists (instead of a resist-masked oxide) for ion implantation masking is providing the ion stopping power. That is

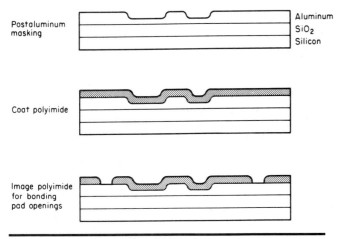

Figure 1.33 Polyimide pad masking.

why a thick positive-resist layer is ideal, and special process techniques have been published to show how this is performed in production. Figure 1.34 shows a resist implant mask sequence.

Photomask etching

IC fabrication requires the production of high-quality photomasks. The photomask etching application is diversified into several different etch materials and their respective etches. It does not include silver halide emulsion masks, which are simply exposed and developed with no etching step. Hard-surface masks are primarily glass-based masks with a coating of some metal or oxide. Chromium, bright and antireflective, is the predominant hard-surface photomask coating material; it provides high resolution, durability, and relative ease of fabrication. Positive resist is used to image and mask the wet or dry etching of the chromium. Chromium can be shiny, black, or antireflective. Since a thin layer is all that is needed to provide image exposure masking and the etchants are very compatible with positive resist, chromium mask etching results in very high quality masks for advanced VLSI and conventional IC fabrication.

Other coatings used on glass to provide hard-surface photomasks include iron oxide, silicon dioxide, and other oxides of metals that can be

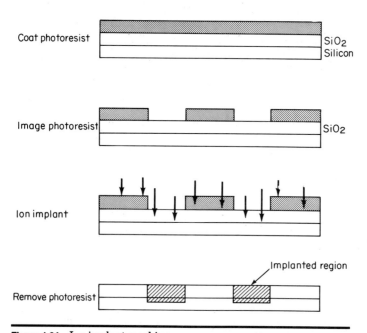

Figure 1.34 Ion implant masking.

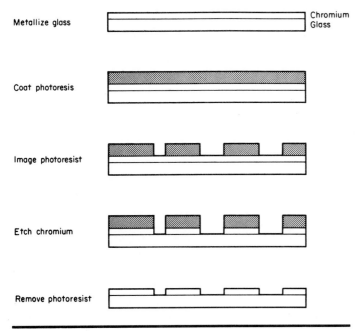

Figure 1.35 Mask etching sequence.

etched with good resolution. Figure 1.35 shows a mask etching sequence.

The use of electron beam resists to image and etch hard-surface photomasks has provided the means for rapid turnaround time on new masks for new device designs. In the future, better e-beam resists will provide higher-resolution capabilities for etched masks. Currently, chromium, with a positive resist as an etch mask, meets the most exacting needs of IC designs for nearly all applications.

Multilevel resist imaging

Multilevel resist imaging has become an accepted optical lithography method for providing increased resolution and greater IC density. There are several ways to utilize two and three separate resist layers directly on top of one another (or separated by a thin oxide or metal layer) for the desired result.

Bipolar devices

The earliest of integrated circuit types, bipolar circuits, still account for a large percentage of all ICs, offering major advantages in speed and power. The basic structure of a bipolar device is shown in Fig. 1.36. The

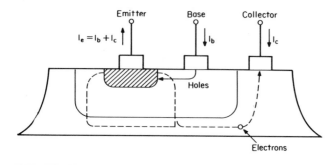

Figure 1.36 Bipolar device structure.

bipole action that gives rise to the device's name is explained by the base and collector current, which, when combined, equal emitter current. Since the current flows in opposite directions across the junction of *p* and *n* regions, with electrons of two types, it is called "bipolar." Bipolar ICs are generally classified as epitaxial or nonepitaxial, depending on the processing used. A typical bipolar circuit requires seven masking steps to fabricate.

MOS devices

The basic patents for metal oxide semiconductor (MOS) ICs were filed in the early 1960s, and this general category of ICs has grown rapidly since then and has gained on the bipolar market share. The structure of a typical MOS transistor is shown in Fig. 1.37.

MOS circuits are basically a unipolar, or field effect, type; they are so called because all the current-carrying electrons are of the same charge. The first field effect transistors were developed into various types, all of them used the same basic unipolar action for electron movement. They include MOSFETs (MOS field effect transistors); silicon gate, metal gate, and junction transistors; and CMOS.

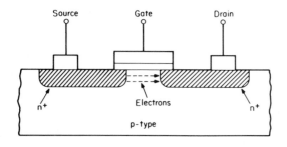

Figure 1.37 MOS device structure.

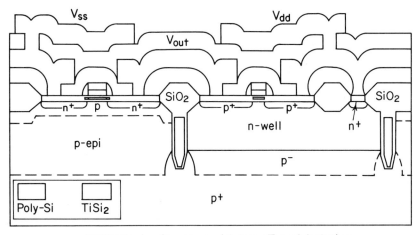

CMOS Device Structure with Deep Trench Isolation

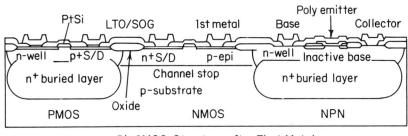

Bi-CMOS Structure after First Metal

Figure 1.38   High-density devices, a cross-sectional view.

One major advantage of MOS devices is that they typically require fewer masking steps than do bipolar types. Since the masking and subsequent etching operations are the major yield determinants, this is a critical difference and is undoubtedly a factor in the rapid growth of MOS devices. Figure 1.38 shows typical CMOS and other high-density device cross sections.

REFERENCES

1. *Basic Technology Course Manual,* Integrated Circuit Engineering, Phoenix, Arizona, 1978.
2. D. J. Leventhal (ed.), "Diffusion System Trends," *Semiconductor Int.,* June 1979, p. 38.
3. W. A. Brown and T. I. Kamins, "An Analysis of LPCVD System Parameters for

Polysilicon, Silicon and Silicon Dioxide Deposition," *Solid State Tech.*, July 1979, p. 51.
4. "Status 80," market report on the integrated circuit industry, Integrated Circuit Engineering, Phoenix, Arizona, 1980.
5. W. C. Till and J. T. Luxon, *Integrated Circuits: Materials, Devices and Fabrication*, Prentice-Hall, Englewood Cliffs, N.J., 1980.
6. R. R. Razouk, L. N. Lie, and B. E. Deal, "Kinetics of High Pressure Oxidation of Silicon in Pyrogenic Steam," *J. Electrochem. Soc.*, vol. 128, 1981, p. 2214.
7. H. F. Wolfe, *Semiconductor Silicon Data*, Pergamon Press, New York, 1969, p. 601.
8. W. A. Pliskin and E. E. Conrad, "Nondestructive Determination of Thickness and Refractive Index of Transparent Films," *IBM J. Res. and Dev.*, vol. 8, 1964, p. 43.
9. F. C. Eversteyn, "Chemical Reaction Engineering in the Semiconductor Industry," *Philips Res. Rept.*, vol. 19, 1974, p. 45.
10. "All Ion Implantation Process for Production of Integrated Circuits," editorial, *Insulation Circuits*, January 1980, p. 39.
11. A. C. Adams, "Electric and Polysilicon Film Deposition," in S. M. Sze (ed.), *VLSI Technology*, McGraw-Hill, New York, 1983, chap. 3.
12. K. Schuegraf, "Low-Temperature Photochemical Vapor Deposition of Silicon Dioxide and Silicon Nitride," *Microelectronic Mfg. and Test.*, March 1983.
13. "Status 1988," Report on the Integrated Circuit Market and Technology, Integrated Circuit Engineering Corporation, Scottsdale, Ariz., 1988.

Chapter

2

The Fabrication Process

Advanced VLSI Process Flow

The advanced VLSI process shown here is a retrograde P-well process with separate optimization of both NMOS and PMOS devices. It can also be characterized as a twin-tub process, since NMOS and PMOS are fabricated in their own tubs, which are made after field oxidation with consecutive implants by using complementary masking, as described in the process flow. Cross-sectional views during different phases of fabrication are given in Figs. 2.1 to 2.15, and the major process parameters are given in Table 2.1.

The process architecture was optimized for high-density, high-performance, and latchup-free operation. These were the driving forces behind the choice of a twin-tub technology with $n/n+$ epitaxial

TABLE 2.1 Major Process Parameters

Substrate resistivity (epitaxial layer)	1 Ω/cm, N
P-well depth	1.4 μm
Field oxide	6500 Å
Gate oxide	250 Å
Polysilicon	4000 Å
Interlevel oxide 1	4000 Å
$n+$ junction depth	0.25 μm
$p+$ junction depth	0.40 μm
Molybdenum/TiW	5800 Å
Capacitor dielectric (Si_3N_4)	1000 Å
Molybdenum (top capacitor plate)	2000 Å
Interlevel oxide 2 (planarized)*	0.4 to 1.4 μm
Alusil	8000 Å
Passivation oxide	5000 Å

*Variable thickness due to planarization

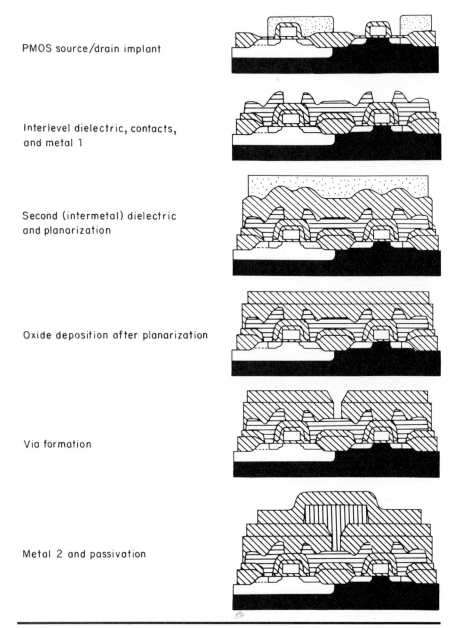

PMOS source/drain implant

Interlevel dielectric, contacts, and metal 1

Second (intermetal) dielectric and planarization

Oxide deposition after planarization

Via formation

Metal 2 and passivation

Figure 2.1   Overview of advanced CMOS process.

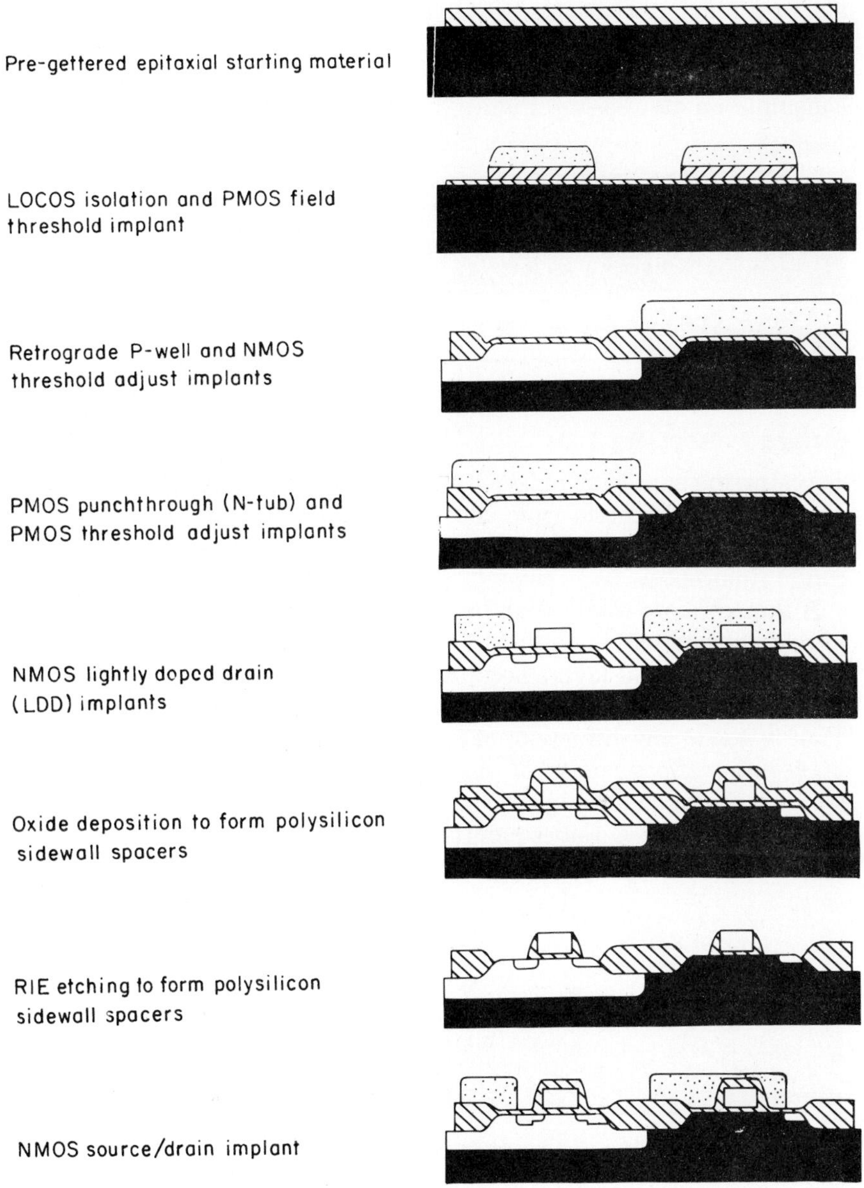

Pre-gettered epitaxial starting material

LOCOS isolation and PMOS field threshold implant

Retrograde P-well and NMOS threshold adjust implants

PMOS punchthrough (N-tub) and PMOS threshold adjust implants

NMOS lightly doped drain (LDD) implants

Oxide deposition to form polysilicon sidewall spacers

RIE etching to form polysilicon sidewall spacers

NMOS source/drain implant

Figure 2.1 (*Continued*)

substrate and retrograde P-well. The twin-tub technology is characterized by separate optimization of both NMOS and PMOS devices, which is an essential requisite for high performance and low power consumption at 1.25 μm. Figure 2.1 shows the process overview.

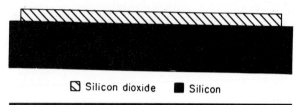

☒ Silicon dioxide   ■ Silicon

Figure 2.2

Starting material (Fig. 2.2)

The starting material consists of $n/n+$ epitaxial wafers. The wafers are purchased directly from the silicon manufacturer, each with its own pregettering treatment for achieving a low and uniform level of contamination.

Although more costly than bulk, epitaxial material offers excellent protection against latch-up. In addition, $n/n+$ substrates allow for shallower epitaxial layers because the $n/n+$ interface is more stationary during high-temperature processing than its $p/p+$ counterpart. That is due to low diffusivity of antimony compared to boron. This improves the effectiveness of the epitaxial material for eliminating latch-up, because the highly conductive substrate can be set closer to the active devices. Consequently, lateral voltage gradients can be minimized and thereby remove the source of latch-up.

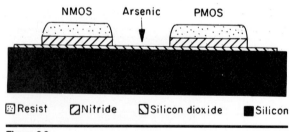

☒ Resist   ⧄ Nitride   ☒ Silicon dioxide   ■ Silicon

Figure 2.3

Isolation and field threshold implant (Fig. 2.3)

The local oxidation of silicon (LOCOS) technique is used to form the active areas. This includes the initial growth of a stress relief oxide, followed by low-pressure chemical vapor deposition of nitride. Then the active area mask is applied for selectively defining the field oxide regions and etching the overlying nitride.

While the active area resist is still in place, the PMOS field threshold is enhanced with a light-dose arsenic implant. Since the P-well retrograde implant has a boron dose about 20 × larger, its concentration overrides, by a large margin, this arsenic implant in the NMOS field region. Hence, its effect on the NMOS field threshold is negligible, although it offers the advantage of saving a PMOS field masking step.

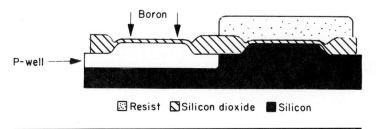

Figure 2.4

Threshold adjust implants (Fig. 2.4)

A retrograde well is formed by high-energy ion implantation after the field oxidation has taken place. The shape of the implant impurity profile is maintained in the following process sequence by avoiding high-temperature steps. Since the peak of the profile is below the surface, the well is called "retrograde" to emphasize the negative slope of the concentration profile from the surface downward. This characteristic is responsible for minimizing the gain of the vertical parasitic bipolar transistor, which is part of the latch-up structure, because of an increase of doping near the bottom of the well. Equally important for reducing latch-up susceptibility is the reduced well sheet resistance provided by the high peak concentration of the retrograde well.

The retrograde well is also beneficial in terms of device packing for density. Since the traditional high-temperature well diffusion is not required, the attendant lateral diffusion also is eliminated. Moreover, in the case of the retrograde P-well, the NMOS field threshold is high (about 20 V), since forming the well after field oxidation prevents boron segregation in the field oxide.

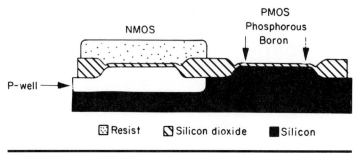

Figure 2.5

N-tub formation and implants (Fig. 2.5)

The P-well dark-field mask is applied to form the retrograde well and to adjust the NMOS threshold voltage by using high and low boron energy implants, respectively. Then the complementary P-well clear-field mask is used to form the PMOS tub and to adjust the PMOS threshold voltage. The deep implant is done with phosphorus and the shallow implant with boron. Since the N-tub is of the same polarity as the substrate, but with higher doping, its major purpose is to eliminate source-drain punchthrough. Hence, the deep implant is specifically called the "punchthrough control implant" in the process flow.

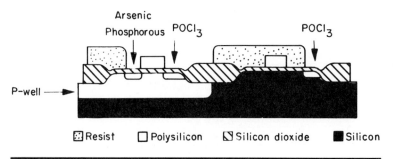

Figure 2.6

Oxide deposition for sidewall spacers (Fig. 2.6)

After the screen oxide is stripped, the gate oxide (250 Å thick) is formed by using TCA as an additive. Next, a polysilicon layer (4000 Å thick) is deposited and later doped with $POCl_3$ to achieve a sheet resistance of about 35 ohms/square.

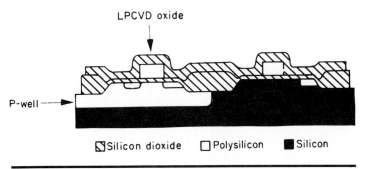

LPCVD oxide

P-well

⊠ Silicon dioxide   ☐ Polysilicon   ■ Silicon

Figure 2.7

RIE etch for sidewall spacers (Fig. 2.7)

After the implants, sidewall oxide "spacers" are formed on the edges of the polysilicon gate by conformal deposition of a CVD oxide layer, followed by reactive ion etching. The purpose of the spacers is to shift outward from the channel edges the source-drain implants in order to reduce the lateral doping gradient of the NMOS junctions by creating lightly doped drain (LDD) extensions. The net effect is a reduction of electric field near the NMOS drain and improvement of the hot electron tolerance for 5-V operation because of reduced impact ionization near the drain.

The PMOS devices also benefit from the spacers because the lateral diffusion of boron from the source-drain regions is only partially overlapped by the gate with an attendant reduction of Müller capacitance and improved speed characteristics.

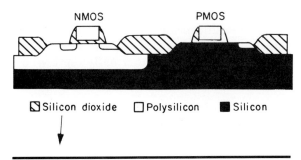

NMOS   PMOS

⊠ Silicon dioxide   ☐ Polysilicon   ■ Silicon

Figure 2.8

NMOS source/drain implants (Fig. 2.8)

In advanced VLSI devices the LDD structures process sequence consists of two consecutive light-dose implants of arsenic and phosphorus in the $n+$ select regions. The energy is as high as possible, within the constraint of avoiding penetration of these ions through the polysilicon gate into the underlying channel. The choice of the buried-graded structure was made on the basis of 2-D device modeling and experimental verification with accelerated hot-electron lifetime testing.

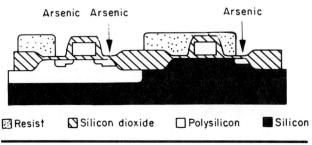

Arsenic   Arsenic   Arsenic

▦ Resist   ◪ Silicon dioxide   ☐ Polysilicon   ■ Silicon

Figure 2.9

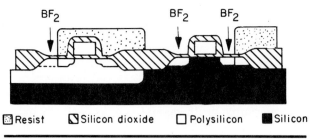

BF_2 BF_2 BF_2

▦ Resist   ◪ Silicon dioxide   ☐ Polysilicon   ■ Silicon

Figure 2.10

PMOS source/drain implants (Figs. 2.9 and 2.10)

Source-drain implants are done with arsenic for $n+$ and BF_2 for $p+$ regions. Both implants are implemented with low energy and high dose through 250 Å of screen oxide, which was thermally regrown on the bare source-drain areas after spacer formation.

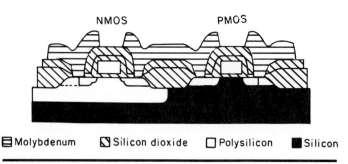

NMOS   PMOS

▤ Molybdenum   ◪ Silicon dioxide   ☐ Polysilicon   ■ Silicon

Figure 2.11

Interlevel dielectric, contacts, and metallization (Fig. 2.11)

The interlevel dielectric is deposited at high temperature to eliminate the need for a separate anneal of the source-drain implants. Contacts are RIE-etched down to silicon. Then the resist is removed and the wafers are RIE-etched again for a short time to clean the bottoms of the contacts. This technique has increased the yield significantly, as measured on contact strings of 2000 and 5000 contacts, compared to a previously used dry or wet etch.

The first metallization is based on molybdenum over a thin layer of TiW. This metallization is resistant to electromigration and makes excellent ohmic contacts to $n+$ silicon. The sheet resistance is about 0.17 Ω/square, which is about two times higher than aluminum-silicon alloy layers of equal thickness. However, it brings the significant advantages of avoiding hillock formation, resisting electromigration, and eliminating spikes through shallow $n+/p$ junctions.

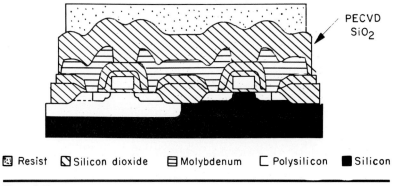

PECVD
SiO$_2$

▦ Resist ◨ Silicon dioxide ⊟ Molybdenum ⊏ Polysilicon ■ Silicon

Figure 2.12

Interlevel dielectric II and planarization (Fig. 2.12)

The second dielectric is a thick oxide layer. The surface relief is minimized with a planarization step, which consists of spinning a resist layer over the deposited oxide and then reflowing it with a postbake. The surface tension of the resist promotes the formation of a smooth surface, which can be replicated into the oxide with an RIE etch-back adjusted for equal resist and oxide etch rates. The end point is characterized by the complete resist removal plus a safety margin.

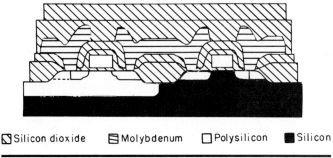

Silicon dioxide   Molybdenum   Polysilicon   ■ Silicon

Figure 2.13

Oxide deposition (Fig. 2.13)

After planarization, another oxide layer is deposited to ensure complete dielectric integrity of the intermetal insulator and to guarantee its minimum thickness.

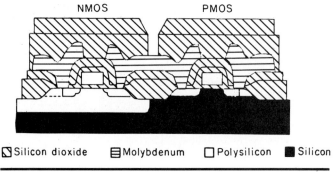

Silicon dioxide   Molybdenum   Polysilicon   ■ Silicon

Figure 2.14

Via formation (Fig. 2.14)

Vias are RIE-etched by utilizing a semitapered process to improve the step coverage of metal 2. This method is based on (1) patterning the resist to provide the desired via sloped-edge profile at the resist edges and (2) a two-step etching process consisting of an anisotropic RIE etch followed by an isotropic tapering etch.

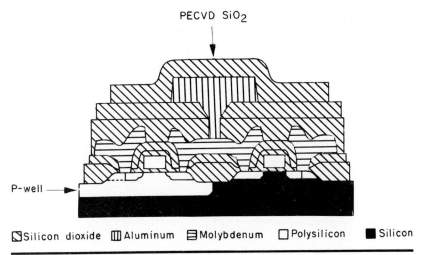

PECVD SiO$_2$

P-well ⟶

⬜Silicon dioxide ⬛Aluminum ⬛Molybdenum ☐Polysilicon ⬛Silicon

Figure 2.15

Metal II and passivation (Fig. 2.15)

Vias step coverage is eased by the tapered sidewalls and by restricting the placement of the vias to metal 1 over field oxide. Metal 2 is RIE-etched.

The passivation sequence is based on low-temperature-deposited oxide to provide mechanical protection during handling and packaging and to seal the devices against contamination.

The chip shown in Fig. 2.16 was made by using the advanced VLSI process described in the preceding sections and illustrations.

Electrical Test and Probe, Die Separation, and Bond

Wafers are now tested against electrical specifications in automatic wafer probe systems in which dies that do not pass are identified, often by having a small dot of magnetic ink placed on them. This will key a magnetic die remover further down the line from the probe.

Automatic wafer scribing, by either diamond or laser, prepares the wafers for separation into individual dies. After separation, during which some dies may not break properly and are removed, each die is individually bonded to the support package. A wide variety of bonding

Figure 2.16   A 70K transistor advanced VLSI chip fabricated in a 1.25-μm CMOS technology with a 6.2- by 7.6-mm die size and featuring data synchronization, pipeline latency compression used in systolic array signal processing. (*Courtesy of Kenneth J. Polasko, General Electric Company, Corporate Research and Development Center, Schenectady, N.Y.*)

techniques are used, all of which must provide good ohmic contact between the die and the portion of the lead frame it contacts. Flexible lead frames (copper or mylar) have permitted a lower-cost form of support package, and die bonding on 35-mm-tape lead frames achieves a higher level of automation than earlier Kovar lead frame packages, since they provide for both the die and pad electrical contacts. Prior to this type of lead frame, individual wires were and still are bonded to each pad area by using ultrasonics. The problem of wire breakage and poor connections is ever-present, since the wires are only 2 to 3 mils thick and can withstand only a 4- to 6-g pull before separating. Another technique, in which all pads are simultaneously connected to the lead frame, is "spider bonding." In spider bonding, a multilegged package makes solderable contact with each of the die's pads in one step.

Encapsulation follows die and pad bonding, in which a plastic is melted over the completed circuit. It is then hermetically sealed in a vacuum. The last step is the final test and identification. Here each circuit must pass an electrical test and then be numbered serially and identified according to company, in case there is a problem later on.

REFERENCES

1. Ken Polasko, General Electric Research and Development Center, Schenectady, N.Y., private communication and text writing, 1988.

Chapter

3

Resist Classification

Resists used to fabricate VLSI devices are highly customized formulations of polymers, usually solvent-based. Organic polymers are used throughout the IC process as etch masks, and occasionally as metallization masks, in a process called "lift-off." Resist materials selected for lithography must meet a large number of physical and functional requirements before being put into a production line. A resist is used for each mask pattern that is etched into the wafer surface. Figure 3.1

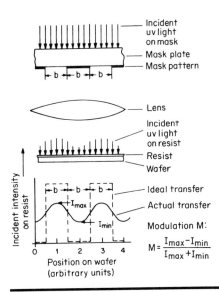

Figure 3.1 Parameters of resist exposure in projection optical lithography.[1]

shows the basic aerial image transfer mechanisms involved in resist imaging. After a resist layer is imaged, the pattern is transferred by etching a layer below the resist. Finally the patterned area is doped with ions that change the conductivity or charge-carrying properties of the silicon. Resists are used as many as 15 to 20 times in these lithography operations, but a typical process has only 7 to 9 resist-imaging steps. A standard resist-imaging and etch sequence is shown in Fig. 1.29.

Resist Requirements for VLSI/ULSI Chemical and Coating Properties

Resist formulations are mixtures of solid and liquid components. The solids content is made up of one or more resin materials and a sensitizer in multicomponent resists. The solid components must be completely dissolved in a carrier solvent, usually an organic solvent mixture. Resists are applied by spin coating, so a key property is the ability to cast highly uniform (± 50-Å) films from these solutions. The resist solutions must be stable over temperature ranges typically encountered in shipping at various times of the year. This means up to 110°F for several hours on a shipping dock in July to several hours at -10°C in the back of a commercial carrier.

The resist solution must also be highly filtered down to 0.1 to 0.2 μm with an absolute membrane filter. Partially dissolved resin and any solid contaminants that could interfere with image formation must be removed. The solvent system used must be optimized to provide extremely smooth coatings and permit uniform film formation after a desolvation bake (softbake). The solvents used cannot be harmful to operators using the resist under normal process conditions. Some organic solvents have been correlated with operator health problems and are no longer permitted in many facilities.

Adhesion properties

Resists are used on many different types of VLSI surfaces, and they must have sufficient adhesion to those surfaces to withstand spray developing and wet- or dry-etching environments. Adhesion is measured as a function of undercutting during etching. Adhesion includes bonding to the following surfaces:

Typical metals and conductors	Typical semiconductors
Aluminum and its alloys (Si, Cu)	Gallium arsenide
Refractory metals and their silicides	Silicon (poly and single-crystal)
Chromium	Silicon dioxide (doped/undoped)
Titanium	Silicon nitride
Copper	Metal oxides
Gold	
Tantalum	

Resolution

Resolution is the overriding parameter in resist material selection, one that defines the length of time a resist can be used for a given IC design rule set. Resolution is measured as a function of the smallest opening or island structure that can be made under a given set of process conditions and as a function of resist contrast. Contrast is described in detail in Chap. 8, but basically it is a measure of the resist response to an aerial image that has an intensity gradient defined by the optical modulation transfer function of the optical imaging system. Contrast plots for a negative and positive resist are shown in Fig. 3.2.

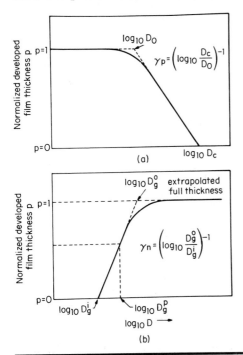

Figure 3.2 Contrast curves for (a) positive and (b) negative photoresists.[2]

In VLSI lithography, it is desirable to have high contrast in a resist material to compensate for light scattering that occurs as an image is projected through an optical system and onto the resist surface. Contrast values of 3 to 5 are needed with resist materials for ULSI imaging, in which 0.5-μm and smaller features are needed to make devices. The MTF values for a printing system for this level of resolution should be 0.5 to 0.6. Final resolution is then a product of the MTF of the optical system, the contrast of the resist material (which may compensate for MTF values below 0.6), and subsequent processing. Resolution is partly controlled by the developing process, in which subtle changes in developer temperature, concentration, or spray pressure will change the resist sidewall angle and final image resolution. New approaches to increasing contrast in lithography include imaging only the top 2000 to 3000 Å of a material and transferring this surface layer image via dry plasma or reactive ion etching.

Sensitivity

Resist material sensitivity is the relative response of the polymer film to the exposing radiation, and it is measured in many different ways. A functional definition used frequently on a production line is related to the amount of exposure time or dose (energy × time, usually integrated automatically) necessary to reproduce exactly the mask geometries on the wafer. This is termed "optimum exposure," and it is a good practical test of sensitivity. A variety of resists are usually compared on this basis.

The actual energy used by the resist to effect a chemical reaction is more specific to real sensitivity. Exposure dose is a function of incident exposure multiplied by the resist absorption efficiency. Resists are formulated to match, as closely as possible, the wavelength of the exposing radiation. Sensitivity is critical to obtaining short exposure times and thus good wafer throughput to meet production needs. The molecular parameters of the resist are optimized for maximum absorbance at the exposing wavelength.

Sensitivity is more scientifically defined as a function of the resist photoefficiency, which is measured quantitatively by the formula:

$$\text{Quantum efficiency} = \frac{\text{number of reacted sensitizer molecules}}{\text{number of absorbed photons}}$$

The sensitizer molecules reacted are measured in standard monochromer studies in which a fixed volume of resist is exposed to a single bandwidth of energy (g-, h-, or i-line, excimer wavelength, e-beam, ion beam or x-ray).

The quantum efficiency of positive-working novolak photoresist (two-component systems) is 0.2 to 0.3, whereas that of negative photoresists is much higher, typically 0.4 to 0.8.

Single-component resists, like poly(methyl methacrylate) (PMMA) and polymethyl isobutyl ketone (PMIPK), are used in deep-uv lithography, and they have relatively low quantum efficiencies (0.3 for PMIPK at 313 nm, for example) as photoresists. Single-component resists are polymers wherein a single polymer unit or chemical species embodies the sensitivity parameters, etch resistance, coating property, and all other functional behavior. This leaves little flexibility, except in molecular weight, viscosity variation, and dye addition, to modify the behavior of the product used on the wafer.

Two- or multicomponent resists (three solvents, one or more resins, one or more photoactive compounds) use the resin to determine the energy absorption, etch resistance, coating behavior, and other physical properties and the sensitizer or photoactive compound (PAC) to determine the photochemical conversion behavior. Most positive and negative resists used in VLSI are two- or multicomponent formulations.

Etch and process resistance

Resist materials must be able to withstand a wide variety of chemically and physically corrosive environments in wafer processing. Some resists are used as direct-implant masks, and they must be able to absorb high-energy ions, electrons, and neutrons. Plasma environments are the most common examples of situations testing the etch resistance, thermal stability, and adhesion of radiation-sensitive polymeric films. Some resist is removed in these processes, and sufficient resist thickness must remain so that etch species do not penetrate and cause pinholes in unwanted areas. Lateral etching or undercutting must also be held to specified limits in order to maintain effective pattern transfer and critical dimension control.

Process resistance includes withstanding developer solutions sufficiently to leave a continuous film of high surface integrity prior to etching. Some resist is lost in developing, and the differential solubility of the resist needs to be large enough to allow for overdeveloping without significant loss of thickness or surface integrity.

Removal of resist

Resists processed through various etch environments are frequently hardened (cross-linked, chemically reacted) to make removal difficult. Resist materials must be clearly removed without adversely affecting the underlying substrate. Complete removal means no trace elements, especially metals, that could be electrically active in the silicon. Specifications for maximum allowable concentrations of metals are established.

Resist Types

Resists are of two main types, negative tone and positive tone. Negative-tone resists increase their thickness as a function of exposure energy; positive-tone resists lose developed thickness as a function of exposure energy. Figure 3.3 shows the main difference in these two basic types of resist. The contrast plots shown in Fig. 3.2 illustrate the exposure versus developed thickness behavior of these types of resist.

Optical negative-working resists

Polymer resist materials that cross-link or harden to form an image when exposed are termed "negative working" resists, simply because they form a negative of the aerial incident image of the mask or reticle. The unexposed resists are soluble in organic solvents, the basis for developing solutions. The exposed areas of a negative resist form cross-linked matrices and photosensitivity is high, since an entire polymer chain is cross-linked in a single exposure event. In the solvent developer, the unexposed polymer is first swollen and then comes apart in molecular chain units. The chemical mechanisms of long-chain exposure and dissolution limit the resolution of negative optical resists. Resist coating thickness is kept to a minimum in order to achieve maximum resolution. A SEM of a negative-working optical resist on an IC device is shown in Fig. 3.4.

The cyclized poly (*cis*-isoprene) type of negative-working photoresist supplied the imaging needs of the IC industry for over 20 years. The

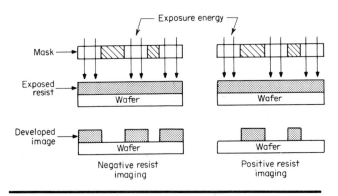

Figure 3.3 Exposure principle for positive and negative resists.

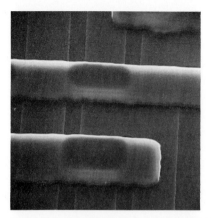

Figure 3.4 Typical negative resist image. (*Courtesy of Mike Nash.*)

chemistry, shown in Fig. 3.5, is functional for resolution levels down to about 3 μm. This system uses a bis-aryldiazide sensitizer. The resin is a synthetic rubber. The partially cyclized resist has good adhesion to semiconductor surfaces and excellent resistance to commonly used wet etchants. The developers used with these resists are nonpolar organic solvents such as xylene, toluene, and halogenated aliphatic hydrocarbons. The spectral absorption of these resists peaks at about 360 nm, so it is ideally matched for mercury i-line (365-nm) lithography. It is not photographically useful at the mercury g-line (436 nm) or h-line (405 nm). Both the swelling and the high-molecular-weight components limit the use of these resists (and any resist with similar

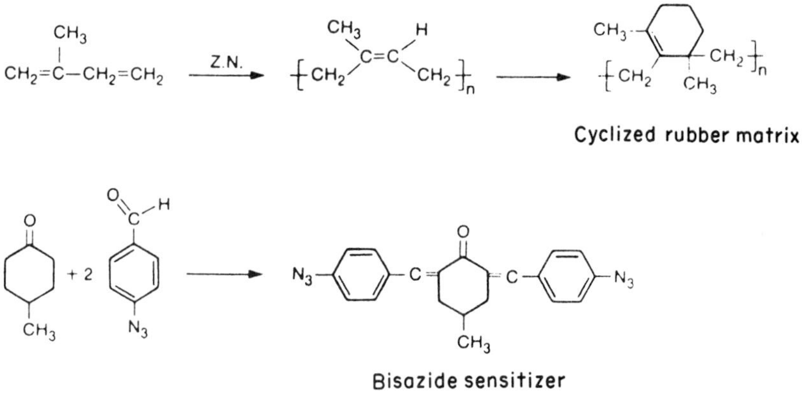

Cyclized rubber matrix

Bisazide sensitizer

Figure 3.5 Poly (*cis*-isoprene) negative resist chemistry.[3]

chemistry) for VLSI lithography in the 2-μm and below resolution regime.

New optical negative-resist chemistries using aqueous-based developers or developers that greatly reduce swelling will extend the use of this class of resists. However, the relative ease of producing negative-tone images by using high-resolution positive photoresists (novolak type) eliminates much of the motivation for producing new high-resolution negative photoresist materials.

Optical positive resists

The increase in IC density as the industry moved from LSI to VLSI created the need for new imaging and etching technology to satisfy the resolution requirements of the new and much denser devices. Isotropic wet etching was slowly displaced by anisotropic dry plasma and reactive ion etching. Contact printing was made obsolete by proximity and projection printing, and optical negative resist was gradually replaced by higher-resolution positive resist.

Optical positive resists have inherently better resolution because their chemistry does not rely on long-chain polymer structures. Instead, they are formulated with low-molecular-weight components. A typical resin system is a base-soluble phenolformaldehyde novolak resin matrix that is base- or alkaline-soluble. These polymers do not form chain groups; instead, they break cleanly, without swelling, as they are dissolved by their aqueous developer. The sensitizer is a diazonaphthaquinone dissolution inhibitor which greatly reduces the alkaline solubility of the resin system. The molecular weight of the sensitizer is about 200.

During exposure, the diazoquinone undergoes Wolff rearrangement, hydrolysis, and conversion to indene carboxylic acid. The chemistry is outlined in Fig. 3.6.

The acid product of exposure is highly soluble in the aqueous alkaline developer. The difference between exposed and unexposed resist solubility is the basis for imaging, and it also gives the resist its wide process latitude properties so critical in VLSI device production.

The sensitivity of optical positive resists is relatively low, mainly because a sizable number of photochemical events are required to expose all of the sensitizer. The resolution is so high (>50 nm) that extra exposure time is a small cost for nanometer-level-resolution sharp line edges and process latitude. Figure 3.7 shows a SEM of positive optical resist covering high steps with good line control and micron-level resolution. The absorption of such resists is strong in the 320- to 420-nm range, and modifications of the sensitizer and resin permit mid-uv (313-nm) and deep-uv (248-nm) wavelengths.

Positive optical resists have good adhesion to most semiconductor

Positive Photoresist Chemistry

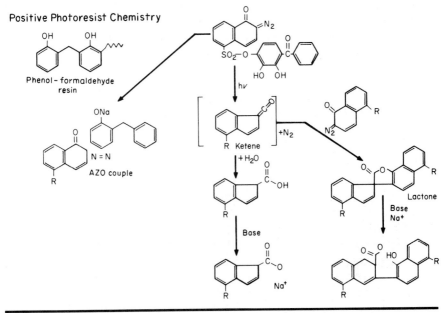

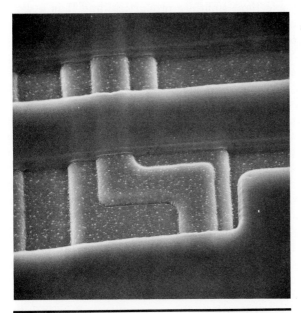

Figure 3.6   Positive novolak-type resist exposure reaction.[4]

Figure 3.7   Positive photoresist step coverage. (*Courtesy of Shipley Company.*)

surfaces, primed and unprimed, and also have satisfactory dry-etch resistance. The addition of one-step postexposure or postdevelopment treatments (deep-uv flooding) affords complete etch process compatibility without high-temperature postbaking.

Electron beam resists

Electron beam lithography is a direct-writing technology that uses a focused beam of electrons to expose a resist pattern. Since the amount of resist area that can be exposed at one time is very small, electron beam exposure systems have relatively low throughput. Second, electron beam energy scatters in resist films to create unwanted lateral exposure. These resolution and throughput constraints, coupled with the nonavailability of plasma-resistant chemistry, have kept electron beam technology in only limited production (critical mask levels) of silicon devices. However, the low throughput requirements and thin resist coating needs of mask fabrication technology have made electron beam imaging the process of choice. Electron beam systems can also provide very rapid turnaround time (compared to optical pattern generators) on prototype chip designs.

A wide variety of resist chemistries have been used for electron beam imaging. Early development work focused on poly(methyl methacrylate), or PMMA. PMMA is produced by the radical-initiated polymerization of methyl methacrylate monomer. The sensitivity of PMMA is relatively low, requiring a 50- to 100-μC/cm^2 exposure dose at 20 kV. PMMA has excellent imaging properties, and it is a very useful material for process development. Chemical modification to improve PMMA's dry etch resistance has led to "terpolymer" formulations and other PMMA analogs with sensitivity improvements along with better plasma resistance. One such analog commercially available is poly(fluorobutyl methacrylate), a resist suitable for both mask and silicon device fabrication.

Since most VLSI/ULSI masks are produced by wet (and dry) etching of thin chromium films, resists not suitable for device processing are acceptable in mask production, as long as the resolution, coating properties, and imaging latitude characteristics are preserved. Poly-(butene-1-sulfone) or PBS is one such material. PBS is a high-sensitivity (1 μC/cm^2 at 10 kV) positive-working copolymer of sulfur dioxide and 1-butene. The chemical structure and decomposition pathway are shown in Fig. 3.8.

PBS is widely used as an electron beam resist for mask making. Electron beam resists with the dry-etch resistance needed for device fabrication have been formulated by using the same novolak resin system used in classical positive optical resists and a sensitizer based on poly(olefin sulfone). The structure of this system (called "NPR" for new positive resist) is shown in Fig. 3.9.

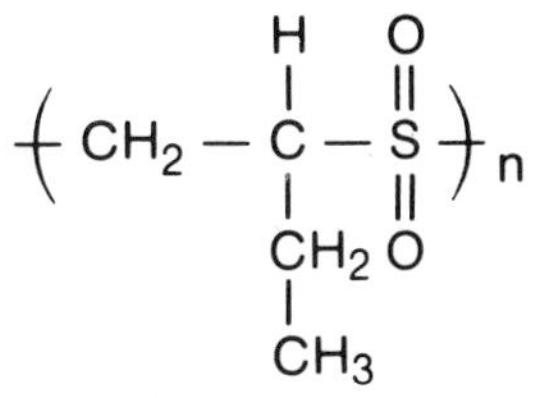

$$\left(CH_2 - \underset{\underset{\underset{CH_3}{|}}{\overset{\displaystyle |}{CH_2}}}{\overset{\overset{\displaystyle H}{|}}{C}} - \underset{\overset{\displaystyle \|}{O}}{\overset{\overset{\displaystyle O}{\|}}{S}} \right)_n$$

$$R - SO_2 - R' \longrightarrow [RSO_2R']^{\overset{\centerdot}{+}} + e^-$$

$$[RSO_2R']^{\overset{\centerdot}{+}} \longrightarrow RSO_2^+ + \cdot R' \longrightarrow R^+ + SO_2$$

Figure 3.8 Poly(butene-1-sulfone) (PBS) exposure mechanism.[5]

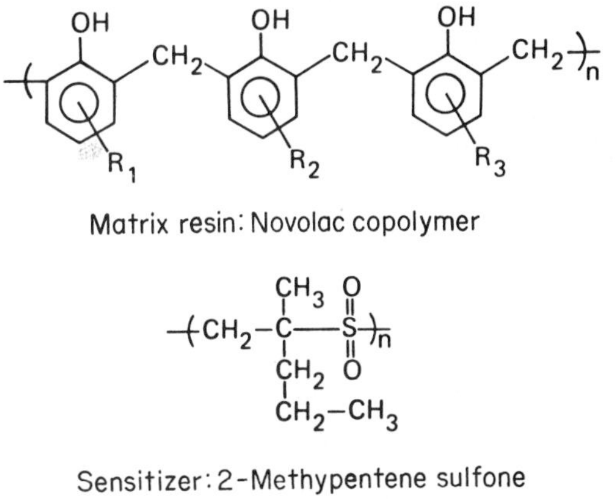

Matrix resin: Novolac copolymer

$$\left(CH_2 - \underset{\underset{CH_2 - CH_3}{\overset{|}{\underset{|}{CH_2}}}}{\overset{\overset{CH_3}{|}}{C}} - \underset{\overset{\|}{O}}{\overset{\overset{O}{\|}}{S}} \right)_n$$

Sensitizer: 2-Methypentene sulfone

Figure 3.9 NPR e-beam resist.[6]

Negative-tone electron beam resists were originally based on a copolymer of glycidal methacrylate and ethyl acrylate called "COP." The exposure mechanism is the same as for optical negative resists (radiation-induced cross-linking and polymer chain linkage with relatively high molecular weight). The same resolution problems occur with COP as with optical negative resists and for the same reasons, including image swelling in solvent developers and thickness-limited resolution a fraction of that possible with optical positive resist. COP, in thin coatings, is still functional as a wet-etch resist for chromium etching of masks. The structure of COP is shown in Fig. 3.10.

Positive- and negative-tone resists are needed in electron beam imaging because of the differences in the amount of circuit area to be exposed. For example, a pattern with only a small number of pixels is much more efficiently written with a positive-tone resist; conversely, direct writing

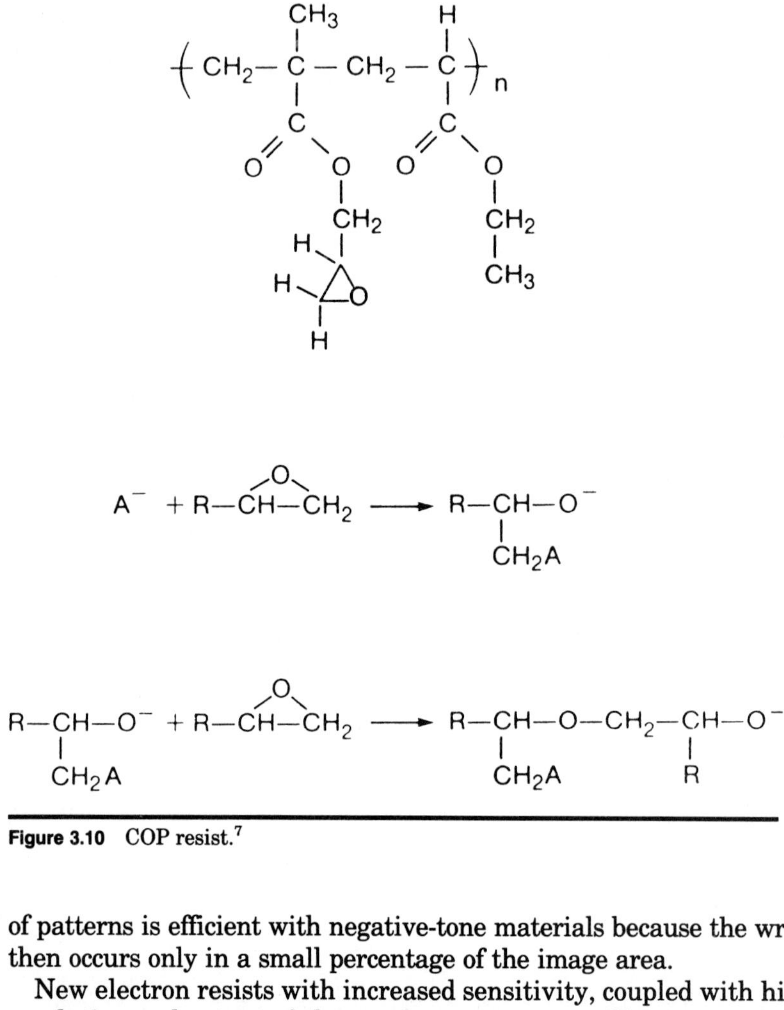

$$\left(CH_2-\underset{\underset{O\diagup\quad\diagdown O}{\overset{\|}{C}}}{\overset{\overset{CH_3}{|}}{C}}-CH_2-\underset{\underset{O\diagup\quad\diagdown O}{\overset{\|}{C}}}{\overset{\overset{H}{|}}{C}}\right)_n$$

CH₃ ... C ... CH₂ ... C (repeat unit with CH₂ and epoxide ring, and CH₂—CH₃)

$$A^- + R-\underset{\diagdown O\diagup}{CH-CH_2} \longrightarrow R-\underset{|}{CH}-O^-$$
$$CH_2A$$

$$R-\underset{|}{CH}-O^- + R-\underset{\diagdown O\diagup}{CH-CH_2} \longrightarrow R-\underset{|}{CH}-O-CH_2-\underset{|}{CH}-O^-$$
$$CH_2A \phantom{{}+ R-CH-CH_2 \longrightarrow R-CH-O-CH_2-}CH_2A \quad R$$

Figure 3.10   COP resist.[7]

of patterns is efficient with negative-tone materials because the writing then occurs only in a small percentage of the image area.

New electron resists with increased sensitivity, coupled with higher resolution and output of electron beam systems, will permit wider use of this technology in device manufacturing. See Chap. 8 for more information on electron beam lithography.

x-Ray resists

Resists specifically for x-ray imaging have not been as actively pursued as other types for two major reasons. The first is that x-ray lithography has not developed as rapidly as optical and electron beam imaging technology, which has kept demand for resist limited to efforts on the research and development scale. Second, resists formulated by electron beam lithography work with x-ray exposure tools also.

The success of optical lithography in producing submicron, production-level image process technology has left x-ray ion beam and electron beam imaging with a small segment of the high-resolution imaging materials market. Optical lithography using excimer lasers can generate 0.3- to 0.5-μm resolution routinely. Process improvements in surface-reacted, bilevel resists are expected to extend resolution down to 0.2 μm levels and possibly below if even deeper uv lasers, operating at 193 nm (argon fluoride) or 152 nm (fluorine), can be made practical exposure systems.

The ability to use electron beam resists for x-ray exposure affords a sensitivity improvement, since organic resists present a greater total cross-sectional area for x-ray radiation (protons) than for electrons. Figure 3.11 shows a SEM of an x-ray resist image.

x-Ray resist sensitivity is derived in much the same way as it is for optical resists: by tuning the resist absorption to the emission lines (palladium, etc.) of the x-ray source. Halogen atom doping is used for this purpose in a number of acrylate-based resist formulations.

x-Ray resists, like e-beam resists, must provide resolution consistent with (and take advantage of) the short wavelengths of beam imaging sources (lasers, ions, electrons, and protons). Also, the full spectrum of resist properties needed in device manufacturing must be available in x-ray resists.

Ion-sensitive resists

Positive and negative resists used in optical and electron beam lithography also find application in ion lithography and often with gains in sensitivity. From the point of view of photomechanics, ions move efficiently through polymer structures, mainly because of their larger atomic number and reduced velocity compared to electrons. Thus,

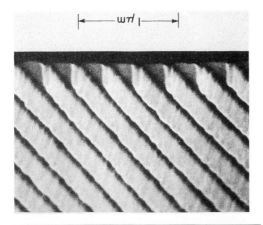

Figure 3.11   x-Ray resist image with 0.1-μm resolution. (*Courtesy of H. Smith, M.I.T. Lincoln Laboratories.*)

relatively fewer ions can, for example, produce a chain scission reaction. This allows the use of resists with lower exposure dose requirements, such as conventional optical positive resists.

Imaging with ion beams can be affected by poor beam or dose uniformity, such as beam position shift, beam shape aberrations, ion scatter, ion spatial arrangement, and dose inconsistency. Calculations show that a dose of about 800 ions/pixel is necessary to maintain pattern dimension control of 10%, 3-σ.

Resists used in ion imaging, whether focused beam or masked lithography, are standard electron and optical resists. If ion lithography becomes more of a production technology, new materials optimized for ion exposure reactions will be needed.

Image reversal resists

Image reversal is accomplished with the well-characterized optical positive resists discussed above. One approach is to add small quantities of monoazoline, imidazole, or triethanolamine to diazoquinone novolak photoresist. The standard process is modified as follows: coat and softbake, expose, and then post-expose bake; follow with uv flood exposure and standard development. Negative-tone images result, and the chemical reaction is as shown in Fig. 3.12.

A more recent image reversal technique with the same resist involves processing exposed images through a vacuum chamber filled with a gaseous amine. Extremely high resolution and negative-sloping resist sidewalls are possible with this technique, in addition to a high degree of control of image size and wall angle depending upon the treatment parameters in the amine environment. See "Image Reversal" in Chap. 8.

Multilayer resists

A multilayer resist process that also accomplishes image reversal involves treating positive optical resists with organometallic agents in a vacuum chamber, again after resist exposure. The organometallic reagent hexamethyldiazane (HMDS) or chlorotrimethylsilane (CTPS) is seeded onto the resist surface. Figure 3.13 shows what is called an IGAS phase-functionalized resist system along with a SEM photo of the image and a typical sequence used to produce the results.

There are a number of American Society for Testing and Materials (ASTM) procedures that show various methods for measuring physical and chemical properties. A partial listing of these methods is given in Table 3.1.

Resists for short-wavelength lithography

Short-wavelength lithography permits finer pattern resolution, but a number of rules must be obeyed to take advantage of the natural gain

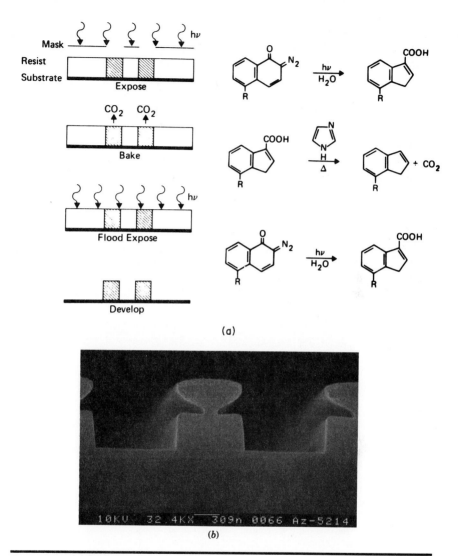

(a)

(b)

Figure 3.12 (a) Image reversal mechanism for positive photoresist and SEM of image.[7] (b) AZ-5214, image-reversal-processed, over etched thermal SiO_2 on silicon. (*Courtesy of I. Higashikawa, Toshiba.*)

in pattern size caused by reduced wavelength. The minimum line width and depth of focus are functions of the numerical aperture (NA) and wavelength of the optical imaging tool. At a wavelength of 250 nm, the NA is limited to 0.5 and the depth of focus is 0.5 μm. The depth of focus should never be less than the resist thickness, or image dimensional variation will occur in the resist. In the resist, the con-

Figure 3.13 Gas-phase functionalized resist system (SEM of actual image and process sequence). (*Courtesy of B. Roland, UCB Electronics.*)

TABLE 3.1   ASTM Photoresist Testing Publications

Method	Publications standard
Testing physical properties of photoresist used in microelectronic fabrication	F66-66T
Nonvolatile content of resin solutions	D1259-61
Viscosity of resin solutions	D1725-62
Refractive index of viscous materials	D1747-62
Specific gravity of industrial aromatic hydrocarbons and related materials	D891-59
Abrasion resistance of coatings of paint, varnish, lacquer, and related products by falling-sand method	D968-51
Viscosity of transparent and opaque liquids—kinematic and dynamic viscosities	D445-69
Measuring and counting particulate contamination in clean rooms and other dust-controlled areas designed for electronic and similar applications	F25-68
Water in volatile solvents—Fischer reagent titration method	D1364-64
Dielectric constant	D150-59T
Volume resistivity	D257-6
Voltage breakdown	D149-61

trast must be high to allow for the formation of high-aspect-ratio images. Most current g-line resists have considerably lower contrast values at 240 nm and below than at 308 nm and above. For example, the Hunt 204 and Shipley 1300 and 1400 resist families and the American Hoechst 1300 and related novolak-based formulations all have 3.4 to 3.6 contrast values in the 310- to 436-

nm range. Their contrast at 240 nm is only 0.6 to 0.7, except Hunt 204, which is 0.85. PMMA at the 240-nm wavelength has a contrast value of 1.7.

In addition to contrast, resists for short-wavelength lithography need to be sensitive to the exposure source and coat without surface striations. PMMA coats without striations, filters out longer wavelengths, and has high contrast and sensitivity to deep-uv sources. Its main drawback is poor dry-process resistance. It is useful as a bottom or planarizing layer in two-layer processing in which the top layer is a thin g-line material such as Hunt 204. The addition of a dye is also used to absorb reflections and calibrate the filtering of unwanted wavelengths.

Other resists for short-wavelength lithography are experimental analogs of diazo-sensitized novolak resin-based positive resists modeled after the 1350-type resist system. These variations are modified to improve the differential solubility or contrast of the system for use as short-exposure wavelengths. One approach is to allow a resist to react at the surface with high sensitivity and produce a chain reaction that chemically breaks bonds down through the resist layer, replacing the conventional exposure mechanism and solving a resist sensitivity problem. Since many resists have high absorption at deep-uv wavelengths and poor transmission, an alternate chemical mechanism to achieve the effect of transmission is a likely path to follow.

Resist Applications

Photoresists are used in an interesting variety of applications within the scope of IC fabrication. Aside from the common use as an etch mask, resists are used for "spike etching," backside coating, dielectric insulation, lift-off, implant masking, and multiple-layer imaging and as an actual optical mask for resist exposure, to name a few other uses. Outside the field of IC fabrication, resists are used for holography, chemical machining, nameplate fabrication, printed-circuit manufacturing, textile patterning, electroforming, ruling fabrication, solar cell production, and numerous specialty areas in which an imaging material of good resolution is required. One of the oldest uses for photoresists is photoengraving; it replaces mechanical milling equipment when precision is required. An obscure and ancient use of what was probably the first photoresist was the Egyptian combination of oil of lavender and Syrian asphalt (which cross-links when placed in the sun) to cure mummy wrappings.

One of the most commonly overlooked aspects of photolithography is matching the proper resist with the application. The variety of resists available makes this no small task. Yet there are some simple and practical criteria to make the job easier (as applied to IC etching applications), as the following list illustrates:

1. What are the surfaces onto which the resist must be applied?
   a. Are the surfaces highly doped or only lightly doped?
   b. Is the surface highly reflective, so as to limit the spacing to prevent bridging with a particular resist?
   c. Does the surface have irregularities that must be covered by the resist?
   d. Does the surface have special cleaning requirements (scrubbing or priming)?
   e. What is the surface flatness?
   f. Is the surface uniform in terms of material integrity, so that uniform etching results independent of resist patterning can be expected?
   g. Will the surface and resist react to form a "memory" that may be hard to remove?

2. What are the step heights to be covered during coating?
   a. The resist should be so applied as to provide at least 2000 Å of protection over the tops of any of the steps.
   b. Have coating techniques to provide optimum coverage without having to add unnecessary resist thickness been explored?

3. What are the etchants to be employed?
   a. What resist thickness is required to provide pinhole resistance?
   b. Is the resist sensitive to a particular component in the etchant, and can that component be substituted or another etchant be used?
   c. Have the optimum postbake time and temperature been derived to provide the desired undercut?
   d. Can plasma reactive ion and sputter etching be employed satisfactorily?
   e. Does the etchant make the resist difficult to remove?
   f. Have the proper etch time, temperature, and concentration been derived?

4. What are the minimum line and space resolution?
   a. Is there sufficient latitude in the resist system to allow for the process variation and the resist resolution and still meet CD objectives?
   b. Have optimization schemes for resist baking, exposure, and development been run?

5. What special equipment considerations are necessary?

a. Are the resists and developers materially compatible with the processing equipment?

b. Is the resist sensitivity adequate to permit the required exposure throughput?

c. Has the resist supplier made equipment recommendations; have studies identified problems with or advantages to particular types of equipment?

d. What is the technological life span of the intended equipment likely to be?

e. Are other people using the resist system and equipment combination that you intend to use? If not, find out why. If so, there are probably good lessons that have already been learned.

6. What are the economic restraints for the resist, developer, primer, thinner, and stripper materials?

a. Imaging material cost is low in proportion to its potential impact on device yield. Therefore, are higher-grade materials (highly filtered) that will enhance imaging performance available?

b. Have the economics of a given imaging system been compared with those of another imaging system for a given mask level? Some studies have been conducted in this area.

7. What are the ecological or disposal requirements?

a. Are any toxic or hazardous materials that cannot be safely handled with your current processing equipment involved?

b. Have you obtained waste treatment recommendations from the manufacturer?

c. Do you know what the current guidelines for legal disposal are?

Answering these questions will probably reduce the field of resists to two or three candidates, and from there a simple test matrix can be designed to arrive at the proper choice for your application.

Resist Process Control

Process control in IC fabrication is a complex task requiring the resources of several departments to manage the fabrication environment and the equipment and materials used in that environment. Even the sources of operator-induced variability must be examined as part of the analysis.

There are many other aspects of IC manufacturing process control, but none has more direct impact on device yields and overall economics than lithography. Lithography is the largest contributor (70 to 75%) to yield in IC manufacturing, measured at wafer probe as a percent of good die. Yields in lithography can drop drastically and ruin the economics of the entire process.

Resist users perform a large number of functional tests to ensure that the lithographic process stays within predetermined design rule specifications. Testing in the wafer fabrication is kept to a minimum by having a separate in-house quality control laboratory perform key spot tests on film thickness that simulate conditions on the production line. The thickness and sensitivity of the resist, along with the "strength" (normality) of the developer, are the major functional parameters governing lithographic performance. The numerous analytical and functional tests to monitor these properties are performed by the supplier or in joint specification programs established by resist manufacturers and users.

Assuming adequate controls are in place to govern the chemical uniformity (sensitizer concentration, resin molecular weight, optical absorption, etc.) of a resist material, the users can focus on testing to ensure consistent resist film thickness as seen by the exposure aligner.

Control of photo- and other radiation-sensitive resists begins with the manufacturer. The main objectives set by resist producers are to reduce product variations by controlling the raw materials used, controlling the manufacturing process, and finally by performing a set of tests to monitor the quality level produced.

The emphasis throughout resist and companion product (developer, removers, etc.) manufacturing is to control aspects of the product that have functional impact on the IC manufacturing line. This requires close and frequent communication between resist manufacturer and user. The result of close communication will be a set of specifications on key resist and developer properties.

The major property to be controlled in resist materials is radiation sensitivity. In the case of photoresist, photospeed or photosensitivity control is the main goal. Most users require photosensitivity control to less than $\pm 5\%$ and advanced processes to less than $\pm 3\%$. In resist manufacturing this requires first identifying the components of variance and then establishing limits on the parameters that permit the required photospeed control. For example, resin viscosity variance and resin photospeed variance are two parameters monitored to meet a photospeed specification.

Many statistical tools are used to bring and keep a resist-manufacturing process under control. They include mathematical modeling for thickness and photospeed (raw materials), hierarchical or nested designs, and full factorial design models.

The result of this approach to resist quality control is more uniform photospeed, to cite one property as an example. Figure 3.14 shows the

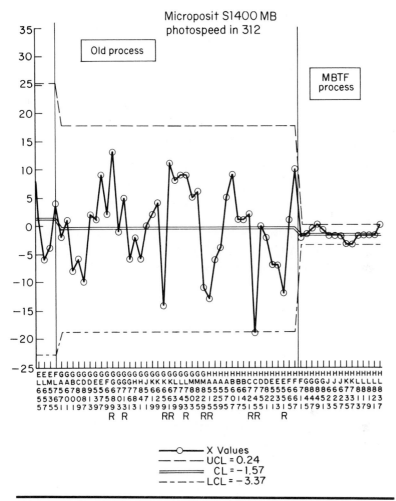

Figure 3.14 Effect of quality and process control on resist photospeed. (*Courtesy of Shipley Company.*)

results of the statistical testing tools described earlier.

The imaging parameters are driven by the resist film thickness along with the aerial image intensity profile. The dry film thickness of the resist is a function of resist solids content and viscosity, along with the coating environment parameters. Dilution models are used to explain the relations among solids control viscosity, uv absorbance, and thickness. The emphasis is on predicting and controlling the photo-response of a given resist at a specific thickness. The ability to control the solids content, viscosity, and uv absorbance allows for predictable and repeatable behavior from a given resist film thickness.

TABLE 3.2   Typical Resist Supplier Specifications

Microposit S1800 Series Photoresist Properties
(For S1805, S1811, S1813, S1815, S1818, S1822)

Film thickness reproducibility	
S1805, S1811, S1813, S1815	± 200 Å with respect to a reference
S1818	± 300 Å with respect to a reference
S1822	± 500 Å with respect to a reference
Filterability constant, n/n_0	0.0075 maximum
Water content	0.5% maximum
Index of refraction	1.64 at 6328 Å, 1.68 at 4360 Å
Na content	<1 ppm
Fe content	<1 ppm
Type of solution	Solvent base:
	Propylene glycol
	Monomethyl ether
	Acetate
Flash point (closed cup), approximate	46°C
TLV rating*	100 ppm

*Rating is for propylene glycol monomethyl ether.
SOURCE: Shipley *Technical Manual.*

TABLE 3.3   Commonly Used Radiation-Sensitive Resists

Type of resist	Designation (manufacturer)
Positive optical	AZ5214 (American Hoechst)
	MicropositS-1400 series (Shipley)
	PR-1024 (Macdermid)
	HPR-200 series (Olin Hunt)
	AZ-4000 series (American Hoechst)
	System 8 resists (Aspect Systems)
	EPR5000 (Dynachem)
	OFPR800 (Dynachem)
	Positive resist 820 (KTI Chemical)
	Selectilux P2100 (EM Industries/E. Merck)
	PR914 (MacDermid)
	MS6000 series (Petrarch Systems/Dynamit Nobel)
	Raycast RG-3900B (Hitachi)
Negative optical	Waycoat HNR (Olin Hunt)
	Waycoat HR/SC (Olin Hunt)
	Waycoat type 31L (Olin Hunt)
	Good Rite NR1000 (BF Goodrich)
	Good Rite NR3000 (BF Goodrich)
	KTI 732/747/752 (KTI Chemicals)
E-beam, positive	EBR-9, FMR E102, CP3, RE-5000P, AZ- 2400, FBM, PBS (generic)
E-beam, negative	OEBR-100, CMS (SS), SEL-N, MES, GMC, COP (generic), GAF, KTI
Photosensitive and nonsensitive polyimide	Ciba Geigy Probimide
	Dupont
	KTI Chemicals
	EM Industries

Part of the control programs to provide minimal photospeed variance are reduced variance in the resin and photoactive compound (PAC or dissolution inhibitor). These activities, used in a planned program of statistical quality control (SQC) and statistical process control (SPC), used in a close and communicative user/supplier relationship, will result in a level of control consistent with the critical dimension (CD) specifications in the IC job. Establishing limits on the various components (resist materials, equipment, environment) of the lithographic step is a fundamental aspect of building overall lithographic control programs. Table 3.2 shows typical resist material specifications, and Table 3.3 shows commonly used radiation-sensitive resist materials.

REFERENCES

1. M. J. Bowden, *J. Electrochem. Soc.* vol. 128, no. 5, 1981, p. 195c.
2. M. J. Bowden, "A Perspective on Resist Materials for Fine Line Lithography," in *Materials for Microlithography,* Advances in Chemistry Series, No. 266, American Chemical Society, Washington, D.C., 1984, chap. 3, pp. 39–117.
3. C. G. Willson, "Organic Resist Materials—Theory and Chemistry," in *Introduction to Microlithography,* ACS Symposium Series, Vol. 219, American Chemical Society, Washington D.C., 1983, chap. 3, fig. 13, p. 108.
4. Ref. 3, fig. 30, p. 127.
5. Ref. 3, fig. 31, p. 128.
6. Ref. 3, fig. 32, p. 129.
7. Ref. 3, fig. 24, p. 121.
8. E. Reichmanis and L. F. Thompson, "Polymer Materials for Microlithography," *Am. Rev. Mater. Sci.,* vol. 17, 1987, pp. 235–271.

Wafer Characterization

IC fabrication takes place on a very specially prepared silicon wafer surface, and since this technology continues to demand more from the integrated circuit, it also demands more from the surface on which it all takes place. This chapter deals with the various ways of producing, measuring, and monitoring silicon wafers and identifies some of the major problem areas associated with wafer surface damage, non-flatness, or other irregularities. In the future, the IC industry will find it necessary to become much more critical of the wafer substrate. In order to capitalize on the large emerging IC markets and the yield-related economies they will demand, IC processing will need to become better characterized and more productive in terms of wafer throughput and good die yield.

Semiconductor Crystals

The basis of all semiconductor device technology is a crystalline material that serves as the carrier for electron charge state movement. Crystal growth technology has developed to meet the increasing needs of finished VLSI circuits. Reduction in the width of IC line geometries means increased sensitivity of the functioning IC to anomalies in the base material, the silicon crystal. Crystal-growing processes, beginning with the selection of raw material, have had to incorporate the needs of increasingly critical performance specifications of finished ICs.

Circuit line widths in ICs have a specific and predictable sensitivity to stacking fault defects, oxygen pockets, crystal impurities, and other nonuniformities occurring within the body of the as-pulled ingot. In the sliced wafer, these nonuniformities may occur directly under the gate of the device and can alter the performance characteristics of the chip. The relationship between all the crystal nonuniformities (that "surface" in the sliced silicon wafers) and the chip is understood and quantifiable. A particular chip will tolerate a given number and size of nonuniformities before device performance is affected. In a well-controlled process, engineers will specify a quality level for all wafers used to produce a given chip, knowing that process yields and probe yields will be adversely affected when wafer quality below the specification level is introduced. It is critical to balance IC design and lithography criteria into a single specification set with corresponding quality level for the starting crystal material.

Crystal behavior

Several types of crystals are used as a basis for fabrication of semiconductor devices. Silicon is by far the most popular. There are several special atomic conditions that make silicon such an excellent medium for IC fabrication. For example, crystals typically have a very large number of atoms sharing the valence electrons. The valence electrons are often called "conduction electrons," and the atomic sharing creates a cohesive bond within the atomic structure. The strength of crystals is to a large degree a function of the electron sharing.

Silicon has another key physical aspect tied directly to its atomic configuration: the energy band structure. The energy levels in a silicon (or any) atom determine the degree of interatomic adhesion in the crystal structure. Energy band structure relates to the energy state of the atom. The properties of an atom differ according to which of the various energy states the atom is in, and that in turn is determined by the positions of the outer electrons. When outside forces (temperature) cause the electrons to change positions or to move to another energy state (or "shell"), energy is either acquired or lost. Silicon, as a group IV element in the periodic table, contains four valence electrons. The electron sharing cited above is such that the four valence electrons of each atom are shared with another four silicon atoms. The result is that a single silicon atom has eight valence electrons outside its nucleus, four "owned" but shared and four "loaned" electrons. The actual crystal structure is a diamond lattice (Fig. 4.1).

The functioning IC conducts electrons because of unoccupied energy states within the silicon crystal. The lower valence energy bands in the silicon atom are typically separated from the upper or outer energy band, called the "conduction band." When the crystal is heated, thermal energy causes the electrons in the lower valence energy bands to move outward, across the energy gap between valence and conduction bands, and into the unoccupied sites in the conduction band. The larger the gap between the two layered energy bands, the more "insulative" the electrical behavior will be.

For example, an unheated or ambient temperature crystal that con-

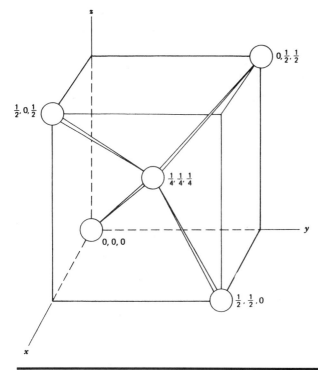

Figure 4.1 Tetrahedral diamond lattice structure.

tains numerous electrons in the outer conduction band will be classified as a semiconductor. If, however, there are just a few electrons (or
no electrons) in the ambient crystal, it will be an insulator. The factor
determining the significant difference between the electrical behavior
of crystals is the width of the energy gap between the inner valence
and outer conduction bands of the atom. In the silicon atom, the energy gap has been measured as 1.1 eV, making it a semiconductor.
The similarly structured diamond crystal, which is an insulator by
definition, has an energy gap of 6.0 eV. In any atom, the amount of
thermal energy available to drive electrons across their respective energy bands is approximately 0.03 eV. When a "threshold energy"
(enough to dislodge it from the valence energy band) is applied to a
valence electron, an electron pair bond is broken, leaving behind a
positive charge (or "hole") as the electron moves outward to the conduction band. The positively charged hole site, or vacant bond, can be
thought of as a positively charged "quasistructure" because of its mov-
ability within the crystal. If, for example, another electron comes
along to occupy the hole, the positively charged "area" will be moved
to another place in the crystal.

Semiconductors are often classed as either intrinsic (undoped) or extrinsic. Extrinsic semiconductors have been doped with impurities from the group V section of the periodic table of elements. Pure or intrinsic semiconductor material is best thought of as the quasi-insulative base or foundation of all integrated circuitry. The electron mobility in material such as silicon is relatively low compared to that of the doped silicon to which boron, phosphorus, or other dopants are added. Once doped, the electron speed is quite high. For example, the mean free path of an electron is about 0.2 μm, meaning that the electron will travel as a ballistic particle for that distance, unobstructed, until it begins to collide with other electrons, the crystal lattice, temperature "waves," or barriers.

Ballistic-effect devices, such as the vertical electron transistor (VET) or the permeable base transistor (PBT), are useful for ultra-high-speed (10-ps) digital circuitry for military and information-processing applications. Ballistic speeds are more easily achieved in materials that permit even greater electron mobility, such as gallium arsenide. Ballistic-effect devices based upon gallium arsenide provide speeds exceeding those achieved in the Josephson junction transistor.

In essence, then, the fabrication of a solid-state electronic device involves chemical modification of a semiconducting material in areas in which increased electron mobility is desired. These "high-speed electron pathways" are first outlined by the circuit designer and, by use of appropriate software and a computer-aided design (CAD) system, taken to digital form. Photolithographic processing then transforms the digital circuit patterns into patterns in resist, etched oxide masks, and, finally, doped areas that are the electron- or charge-carrying pathways that constitute solid-state circuit behavior.

Crystal impurities

Impurities have a strong influence on both the physical and chemical properties of finished silicon or gallium arsenide wafers and finished ICs. The two primary impurities in silicon crystals are oxygen and carbon, which act as dopants in modifying the charge-carrying properties of the crystal. The amount of each of these impurities in the crystal is determined by the maximum solubility of each in silicon, as follows:

Maximum Solubility of Oxygen and Carbon Impurities in Silicon Crystals	
Carbon	5.0×10^{17} atoms/cm^3
Oxygen	2.0×10^{18} atoms/cm^3

Oxygen and carbon impurities, as unintentional dopants, enter the crystal during the crystal growth phase. Since the Czochralski method is used for growing the majority of silicon crystals, we will use it as an example of impurity sourcing. Figure 4.2 shows the quartz crucible, overall growth dynamics, and graphite heaters. The oxygen comes from the reaction wherein quartz is dissolved from the sides of the crucible as follows:

$$SiO_2 \text{ (crucible)} + Si \text{ (melt)} \rightarrow 2SiO$$

Several parts per million of oxygen are present in the ambient growth gas, which is generally argon or helium. The graphite heaters react with this gas, resulting in carbon monoxide ($2C + O_2 \rightarrow 2CO^+$) and carbon dioxide [C (heaters) + O_2 (contaminant gas) $\rightarrow CO_2$]. In this way, carbon enters the melt and, eventually, the crystal.

Oxygen bonding in the crystal lattice, in which it occupies interstitial sites, gives the bonding configuration of Si ˘ O ˘ Si or SiO. During subsequent heat treatment, the oxygen may rearrange in the crystal and take other forms such as SiO_2 or SiO_4.

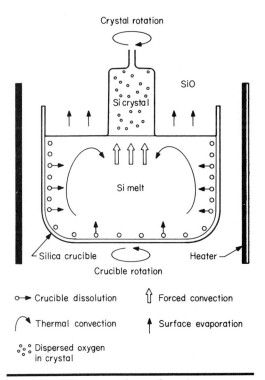

Figure 4.2 Silicon crystal growth environment.

Carbon, on the other hand, takes to the substitution sites in the silicon lattice by covalent bonding with silicon. Fortunately, the carbon is not electrically active in the crystal as oxygen is, especially since it cannot be removed from the lattice by heat treating, whereas oxygen can be. Figure 4.3 shows oxygen and carbon concentration across a crystal diameter.

Heat treatment.   Thermal annealing is used to lower the solid-soluble oxygen content in crystals. Since the absorption coefficient is a function of temperature, various levels of oxygen occur as functions of anneal parameters. Isochromal anneals of samples with different initial oxygen levels show that the anneal step and final oxygen concentration are functions of initial oxygen levels. For example, a crystal that has a medium concentration of oxygen and is annealed for 2 hr will undergo little change in oxygen level. A 24-hr anneal of the same crystal will result in the coefficient approaching zero at 800°C anneal temperature. In most cases, minimum oxygen absorption occurs at the 800 to 900°C anneal temperature level.

Effects of oxygen and carbon: Summary.   Oxygen and carbon levels in a silicon crystal can have good and bad effects, depending upon where and when they occur in the wafer fabrication process. Oxygen donors, for example, are created by heat treatment at 450°C and make it difficult to interpret resistivity readings and establish the dopant level accurately. Oxygen donors also determine or at least affect the breakdown voltage, a key electrical parameter in any *pn* junction device. Other undesirable effects of oxygen in crystals include precipitates (SiO_2) from wafers that were heat-treated at 700 to 1000°C, swirls, and stacking faults. These effects are associated with oxygen levels greater than 25 ppm.

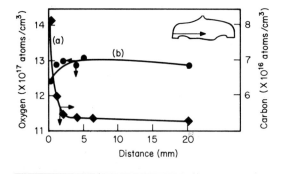

Figure 4.3 Oxygen and carbon concentration across the crystal diameter.

Oxygen is responsible for positive effects in crystals, such as an improvement in the epitaxial quality caused by high oxygen levels. Oxygen can also prevent dislocation multiplication in wafers during thermal cycling. The precipitation of SiO_2 associated with high oxygen concentration in turn acts as a gettering sink for epitaxially induced crystal defects or impurities.

Carbon, by comparison, is not as critical an impurity as oxygen, partly because of its relatively lower concentration level. Carbon has been blamed for being the nucleation site for crystalline defect formation as well as for being a factor in the degree of swirl density in a crystal.

Oxygen and carbon effects, good or bad, are unavoidable, since the impurities will always be present in Czochralski silicon crystals. The oxygen levels (1.0 to 1.5×10^{18} atoms/cm^3) and carbon levels [1.0×10^{17} (tang); 2.0×10^{16} (seed)] can be varied by changing the crystal growth parameters, including growth ambient pressure, rotation rate, pull rate, and crystal/crucible diameter ratios. Careful measurement of the impurities is made by infrared absorption in 9- and 16-μm bands. The sensitivity of the measurement increases with lower temperature. The instrument used to generate this information was a double-beam infrared grating spectrophotometer. Infrared spectra information is computer-processed for direct calculation of absorption coefficients and other data interpretation.

Crystal defects

Defects in semiconductor silicon crystals are often divided into two basic types, intrinsic and extrinsic, depending upon their point of origin.

Intrinsic. Intrinsic defects are those that are introduced before the crystal becomes a wafer. The types of intrinsic defects include:

Stacking faults

Point defects

Oxygen, carbon, and other grown-in impurities

Dislocation in the crystal

Interstitial vacancy clusters

Intrinsic defects are removable to a certain extent by applying thermal cycling (denuding) and gettering techniques. The defects that remain after these treatments are possible sources of electrical failure in the IC device after fabrication. Since the crystal-purifying techniques to remove intrinsic defects are costly, time-consuming, and not entirely effective, major emphasis has been placed on crystal-growing processes that minimize the formation of intrinsic defects.

Point defects. Many types of defects can interrupt the regularity of a nearly perfect structure of single-crystal silicon. A major intrinsic defect is the point defect, which can be one of the following: vacancy in the crystal lattice, interstitial atoms, interstitial impurity atoms, or simply dopant or impurity atoms replacing the original atoms in the lattice.

Vacancies, one of the most fundamental of point defects, are caused by an atom moving out of its normal lattice site and into the surface of the crystal. The atomic movement and resultant vacancy site are caused by thermal fluctuations in the crystal. The result is also called a "Schottky defect," especially if the atom is transported far from its original lattice site. In some cases the dislodged atom is moved only a short distance and remains near the vacancy as an interstitial impurity. In that event, the defect has lower activation energy and is called a "Frenkel defect." The energy required to create vacancies, either Schottky or Frenkel defects, is relatively large, being about 2.3 eV. Thermal processes such as annealing may be the source of defects of these types. Probability plays a role, since the atom must be first freed from its host site and then caught in an interstitial site. Calculation of this probability involves first figuring the breaking of four covalent bonds. If a vacancy occurs next to an existing vacancy (called a "divacancy"), only two bonds need breaking. Divacancies in silicon crystals are common.

An interstitial point defect is an atom that occupies a natural void in the crystal lattice. The diamond lattice presents five such natural voids per unit cell, and since the energy of formation of an interstitial atom is nearly equivalent to that of a vacancy, interstitial defects are very common. Each void is large enough for a silicon atom, and whereas the Schottky defect has a formation energy of 2.3 eV in silicon, the energy of formation of an interstitial atom (vacancy interstitial or Frenkel defect) is 0.5 to 1.0 eV. That is to be expected, since the Schottky defect requires movement of the atom to the crystal surface, a much greater distance than to a nearby interstitial void.

Impurity atoms as point defects also are common occurrences in silicon crystals. Typical impurities found in the silicon lattice include copper, iron, cobalt, manganese, nickel, and zinc. They are generally found in the interstitial sites or voids. Impurities that occur in place of silicon atoms are called "substitutional"; they include gold and such elements from groups III and V of the periodic table as phosphorus, arsenic, boron, aluminum, indium, gallium, and antimony. The fact that silicon crystals have the ability to accommodate such a wide range of impurities makes possible the wide range of electrical properties. Impurity atoms are the determinants of conductivity in an integrated circuit, the factors which establish the movement of electrons or holes (charge states) through the crystal in a predetermined manner.

Silicon atom replacement by an impurity atom can create stresses and distortion, and stress created by substituted dopant atoms is a function of the diameters of these atoms. Intralattice "space" is calculated to be about 2.35 Å, and dopant impurity atoms that replace silicon atoms in the lattice need to be 2.35 Å in diameter or smaller to prevent stress and distortion of the crystal. The diameters of dopant atoms typically used are as tabulated here.

Atom	Diameter, Å	Atom	Diameter, Å
Al	2.52	In	2.88
As	2.36	P	2.20
B	1.76	Sb	2.72
Ga	2.52		

Extrinsic.   The other primary category of defects is extrinsic, and extrinsic defect types include:

Design rule violations

Wafer surface impurities or physical defects from sawing or lapping

Dielectric definition

Wafer warpage, poor metal coverage, and process-included effects

Extrinsic defects include all the problems arising from photolithography, including mask misalignment, residual sodium and resist developers, resist monolayers that become entrapped between oxides, over- or underetched structures, and metal bridging, to name a few. Photolithographic defects may be caused by poor attention to device design rules.

Crystal structure

Silicon is one of many elemental semiconductors; both elemental and compound semiconductors are available for semiconductor manufacturing. Gallium arsenide (GaAs), a compound semiconductor, is widely used for metal oxide semiconductor (MOS) integrated circuit manufacturing. Germaniun, one of the earliest types of semiconductor used for transistor fabrication, is still produced in crystal for power transistor production. Compound semiconductors such as gallium phosphide (GaP) and lead telluride (PbTe) are used in optoelectronics for light-emitting devices, photosensors, and similar devices.

Silicon is an ideal material for semiconductor device manufacturing for many reasons, including its relatively low cost and raw material abundance (sand), ease of generating passive oxide layers on the surface, ability to readily bond resist or other patterning layers to it or its oxides, high-temperature stability as a finished device, and relative ease of working (cutting, polishing, etching) structures needed for finished ICs.

Molten silicon, when cooled, crystallizes into a diamond lattice structure. The silicon crystal is represented by two interpenetrating face-centered cubic crystal lattices. Each atom has four adjacent atoms to which it is covalently bonded, with a lattice constant of 5.43 Å. The crystal unit of the silicon crystal lattice is a tetrahedron. One silicon atom in this structure is positioned centrally with four atoms surrounding it, each at a distance of 2.35 Å (interlattice distance). The shape of this basic tetrahedron diamond is shown in Fig. 4.1. The points at which various silicon atoms are located are called "lattice sites." This structure belies the shape of all crystals in the "cubic" class. The cube-edge distance in silicon is 5.428 Å, and the distance to the next silicon "cube," or crystal unit, is 1.18 Å, with eight atoms per unit. The face-centered cubic (FCC) lattice with eight atoms is shown in Fig. 4.4. In this structure, the interpenetrating corner of the second lattice occurs one-fourth the distance into a major diagonal of the first FCC lattice. In this structure the density (or number of atoms divided by the number of unit cells) is 34%. Dimensionally, one silicon atom can be imagined as being placed in the center of a cube, and adjoining silicon atoms reside at each corner of the cube.

Crystals are described and characterized physically according to the number of times crystal planes intercept crystal axes in a single crys-

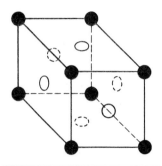

Figure 4.4   Face-centered cubic lattice.

tal unit cell. This description is given by equations and referred to as "Miller indices." For example, for a crystal whose planes intercept the *a* axis once, the *b* axis once, and the *c* axis once, the notation of (111) would be used. Miller indices are generally derived by taking the crystal plane intercepts, inverting them, and expressing them in the smallest integers. The Miller indices of several common planes in a silicon crystal are shown in Fig. 4.5.

Semiconductor silicon crystals are grown in different crystal planes, usually either (111) or (100). The crystal plane chosen will determine where on the ingot the flat will be cut. In wafer fabrication later on, the fault orients the wafer so that scribing the dicing will occur along natural crystal planes and the wafer will cleave easily into individual dies. The crystal orientation also determines the position of lattice

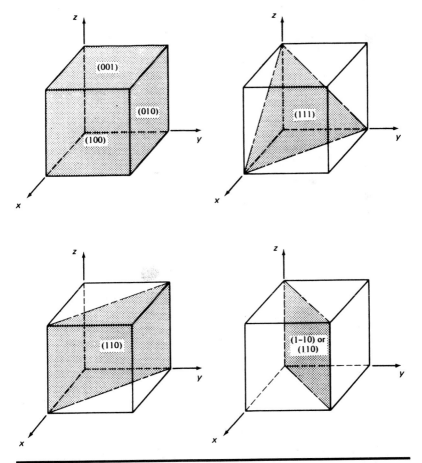

Figure 4.5 Miller indices for common semiconductor crystallographic planes.

sites that will accept donor or dopant atoms in ion implantation or diffusion. The crystal surface angles exposed to the doping source will in part determine rate of acceptance of impurity atoms.

Crystal manufacturing

Silicon crystals are made by pulling ingots from a melt. This technique has been used for years, and the major change has been in the size of crystals and corresponding increase in melt size. For example, the diameter of silicon crystals has increased from less than 20 mm to more than 150 mm in less than 20 years, with melt sizes quadrupling in the same time frame. A 150-mm crystal could have a melt size approaching 60 kg, double what it was in the early 1980s.

In the late 1960s, automation came to crystal manufacturing in the form of automatic crystal diameter control. The need arose for dislocation-free ingot growth, which in turn increased production with reduced cost per wafer. Digital computers revolutionized the quality of crystals produced, because all key functions from initial growth to final shutdown are highly regulated. Once high-volume production of sizable crystals was achieved, the semiconductor industry pushed hard for more homogeneous crystals. That led to changes in manufacturing equipment (from batch to continuous feed) and technology (nonmagnetic to magnetic).

The primary technique used in the 1980s is the Czochralski (Cz) batch process. The actual fluid dynamics in silicon crystal growth are not totally understood. As the seed crystal is pulled from the melt, a series of slow patterns emerge; they are caused by convective currents. The molten silicon is moving, in either "free form" or pressure-forced, up the container walls and down toward the center. Figure 4.6 shows a Cz crystal being pulled.

The free convection currents in the melt are a result of temperature gradients and solute concentrations in the molten silicon. Variations in crystal structure occur partly because of variations in solidification rates caused by convective movement and inhomogeneities in the molten material. Rotation of the crystal helps introduce controlled convection, which results in fluid dynamics in the melt. The fluid movement during crystal formation also determines the distribution of impurities, such as oxygen, carbon, and dopant, and other defects. Some of the physical defects occurring in crystals are caused by temperature gradients. For example, backmelting at the seed-crystal-melt interface is the cause of the solidification, melting, and resolidification processes that can cause defects. The crystal-growing variables include pulling speed, rotation speed and direction, and configuration of the

(a)

(b)

Figure 4.6 (a) Cz crystal growth and ingot. (b) Gallium arsenide crystals.

hot zone. The most effective technique for controlling convective flow intensity is an induced magnetic field.

The basic principle governing the use of this technique is current induction, a result of an electrical conductor crossing a magnetic field. The fluid motion of the melt is resisted or opposed by the magnetic field created around and through the molten silicon. The various forces generated in the melt by convection and magnetism keep the melt in motion around the crucible axis. The axial flow patterns increase the rate of flow. The magnetic field can be rotated to offset the excess silicon melt movement; magnetic field rotation opposite to the crucible rotation results in very high temperature uniformity in the melt; and reduction of free convection in the melt subsequently reduces oxygen concentration in the as-grown crystal.

Magnetic fields can be applied to silicon melts internally with heaters or externally with electromagnets. Powerful external magnets are quite effective in suppressing convective flow in the melt, and magnet size and power usually track with melt size. In general, the temperature patterns in the melt are greatly modified by the magnetic field, including the melt-crystal interface.

Continuous liquid feed crystal growth

One way to eliminate nonuniformities in silicon crystals is to keep the melt size constant by continuously feeding the molten silicon into the crucible. The outline of a continuous liquid feed furnace is shown in Fig. 4.7. The meltdown chamber acts as a supply to the main growth chamber to keep the melt level constant and permit smaller melt volumes. Once polysilicon is melted, the liquid transfer system moves it into the main melt, where it is controlled by the automatic melt level control system. Raw polysilicon for melting in the meltdown chamber can be delivered in rod, powder, or chunk form. While the furnace is in operation, ingots can be removed and raw polysilicon added. A helium-neon laser is part of the melt level–sensing system, in which reflected light from the melt surface is sensed and recorded to a resolution of 0.02 in. A quartz tube, shrouded by a heater, transfers the molten silicon to the main melt.

Thus, the Cz method of batch crystal growth is giving way to more advanced methods, such as continuous liquid feed (CLF) techniques,

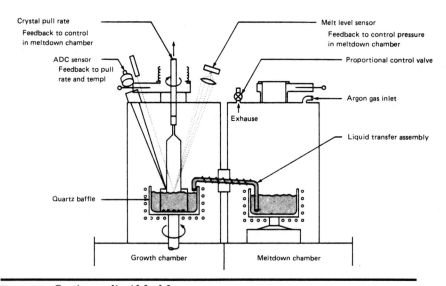

Figure 4.7 Continuous liquid feed furnace.

that permit growth of more nearly perfect and more uniform crystals needed in the fabrication of advanced ICs.

Float zone crystal growth.   The float zone (FZ) crystal growth technique is used to produce a small amount of the wafers used in IC manufacturing. Float zone crystals are grown in an inert atmosphere, usually argon. The starting raw material, a polysilicon rod, is tightly wound with an rf heating coil. Current applied to the coil creates a molten zone, 1 to 3 cm long, in the rod. This zone is then moved throughout the length of the rod by moving rod or coil or both until the melting and subsequent recrystallization reorients the structure of the silicon to a crystalline state.

The process begins in the same way as a Cz crystal growth process, by touching a molten drop of silicon to a seed crystal (which establishes crystal orientation) and then slowly withdrawing the seed as more molten silicon is drawn to the seed, cools, and forms a crystal ingot. In the FZ process, the molten zone may be rotated, just as the crucible is in the Cz processes, to promote crystal growth uniformity. Crystal diameter is controlled by the rate of pull and the rate at which the polysilicon rod is fed into the coil. The molten silicon zone is rf-heated. The levitation effect from rf heating permits more of the melt to be supported than would be by surface tension effects. The result is a capability to grow larger-diameter crystals.

Float zone crystals are produced mainly for less critical applications, such as silicon power devices. The equipment for growing FZ crystals is less critically controlled. Even so, dislocations are kept at a minimum by a technique (used in Cz crystal growth also) called "necking." Necking is the reduction of seed diameter, just after the seed initially touches the melt, that allows dislocations to grow out of the crystal. It is followed shortly by a return to the starting crystal diameter, which leaves behind in the seed crystal a zero-defect zone from which the likelihood of future dislocation is greatly reduced.

Float zone crystals are subject to the same clustered defects (swirls) as Cz crystals; in both cases the defects arise from interstitials or oxygen vacancy combinations. In FZ manufacturing, however, the greater ability to change the rate of growth permits generally lower swirl-related defects. Float zone crystals are also much lower (100 times) in oxygen concentration than Cz crystals. That is mainly because they avoid the quartz crucible dissolution problem of the Cz process, along with the likelihood of the other trace impurities that occur in a crucible melt. Both FZ and Cz crystals can be made in various crystal orientations, except that Cz techniques make larger-diameter crystals possible.

Wafer Production

High-volume production of silicon wafers to exacting flatness specifications requires considerable manufacturing process control of crystal-slicing machinery. The process of taking a large crystal, or boule, and slicing out thin wafer disks is called "wafering." The major problem or goal of wafering is to get *very* flat wafers out of the crystal. Flatness is a requirement simply because the optical (and some nonoptical) patterning equipment can focus only in one plane and will not readily refocus microimages to adjust for any nonflatness of the wafer substrate. The depth of focus of an optical printer then becomes the limiting parameter.

Flatness requirements are always changing as new crystal sizes are employed and new crystal types, with different expansion coefficients, are used. Flatness is described by class and specified by the Semiconductor Equipment and Materials Institute (SEMI) as shown in Tables 4.1 to 4.3. One development that is likely to change the stringent flatness requirement is aligners that focus automatically on each die. Many of the step-and-repeat aligners focus one die at a time, thereby permitting the exposure over topography on the wafer without dis-

TABLE 4.1   SEMI Specifications for 80-, 90-, and 100-mm Wafers

	Wafer size, mm (in)		
	80 (3.150)	90 (3.543)	100 (3.937)
Diameter range (tolerance), mm (in)	± 1 (± 0.039)	± 1 (± 0.039)	± 2 (± 0.79)
Primary flat length, mm (in)	19–25 (0.748–0.984)	24–30 (0.945–1.181)	30–35 (1.181–1.378)
Primary flat location		(110) ± 1°	
Secondary flat length, mm (in)	10–13 (0.393–0.511)	12–15 (0.472–0.590)	16–20 (0.6–0.8)
Secondary flat location (for material identification)	(111) *p*-type primary flat only (100) *p*-type primary flat; secondary flat at 90° in either direction (111) *n*-type primary flat; secondary flat at 45° in either direction (100) *n*-type primary flat; secondary flat at 180°		
Surface orientation	On orientation for (111) and (100) ± 1° Off orientation for (111) toward nearest (110) 3 ± 1°		
Thickness, μm (mils)	400 ± 25 (15.75 ± 1)	475 ± 25 (18.80 ± 1)	625 ± 25 (24.6 ± 1)
Orthogonal misorientation			
Bow, μm (mils)	50 (2)	55 (2.2)	60 (2.4)
Taper (parallelism), μm (mils)	40 (1.6)	45 (1.8)	50 (2)

SOURCE: Semiconductor Equipment and Materials Institute (SEMI).

TABLE 4.2   SEMI Specifications for 150-mm Wafers

Property	Requirement
	Dimension and Tolerance Requirements
Diameter	149.50–150.50 mm
Thickness, center point	650–700 μm
Primary flat length	55–60 mm
Secondary flat length	35–40 mm
Bow	60 μm
Warp	60 μm
Total thickness variation	25 μm
Edge profile coordinate Cy	25 μm
	Orientation and Flat-Location Requirements
Primary flat orientation*	100 ± 1°
Secondary flat orientation	
(119) *p*-type	No secondary flat
(100) *p*-type	90 ± 5° clockwise from primary flat
(111) *n*-type	45 ± 5° clockwise from primary flat
(100) *n*-type	180 ± 5° clockwise from primary flat
Surface orientation	
On orientation for (111) and (100)	± 1°
Off orientation [111] toward nearest	2.5 ± 0.5°
⟨110⟩ on a plane parallel to primary flat	4.0 ± 0.5°
Orthogonal misorientation†	± 5° max

*For (111) slices, $(1\bar{1}0)$, $(01\bar{1})$, and $(\bar{1}01)$ planes are equivalent, allowable planes. For (100) slices, the allowable equivalent [110] planes are $(01\bar{1})$, (011), $(01\bar{1})$, and $(0\bar{1}\bar{1})$.

†The contribution of 5° of orthogonal misorientation to the total off-orientation angle will be less than 0.5°.

rupting pattern resolution. Also, nonflatness has become more tolerable because wafer chucks with vacuum drawdown and special ring sealers and ridges can draw a nonflat wafer down while it is in the exposure stage. The problem with this sort of fixturing is that, when the operating is complete and the wafer is released, the original distortion returns, and, of course, the resist pattern distorts along with it.

Thick silicon slices distort less than thin ones. Advanced lithography processes sometimes use wafers several mils thicker than average in order to reduce wafer warpage and bow. Thick wafers may pose problems in standard wafer-handling systems. Thicker wafers also mean fewer wafers per boule, which increases wafer cost.

Silicon crystals have been increasing in diameter ever since crystals were first pulled. Larger crystals, or boules, are more difficult to slice or saw, and major changes in equipment design have been required as wafers have gone from the 1-in-diameter slices of the early 1960s to 10- to 12-in slices.

TABLE 4.3   SEMI Specifications for 200-mm Wafers

Property	Requirement
Dimension and Tolerance Requirements	
Diameter	199.5–200.5mm
Thickness, center point	700–750 μm
Notch depth	
Depth	1.00–1.25 mm
Angle	89–95°
Bow	65 μm
Warp	75 μm
Total thickness variation	75 μm
Edge profile coordinate Cy	242 μm
Orientation Requirement	
Orientation*	{110} ± 1°
Surface orientation	
On orientation for (111) and (100)	± 1°
Off orientation [111] toward nearest	2.5 ± 0.5°
⟨110⟩ on a plane parallel to notch	4.0 ± 0.5°
Orthogonal misorientation†	± 5°max

*For (111) slices (1$\bar{1}$0), (01$\bar{1}$), and ($\bar{1}$01) planes are equivalent, allowable planes. For (100) slices, the allowable equivalent {110} planes are (01$\bar{1}$), (010), (0$\bar{1}$1), and (0$\bar{1}\bar{1}$).

†The contribution of 5° of orthogonal misorientation to the total off-orientation angle will be less than 0.5°.

Crystal sawing

The technique employed in crystal sawing involves an angularly shaped cutting blade which has tension exerted around its outer diameter. The inner diameter serves as the cutting surface. This method is used to minimize the amount of silicon lost in the sawing process, referred to as "kerf loss." The other major requirement is to cut the crystal with precision and render very flat wafers. The inner-diameter cutting surface affords much more cutting stability than cutting with the outer diameter would. The average kerf loss ranges from 30 to 40% depending upon the condition of the equipment and the blade and the overall cutting parameters. Wafering or sawing is really not an efficient operation, and new equipment designs are being studied to reduce the large cutting loss.

Wafering is similar to lapidary in the sense that a smooth first cut results in much shorter lapping and polishing steps downstream. Also, stress must be minimized during the sawing step in order to minimize wafer breakage later. Figure 4.8 shows a wafering saw. The horizontal position of the ingot, when a vertical cut is being made, is intended to keep vibrations minimized. A smooth cut through the crystal is important, since any vibrations can cause ridges in the cut which will

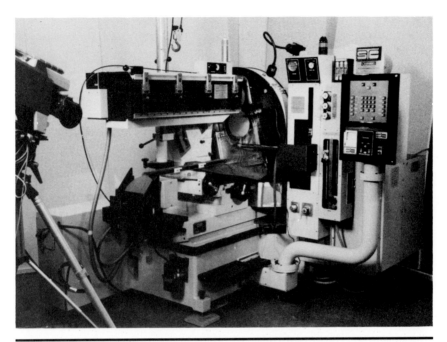

Figure 4.8 Wafer-slicing saw used in production of 200-mm wafers. (*Courtesy of Silicon Technology Corporation.*)

deflect the blade on its return from the cut. The larger the crystal, the more likely the chance such vibrations will occur. A ridge in the crystal from a nonsmooth cut is extremely difficult to remove, since it will have to be completely lapped off before the lapping of the general surface begins. In order to eliminate vibration, wafer saws employ air bearings which support the inner-diameter blade. Air pressure is exerted onto the shaft which holds the blade in tension.

One feature important to wafer manufacturers is real-time recording of the cutting parameters. Wafering saws use monitors which track blade and cutting parameters and record the operation in progress and the data from the process. These saws, like many used in the industry, will cut through crystals with a diameter of over 7 in. A well-engineered saw will cut through crystals that size and have no more than 5 to 7 μm of runout. Most of the runout is corrected in the lapping process.

Cutting blades. The blades used in crystal slicing are industrial-diamond-edged and require careful maintenance. Many saws automatically feed and operate, but a fair amount of "art" remains in running a large wafering machine. Wafers are automatically loaded into cassettes after being sliced off the ingot, but at several stages in the process it is necessary to check the diamond blade to make sure that its core is not rubbing against the ingot. The diamonds will typically last until they become dislodged; diamond wear is not the cause of a new blade being put on a slicing saw. Poor tracking and rubbing of the saw on the ingot will distort the blade, reduce its rigidity or tension, and result in nonuniform sawing.

The amount of friction caused by the blade against the ingot generates considerable heat, which is minimized by the introduction of a coolant between the two surfaces. If the coolant cannot reach in between the two surfaces, however, a problem because of the air velocity caused by high blade rotation results in coolant starvation, and heat is rapidly generated. If that occurs, even for a short portion of the cutting process, considerable stress can be generated in the wafer, as can nonflatness. Heat is also generated by silicon particles which get between the blade and the ingot, where they burnish and gall the wafer. Silicon particles "load" themselves between diamonds on the blade's edge and need to be cleaned out. Selection of the proper grit size and coolant is especially important in keeping heat at a minimum during wafer slicing. The pressure to cut slices faster to meet the production needs of the industry forces this specific operation to be more productive and efficient.

Strategies for blade design present dilemmas of their own. For example, producers can use a thinner blade to reduce kerf loss, but blade life will be reduced and quality sacrificed. Another option is to go the other way with thicker blades to produce higher-quality wafers which carry higher price tags, in the hope that the added income from higher-quality wafers will more than offset the kerf loss. In any case, the industry requires better wafers for advanced imaging processes, and consumers of wafer substrates will no doubt pay the price necessary to obtain the minimum quality necessary to produce advanced VLSI devices.

Wafering trends.   The main trends in wafer slicing are to improve wafer quality and production rates. Wafer quality is improved by many of the new monitoring techniques used to give real-time control to slicing saws. Single-slice recovery (SSR) allows for each wafer to be removed from the saw and loaded into a cassette, thereby reducing blade drag. The overall emphasis is to reduce bow, increase planarity, and reduce work damage. Bow is controlled by measurement of blade deflection, as cited earlier. Since wafer slicing is the first critical step in the fabrication of an integrated circuit from a grown crystal, continued emphasis on improving both quality and productivity is expected. Multiple-band saws and better-quality cutting blades are providing the means to cut more wafers per hour from a given ingot.

Lapping

The as-cut wafer slice from an internal-diameter saw will not be flat. Lapping is a free-abrasive machining process to make the slice flatter and thereby meet industry standards. It planarizes both wafer surfaces, generally in a double-sided lapping machine that has rotating metal lapping plates. These plates move with planetary motion to remove between 10 to 50 μm of substrate material. Lapping machines hold up to 50 wafers and typically require a 5- to 15-min cycle.

Edge profiling

The slicing operation often leaves microcracks which are removed in the edge-profiling operation. Edge contouring helps maintain good device yields by eliminating the source in silicon chips of cracks that emerge to cause problems later on in the process. Advantages of edge profiling include:

- Wafer is more fracture-resistant.
- Line contamination from flakes and particles is reduced.
- Quartz boat loading is easier.
- Mask life in proximity and/or contact printing is improved.
- Photoresist edge buildup is minimized.
- Epitaxial crown effect is reduced.

Several types of equipment are used for edge profiling, including edge grinders, abrasive blasting, and abrasive disk polishing or grinding. The wafer edge is typically ground on the top, bottom, and periphery or outside edge to ensure a uniform and accurate wafer-to-wafer result. As wafer fabrication lines continue to automate, a high degree of consistency in wafer size becomes increasingly important. Eventually, it is possible to imagine perimeter control at a level that will allow the elimination of global alignment because the wafer will be so repeatedly uniform in diameter.

Wafer surface etching

Wet etching of the polished wafer is performed to remove surface damage caused by previous mechanical lapping steps. Caustic and acidic etchants are used to remove 10 to 30 μm of silicon wafer from both sides. Wet acid etches are composed of hydrofluoric, acetic, nitric acid mixtures, but noxious gases and nonuniform etch rates are problems. The caustic etches are sodium or potassium hydroxide–based solutions, used at about 100°C, that etch more uniformly. In either case, automated etch process equipment is used, followed by multiple deionized water rinses.

Mechanical and thermal gettering

Intentional work damage on the backsides of wafers is used to getter, or absorb impurities, during wafer processing. These damage or strain sites in the crystal lattice keep impurities from migrating up to the active area of the wafer. The induced gettering process can be mechanical or thermal. Mechanical gettering sites are created by using one of several techniques that include liquid honing, in which a stream of water containing abrasive particles is directed across the wafer backside. Direct surface abrasion also is used, along with a type of sandblasting in which alumina particles are directed at the wafer. A more recent technique involves using a laser beam to melt or remove part of the wafer backside. Several laser pulses are directed in a pattern that establishes the gettering sites.

The main emphasis in backside gettering is to avoid residual contamination and to create uniform damage depth. Thermally and chemically induced gettering offer some advantages over mechanical gettering. The use of oxygen gradients in the wafer is a common way to achieve a gettering mechanism. The gradients are produced ideally in the manufacturing process so as to avoid an extra step. In-process impurities are a well-known source of defect, and the use of gettering sites is well established as a means of improving yield.

Inspection and marking

Wafers that have been sliced from ingots, lapped, and etched are ready for a prepolish inspection. This inspection and the subsequent marking step serve to identify and categorize wafers according to the following parameters:

- Visual surface appearance
- Thickness
- Bow
- Resistivity
- Cosmetic uniformity
- Edge contour

- Chips and flakes
- Etch stains
- Residual saw marks

Many of the inspection and marking operations are being automated to eliminate operator handling and reduce operator-related defects. In many cases, the wafers are automatically handled between cassettes and fed into various types of inspection devices with which an operator can take several measurements and, with the push of a button, index the next wafer into position. At the completion of the inspection step, wafers are marked to reflect resistivity and other vital parameters. After the sorting stage, the wafer resistivity is known, and this provides a logical point at which to mark the slice. The marking is often performed prior to the wafer etch to eliminate possible surface damage caused by the laser or similar marking device.

Wafer polishing

The wafer polish step is vital to good lithography, since the wafer must be completely free of microcracks, debris, and anomalies that optically or electronically interfere with the transfer of images from masks to the wafer surface. The polishing step involves two main operations. The first step in polishing is bulk removal of material with a colloidal silica slurry applied to a polishing pad that is moved across the wafer surface. This step will remove 10 to 35 μm of the surface and at the same time flatten out the wafer surface considerably. The slurry is alkaline and etches the wafer chemically while the abrasive removes material mechanically.

Wafer-polishing machines need to control many of the various parameters at work in this final surface preparation stage. The parameters include the slurry temperature, composition, flow rate, distribution, and chemical composition. The wafer-polishing equipment must maintain uniform temperature where it contacts the wafers, along with uniform pressure radially and along the circumference. The wafer-polishing pad must also maintain a high degree of uniformity of pressure, composition, and slurry concentration. The wafers must be processed within closely controlled cycle times and rinsed well to remove debris. The total polishing environment must be well controlled in terms of humidity, temperature, and cleanliness.

Postpolish cleaning

Wafers are generally cleaned after polishing in one of several types of solutions, often in multiple tank operations. Detergent-type cleaners are used first to take off the remaining residues of slurry from polish-

Figure 4.9 Topographical and isometric wafer plots. (*Courtesy of Tropel.*)

ing. This step can be followed by a double rinse, and then wafers are taken into a solvent rinse to remove any solvent-soluble soils. A final dip in a mild etch, giving the wafer surface a microetch finish, is a desirable last step in postpolish cleaning operations. These techniques are varied considerably, depending upon the needs of a given process.

The postpolish clean step is typically followed by a final cleaning step that occurs before photo or electron lithography steps begin. The wafers are often mapped for flatness before imaging steps begin. Figure 4.9 shows a standard topographical and isometric plot of a wafer before imaging. Each contour line represents 1 μm of nonflatness.

Flatness

Definition

As integrated circuit geometries become smaller, the effect of the wafer on the lithography process becomes increasingly critical. In many cases, defects encountered in photoresist imaging or etching can be attributed to physical discontinuities in the wafer. The most common wafer problems that lead to defects are

1. Warpage
2. Nonflatness

3. Surface damage

4. Crystal plane slippage

Figures 4.10 and 4.11 illustrate the warpage and nonflatness problems. Wafer "warpage" is deviation of the wafer surface resulting in a concave or convex shape that equals the wafer diameter. Wafer "nonflatness" is simply deviation from the same plane, but by an amount less than the diameter of the wafer.

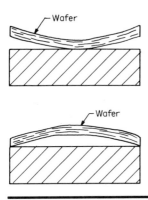

Figure 4.10 Wafer warpage.

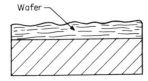

Figure 4.11 Wafer nonflatness.

Surface damage

"Surface damage" is usually associated with a previous mechanical lapping or polishing operation, and it can vary considerably in width and depth. Chemical polishing is designed to remove any final traces of surface damage, but it seldom corrects the damage caused by a mechanical polishing or lapping operation. Wafers with this problem must be culled out; otherwise, that portion of the wafer will be defective. Figure 4.12 shows an example of wafer surface damage.

Flatness versus resolution

Problems with the wafer as a substrate have always existed, but their contribution to device yield has increased measurably as wafer diam-

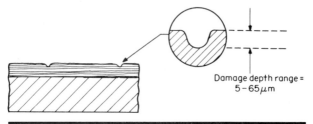

Figure 4.12   Wafer damage.

eter has increased. As device geometries have gone below 1 μm, projection printing has become widespread. The results of projecting 0.5-μm images onto wafers that have significant nonflatness or warp are line width variation and loss of critical dimension (CD) control. All the projection aligners used in IC manufacturing have limited depth of focus. When high-resolution projection systems are used, the depth of focus of the image plane is relatively shallow and must be considered along with substrate flatness and resist thickness. The need for careful and complete wafer characterization becomes obvious when we consider the restrictions placed on imaging by high-resolution geometries, the depth-of-focus limitations of exposure equipment, and out-of-plane wafer distortions.

SEMI wafer specifications[2]

The Semiconductor Equipment and Materials Institute (SEMI) develops standard specifications for silicon wafers, and those specifications are shown in Table 4.1. This type of information has helped suppliers, manufacturers, and users to standardize and control IC process materials and equipment to tighter specifications in order to produce increasingly complex circuits at decreasing costs.

Measurement Techniques

Optical flat

Measuring wafers for flatness can be done optically by fringe counting: placing an optical flat over the wafer and viewing it under monochromatic (sodium) light. This simple technique is easy to set up, since Edmund Scientific supplies both high-quality optical flats and monochromatic sources. The approach is useful for research or and when only a small number of wafers needs to be "surface mapped."

Laser interferometer

A more complex, production-oriented method employs a helium-neon (He-Ne) laser interferometer that generates a topographic contour map of the wafer surface. The evaluation of these interferograms is shown in Fig. 4.13. Note that the instrument has, in this case, been preset to give 1-μm fringes, but the setting can be adjusted by the user.

Ellipsometer

Ellipsometry is a highly accurate method for checking wafer flatness. Figure 4.14 shows a surface topographic map produced by an ellipsometer made by Tropel. The technique is based on optical interference.

Wafer flatness issues

The vacuum clamping that occurs during the wafer fabrication process causes internal stresses that add further to the problem of characterizing changes in wafer surface topography to better control line geometry reproduction from wafer to wafer. Wafers selected for use must be well within the depth-of-focus tolerances of the resist imaging system.

A study of the ingots from which various batches of wafers come shows that some ingots typically produce less-flat wafers than others. This information correlates well with the claim that wafer nonflatness can be induced during the ingot-slicing operation. This would probably mean that all wafers from a single ingot would be affected by a single slicing operation. The slicing problems include contours and nonuniformities caused by slicing blade distortion. Referencing all

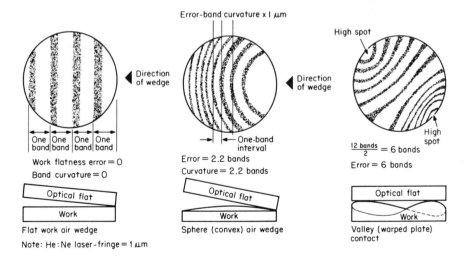

Figure 4.13 Flatness interferometer principle.[3] (*Courtesy of Intop.*)

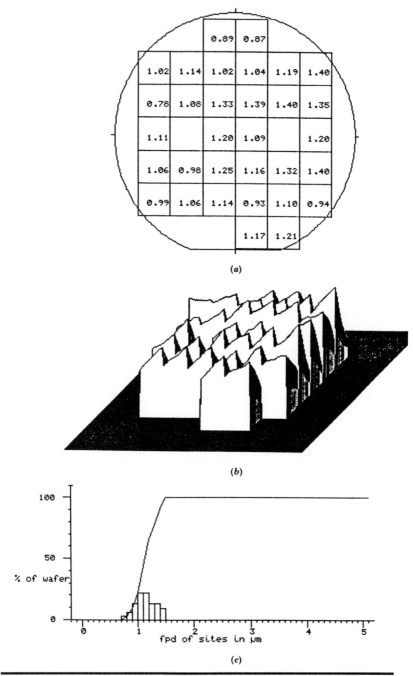

Figure 4.14 Wafer surface mapping. (*a*) Die site map and flatness figures. (*b*) Contour map and scale of deviation. (*c*) Die sort versus yield. (*Courtesy of L. Denis, GCA/Tropel.*)

wafers according to the ingots from which they were sliced would therefore help in the process of wafer flatness characterization. Figure 4.15 shows some flatness values according to the ingot from which they were sliced.

Wafer homogeneity testing

The continual reduction of geometries places greater significance on localized lack of homogeneity within the silicon wafer or in layers of semiconductive material over the silicon. Nondestructive testing of wafer homogeneity has been developed by using a variable-frequency absorption scanner. The device measures local spectral variations in infrared transmission and produces a written scannogram. The immediate application of this device to IC fabrication would be as a quality control tool for checking crystal homogeneity.

High-Temperature Effects

Warpage

After profiling wafers prior to imaging, tests must be run to determine wafer flatness and defect levels, particularly after any high-temperature operations. Multiple high-temperature operations can add up to 2.0 µm of distortion beyond that caused by the first high-temperature step. The greatest changes occur after the first high-temperature step, and they cause a wide variation in wafer flatness.

Plastic flow

While the effects of high-temperature processing on overall flatness are rather easily detected, plastic flow and other process-induced dimensional distortions are more difficult to measure and add to the problem of achieving submicron geometries with 0.1-µm tolerances. For example, above 600°C, plastic deformation occurs in the form of

Ingot number	1	2	3	4	
≤ 4 µm	48%	82%	98%	88%	
≤ 8 µm	17%	15%	2%	12%	
> 8 µm	35%	3%	0%	0%	
Number of wafers	23	88	117	80	
Number of readings	46	176	234	160	(With vacuum clamping only)

Figure 4.15   Wafer flatness uniformity by ingot.

slippage in the crystal planes. That causes dislocations that are to be distinguished from those resulting from the initial crystal-growing process or from stress relief after slicing or lapping and polishing operations. Slippage within the wafer or throughout the wafer is not an uncommon problem. The slippage occurring within active portions of a device will necessarily change the electrical performance and could adversely affect device yield. The size and frequency of process-related slippage and crystal dislocations are not attributable to a single source; they are dependent on numerous parameters, including wafer and oxide thickness, pattern density, doping deposition, and oxide growth cycles.

The many studies conducted to date show that process-induced distortions can be up to 0.5 μm on an individual site. This amount of distortion is not highly critical to IC designs with 3- to 5-μm geometries, but it *is* a factor for VLSI circuits with submicron design rules.

Equipment

Vacuum chuck effects

The increased attention that has been given to wafer flatness and its relationship to resolution in imaging small geometries is causing manufacturers to look to printing equipment flatness factors. The various types of wafer chucks used on aligners affect flatness differently. Chucks with a series of radial pinholes produce a different wafer flatness under a vacuum than concentric-grooved chucks produce. Test results show that total flatness deviation on a pinhole-type chuck can reach 1.25 μm, whereas grooved chucks typically distort a wafer 1.75 μm. Tests on a number of different wafer chucks show a variation among chucks of about 2 μm. Chucks that can be made of materials like aluminum oxide can be optically polished for high flatness.

A more significant difference in wafer flatness is caused by varying the chuck. The relationship between the amount of vacuum and wafer flatness should be measured. The use of additional vacuum pressure adds to exposure throughput time, but it also improves flatness and resolution. Thick wafers will reduce vacuum chuck and thermal effects on flatness.

The effect of high vacuum on wafer flatness during exposure of photoresist raises the question of contamination *between* wafer and vacuum chuck. Entrapment of a 1-μm or larger particle between the wafer and the vacuum chuck is another cause of wafer nonflatness and reemphasizes the need for cleanliness in the imaging area. The front sides of chucks should be given periodic cleaning to remove resist particles or other particulate contamination. The backsides of wafers must be kept free of particles large enough to distort the wafer during vacuum clamping in exposure aligners.

Wafer testing

The importance of checking wafers for flatness has increased the number of available flatness-testing systems.

Economical manufacturing of VLSI devices necessitates complete characterization of the silicon wafers before and during the IC process. Characterization tests should identify warpage, nonflatness, surface damage (including dislocations), and in-process temperature-induced distortions (plastic flow and crystal slippage).

Checking oxide thicknesses is another aspect of wafer characterization. In many cases it will be inconvenient to take actual measurements of the oxide or other film as deposited. Oxides have characteristic and predictable colors (reflection) according to specific thicknesses. A thermal oxide color chart is included in the *IBM Journal of Research and Development,* vol. 8, 1964, p. 43. The chart applies to thermal oxide, Vapox, doped (boron and phosphorus), and sputtered oxides. Table 4.4 is a thermal oxide color chart.

A practical way to identify oxide thickness quickly is to generate several oxide layers of varying thicknesses and use them as a reference. The oxide thicknesses on the reference wafers could correspond exactly with the thicknesses typically used in the process, and this would give line operators a useful in-process control tool. Closer and more absolute control is maintained through the use of beta ray or other nondestructive film thickness devices.

Destructive techniques for measuring deposited or grown layers of metals, nitride, and oxide include Tally Surf, Dektac, and similar stylist-based systems, as well as cross-sectioning followed by SEM measurement. Since IC fabrication is the result of an accumulative layering and structuring of thin films, close quality control of film thicknesses throughout the process is essential to obtaining good device yields.

TABLE 4.4   Thermal Oxide Color Chart*

Film thickness, μm	Order, 5450 Å	Color and comments	Film thickness, μm	Order, 5450 Å	Color and comments
0.050		Tan	0.585		Light orange or yellow to pink borderline
0.075		Brown			
0.100		Dark violet to red violet	0.60		Carnation pink
0.125		Royal blue	0.63		Violet-red
0.150		Light blue to metallic blue	0.68		"Bluish"‡
0.175	I	Metallic to very light yellow-green	0.72	IV	Blue-green to green (quite broad)
			0.77		"Yellowish"
0.200		Light gold or yellow— slightly metallic	0.80		Orange (rather broad for orange)
0.225		Gold with slight yellow orange	0.82		Salmon
			0.85		Dull, light red-violet
0.250		Orange to melon	0.86		Violet
0.275		Red-violet	0.87		Blue-violet
0.300		Blue to violet-blue	0.89		Blue
0.310		Blue	0.92	V	Blue-green
0.325		Blue to blue-green	0.95		Dull yellow-green
0.345		Light green	0.97		Yellow to "yellowish"
0.350		Green to yellow-green	0.99		Orange
0.365	II	Yellow-green	1.00		Carnation pink
0.375		Green-yellow	1.02		Violet-red
0.390		Yellow	1.05		Red-violet
0.412		Light orange	1.06		Violet
0.426		Carnation pink	1.07		Blue-violet
0.443		Violet-red	1.10		Green
0.465		Red-violet	1.11		Yellow-green
0.476		Violet	1.12	VI	Green
0.480		Blue-violet	1.18		Violet
0.493		Blue	1.19		Red-violet
0.502		Blue-green	1.21		Violet-red
0.520		Green (broad)	1.24		Carnation pink to salmon
0.540		Yellow-green			
0.560	III	Green-yellow	1.25		Orange
0.574		Yellow to "yellowish"†	1.28		"Yellowish"

TABLE 4.4 Thermal Oxide Color Chart *(Continued)*

Film thickness, μm	Order, 5450 Å	Color and comments	Film thickness, μm	Order, 5450 Å	Color and comments
1.32	VII	Sky blue to green-blue	1.46		Blue-violet
			1.50	VIII	Blue
1.40		Orange	1.54		Dull yellow-green
1.45		Violet			

REFERENCES

1. D. J. Elliott and M. A. Hockey; "One Micron Range Photoresist Imaging: A Practical Approach," Society of Photo-optical Instrumentation Engineers, *SPIE,* vol. 135, *Developments in Semiconductor Microlithography III,* 1979, pp. 130–146.
2. Semiconductor Equipment and Materials Institute (SEMI), "Tentative Specifications and Standards for Silicon Wafers," 1980.
3. P. Langenbeck, Intop Technical Literature on Flatness Interferometer, Intop Entwicklungen, 1979.
4. Tropel Technical Literature on "Autosort" Flatness Determination Equipment, 1980.
5. Y. F. Chiu, W. H. White, and R. C. Guggenheim, "Semiconductor Wafer Mask Flatness and Warpage," *Proc. Conf. Microlitho.,* Paris, 1977, pp. 317–322.
6. R. A. Colclaser, *Microelectronics: Processing and Device Design,* Wiley, New York, 1980.
7. H. R. Huff, "Chemical Impurities and Structural Imperfections in Semiconductor Silicon," *Solid State Tech.,* February 1983.
8. H. M. Liaw, "Oxygen and Carbon in Silicon Crystals," *Semiconductor Int.,* October 1979.
9. W. C. Till and J. T. Luxon, *Integrated Circuits: Materials, Devices, and Fabrication,* Prentice-Hall, Englewood Cliffs, N.J., 1980.
10. Ref. 9.
11. G. Fiegl, "Recent Advances and Future Directions in CZ-Silicon Crystal Growth Technology," *Solid State Tech.,* August 1983.

5

Surface Preparation

In IC fabrication processes, wafer cleaning is a critical step because the sources and types of contamination are numerous and have a direct bearing on chip yield. For example, devices are so sensitive that a monolayer of sodium ion contamination equal to 1×10^{-4} ions inverts the surface of a 1-Ω/cm silicon device. MOS devices, which rely more upon surface state operation than bipolar devices do, are especially sensitive to contamination. We will consider the various types of wafer contamination and their sources and then discuss detection methods and equipment as well as preventive measures. Methods for removing contamination at various steps in the IC process will be discussed, followed by a review of wafer-cleaning equipment.

Contamination

There is no question about the importance of cleaning in wafer fabrication. The effects of contamination on device yields can be economically disastrous, and with geometries getting smaller every year, the need becomes more acute. The tolerable size and number of contaminants for a given wafer are in proportion to the critical geometry sizes, and the number for a given process is smaller as technology pushes in the direction of larger chip sizes and smaller circuits on those chips. Figure 5.1 gives a relative comparison of contaminant size.

The various types of wafer contamination are usually broken into two general categories: particulates and films.

Particulates

"Particulates" can be defined as any detectable shape of material on the surface of a wafer that has readily definable boundaries. Particulates are generally more easily detected than residues, films, or

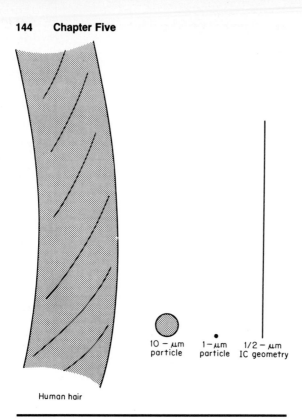

Human hair

Figure 5.1   Relative size of clean-room contaminants.

monolayers because of their distinct three-dimensional nature. The various types of particulates commonly found on wafer surfaces are

1. Silicon dust

2. Atmospheric dust

3. Abrasive particles (transport boxes, metal corrosion, equipment surfaces, cassettes)

4. Lint

5. Organic particles, such as photoresist or bacteria, and organic solvent residues

6. Gold, copper, or iron (atomic layers)

7. Inorganic ions of sodium, potassium, or calcium

8. Fatty acids (from fingerprints)

9. Synthetic waxes

Silicon dust.   Silicon dust is associated with the wafer-polishing and scribing operations. This type of contamination is prevented by coating the wafer with a protective layer of organic material, such as photoresist. A flexible negative optical resist is well suited for this purpose, because it does not require exposure in order to form a more resistant layer prior to scribing. The wafer is then simply coated, softbaked, and sent to the scribing operation. Since the photosensitive properties of either a negative or a positive resist are not required, a nonsensitized coating material may be used. Photoresist-related products are preferred because they are formulated to have the rheologic properties necessary for complete step coverage and excellent coating uniformity.

Silicon particles also may be introduced by way of the slicing operation, but this is much less likely because of the chemical and mechanical etching and cleaning operations that typically follow wafer slicing.

Atmospheric dust.   Airborne material, the most common source of particle contamination, is often classified as "atmospheric dust" to differentiate it from other types of particles that also may be airborne. Airborne dust particles are controlled by a variety of methods, including air locks in clean rooms, Hepa filters, clean-room furniture and clothing, and antistatic or deionizing devices.

The operators in a clean room are the largest single source of airborne particles of any description. For example, after taking a "smoke break," clean-room operators, upon reentering a clean room, can dispel as many as 2 million particles from their bodies. Nonsmokers are much cleaner, they give off as few as 10,000 particles/hr. Cosmetics are another source of contamination, as are all particle-generating materials, such as pencils.

Abrasive particles.   Abrasive particles or inorganic particulates are introduced onto wafer surfaces from several sources, including the polishing compounds used in wafer surface preparation. Most of the chemicals used in wafer fabrication are suspect for the introduction of inorganic particles, including water, resist developers, and strippers and etchants. Abrasive particles from wafer polishing are generally checked by complete preoxidation cleaning techniques, described in the "Cleaning Methods" section of this chapter. Abrasives from the chemicals just mentioned are best controlled by submicron filtration through an absolutely retentive membrane filter.

Lint.   Lint particles are a common source of particulate contamination. They generally come from operators in the clean room, either from their clothing or skin. Even with clean-room garments (hats, boots, coats, and gloves), operators will give off a considerable number of lint particles that are attached by static changes or simply escape around their protective clothing. The exclusion of all padded chairs or furniture with shedding materials is essential in eliminating lint particles.

Occasionally, lintlike particles are introduced from the release of natural fiber filter media. Natural fibers, with uneven strand lengths, release fiber ends when filter pressures become excessive, usually about 20 lb/in^2. Frequent filter changes and the use of synthetic filter materials are advised for this reason. Since cotton prefilters are often desirable for cost reasons, a synthetic membrane absolute filter should be used downstream to catch particles from "media release."

Organic particles: Photoresist. Photoresist "chunks" are often a source of defects and a common and obvious contamination source. They arise from any mechanical abrasion of resist coatings, as when wafers are dropped or come into hard contact with other surfaces. They also result from wafer tweezers gripping the resist coatings on the edges of wafers.

In proximity printing, there is still occasional mask-wafer contact due to close working distances, and resist particles break off and redeposit on wafer surfaces. Resist particles falling onto an image area will result in a defect.

Positive photoresists are particularly brittle (compared with negative resists), and even a slight touch with wafer tweezers can cause localized shattering of the coating. One way to minimize embrittlement of positive-resist coatings is to use longer spread cycles and slower ramps in coating, thereby allowing the fluid to "relax" and reach equilibrium before final acceleration. In this way, resident stress in the film is reduced.

Organic particles: Bacteria. Bacterial colonies and other organic structures grow in water supplies and may end up on a wafer surface. In-line "black light" sources are used to kill bacteria in water lines before they reach the filters. This preventive measure, coupled with good absolute filtration to submicron levels, should eliminate bacterial particulates.

Films

A "film" can be defined as a detectable layer, often random in character, of foreign material. Films, in contrast to particles, have less distinct geometric shapes; yet portions of films may break loose and become particles, as often happens with photoresist scums. Examples of wafer-contaminating films are

1. Solvent residues

2. Photoresist developer residues

3. Oil films

4. Silicone films

5. Metallic films

6. Water stains

Solvent residues. Several solvents are used in IC manufacturing, including acetone, methyl ethyl ketone, trichloroethylene, isopropyl alcohol, xylene, Freon TF, and methyl alcohol. Solvent baths that are not changed often enough may redeposit contamination on wafer surfaces. Moreover, if not applied or dried uniformly, these solvents will stain wafer surfaces, and each solvent has a distinct staining pattern (see Fig. 5.3). Wafer contamination from solvent staining may lead to poor resist adhesion and increased undercutting during etching.

Photoresist developer residues. During development of photoresist films, the resist is dissolved in the developer solution and removed in a subsequent developer rinse operation. In many cases, the developer may contain an overconcentration of dissolved photoresist, in which case a film will be deposited. The developer must be changed periodically according to a set number of wafers or on a predetermined time schedule or increased in concentration for in-line puddle or spray. Added spray rinsing may be needed in these cases. Suppliers of resist developers may provide developer yield data to help establish such a schedule. Another way in which photoresist developer films are deposited onto wafer surfaces is from inadequate postdevelopment rinsing. Spray rinsing is suggested because it provides a mechanical surface-scrubbing action to help remove dissolved resist.

Some positive-resist developers are rendered less active by exposure to air because the active ingredient is neutralized by reaction with carbon dioxide in air. These developers will not completely remove exposed positive resist if they are nearly "spent" by being left exposed to air too long. Periodic analysis for developer bath activity, both in immersion and in spray modes, is suggested to maintain process control of developer activity. Figure 5.2 shows an example of a resist residue. Negative-resist developers also are subject to problems of solvent evaporation at nonuniform rates, which causes a different ratio of solvents with different developing actions or increased concentration of dissolved resist.

The developer rinse step is important in removing resist "scum," as it is often called. Some processes incorporate a two-step developer to prevent excess loading of dissolved resist in a single developer solution. Care must be taken not to overdevelop the image, which would add more dissolved resist to developer solutions and to wafer surfaces.

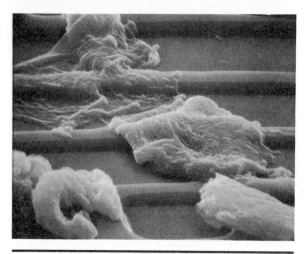

Figure 5.2   Resist residue (SEM at 1000 ×).

Another common cause of excess dissolved resist in developer baths is inadequate softbaking. Excess solvent remaining after softbaking leaves the resist easily attacked by developers, thereby adding resist material to the developer.

Oil films.   Oil is introduced onto wafers through gas or air lines that are not properly filtered. Nitrogen, air, and other gas lines should have in-line cannister filters to catch any oil before it is blown onto the wafers. Oil may also come from the water supply, which, again, should be highly microfiltered to remove both chemical and particulate contamination.

Silicone films.   Silicone films will cause resist wetting and adhesion problems even in parts-per-million concentration levels. Silicones commonly come from operator hand creams or from lubricants used in wafer-handling equipment. All operators should be advised about chemicals that, if present in wafer-processing areas, may find their way onto wafer surfaces. All lubricants that generally contain organic compounds, including surfactants, should be kept out of wafer-processing areas. Severe photoresist dewetting is an indication of the presence of silicone contamination in resist or on wafer surfaces.

The organic films typically exhibit weak electrostatic binding, but polar molecules will bond strongly to the wafer surface. The organic contamination sources include cutting oils, coolants, lubricants, human skin oils, particulates, detergents, and solvent films. These materials give the wafer a hydrophobic surface which resists aqueous cleaning. Detection of these residuals is made by atomizing a layer of water on the surface and checking for uniformity. The contaminated areas will show up as nonuniform streaks, blotches, or islands, and atomizing will detect amounts as small as 0.16×10 g/cm^2 or down to submonolayer thicknesses. Also, cold plate condensation, in which a layer of moisture is formed by placing a warm wafer on a cold plate or by passing warm air over the chilled wafer, is useful for residue detection.

Metallic films. There are several sources of metallic films deposited onto wafers in process. Active silicon, for example, will cause replacement plating of ions onto its surface from any solution containing metals or metal ions. This is most commonly found in etchant or resist stripper baths, both of which usually contain metal ions and free metal in solution. Resist strippers and etchants often contain varying quantities of gold, copper, tin, silver, and other metals. In developer baths, mobile sodium ions are readily entrapped or absorbed by wafers, causing electrical problems later. The best preventive measure is to use metal-ion-free strippers and developers and to check these solutions periodically for metal content.

Good rinsing is a simple technique for removing metal contaminants from wafer surfaces. A triple-cascade deionized rinse followed by a spray rinse is recommended. This will *not* remove metal introduced by replacement plating onto silicon; that can be avoided only by changing the process step at which the metal was introduced.

Water stains. Water stains are caused by impurities in rinse water and can cause resist adhesion problems. Water should be carbon-filtered to remove organics. See "Quality Control."

Detection and Analysis of Contamination

Electron microscopy

The scanning electron microscope is a standard tool for detecting a variety of wafer surface contaminants. SEM analysis is useful for detection and identification of photoresist residues, metallization cracks, metal step coverage, particulates and their possible sources, pits, scratches, and other defects. The SEM can be fitted with an x-ray analyzer that will identify elements of a contaminant. Transmission electron microscopy (TEM) is particularly useful for surface mapping of very delicate structures. In this case, a replica of the surface is made with a surface-conformable plastic. The replica is peeled off the sample and looked at with a TEM.

Another application for electron microscopy in wafer quality control is inspection of wafer surfaces before photoresist imaging operations. A variety of wafer defects, including saw marks, scratches, and dislocations, can be readily observed.

Contamination in photoresist coatings also is readily observable with SEM and TEM analyses. Any dimensional structures on the wafer surface that might be difficult to observe with optical microscopy may be observable with electron microscopy. For example, undesirable oxide structures left after etching and caused by resist residues might not be seen with an optical microscope but will show up clearly in an SEM.

The most difficult forms of contamination to detect are residual layers or films left on wafer surfaces. The SEM will show these contaminants when an optical microscope will not, since its resolution (50 to 100 Å) keys in to the surface texture of various materials, and the contaminant, even if it is a thin uniform film, will display a foreign surface texture. Cross-sectional analysis of a wafer with SEM up to $100,000 \times$ will reveal monolayers not detectable with optical microscopes. The microphotographs in Fig. 5.3 show how various surface contaminants can be identified. The only major disadvantage of

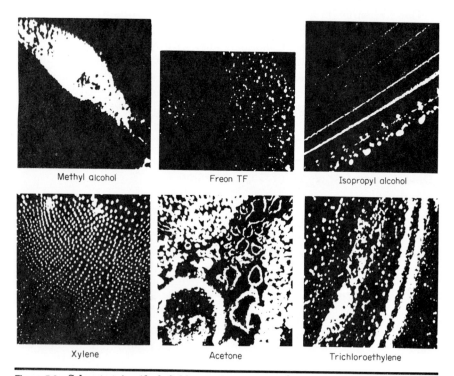

Methyl alcohol   Freon TF   Isopropyl alcohol

Xylene   Acetone   Trichloroethylene

Figure 5.3 Solvent stains (dark-field optical microscopy).[2]

electron microscopy for contamination analysis is that it is often a destructive technique unless the entire wafer is used in the chamber.

Optical microscopy

Several types of optical microscopy are used to detect wafer contamination. The most common is "clear field" or "light field" inspection, whereby the entire sample is flooded with light. A picture taken with a Polaroid camera on an optical microscope can show a considerable level of detail and provide a high-quality image. Figure 5.4 shows the principle of bright- and dark-field optical microscopy.

Attaching an interference objective to the microscope permits highlighting of layer details, and several types of phase contrast and interference devices and lenses are available to achieve this type of analysis. A Nomarski objective lens attached to the microscope is an example of one type used.

Highlighting surface irregularities is an ideal application for dark-field optical microscopy. A dark-field objective reflects back into the objective only rays from surface irregularities; the rest of the field is dark. Dark-field microscopy is useful for detecting contamination in resist coatings, particulates of all types on the wafer surface down to 1 to 2 μm in diameter, surface scratches, and solvent residues.

The main advantages of optical microscopy are relative ease of use,

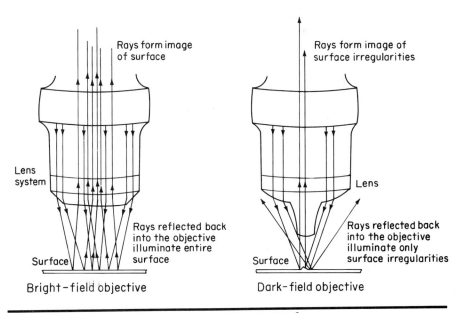

Figure 5.4 Principles of bright- and dark-field illumination.[2]

relatively low cost, and nondestructive nature. The main drawbacks are limited resolution and an inability, in many cases, to identify the exact nature of contamination. One technique for extending the resolution is to use "oil immersion" microscope objectives: A drop of oil is placed between the end of the lens and the sample. When the oil makes contact with both surfaces, it acts as a light collimator, and images that would normally be fuzzy or unresolved because of the reflection of angular light rays become clear and well resolved. Oil immersion is capable of producing sharp images of photoresist structures at $1000 \times$, whereas the same magnification without oil immersion yields "fuzzy" and largely unresolved images.

Other analytical methods

Mass spectrometry is a useful tool for detecting sodium levels in silicon films. Typically, a spark source mass spectrometer is used to ionize the sample. By this method, the detection capability is approximately 1 ppm for sodium in silicon and as low as 10 ppb for selected contaminants.

Auger analysis is particularly useful for identifying the type of contamination when a film has been detected by other means. It works by sputtering the film off the surface and measuring the spectra in the process. It has become a primary tool for identifying all types of contaminants, with the exception of photoresist, which is primarily oxygen and carbon and generally shows up regardless of whether resist is present.

A condensation, or "fog," test is an old and simple method for detecting filmlike wafer contamination. When moisture condenses on a wafer, it forms a uniform layer *if* the wafer is clean. If the wafer has a solvent residue or resist scum monolayer, the condensed moisture layer acts as an optical film and reflects back an image of the irregularity. Moisture condensation can be created by placing wafers in steam or simply by blowing on them. Moisture condensation analysis will prove useful for detecting most types of films and residues that occur nonuniformly on wafer surfaces. In some cases, it will detect films not seen with optical microscopy.

Radiotracer techniques, using radioactive iodine, are used because the tracer reacts with organic materials. The development and use of the other types of analytical equipment have reduced the need for radiotracer analysis, and its desirability is obviously less because of the need for radioactive compounds.

Chemical vapor deposition of silicon dioxide also will detect irregularities on the surface of wafers. Operating in a way similar to moisture condensation, the thin oxide layer "mirrors" back the topography it conforms to via light interference. This is a less desirable test than others because of its destructive nature.

Particle counters. Measuring the density of particulate matter in a clean room with a particle counter is a useful preventive of wafer contamination. Particle counters are available in a variety of sizes and prices, depending on the accuracy and detection limit desired. This tool should be used to identify major sources of contamination so they can be systematically eliminated or controlled to acceptable levels.

Several instruments are used to map the distribution and density of particles. One is the Surfscan from Tencor Instruments, the operating principle of which is shown in Fig. 5.5. The Surfscan will detect not only particles but also fingerprints, haze, and scratches in every step of the process, thereby enabling operators to know where particle contamination is originating. The system uses a laser with automatic calibration to detect surface abnormalities. The resolution of the system is well into the submicron area, and the monitoring rate is good. It also sorts wafers as "good" or "bad" and can process with good cassette-to-cassette throughout.

The hard copy printer logs in all of the data and delivers a printout, as well as giving a live picture on the video screen. A belt drive conveys the wafer under a focused helium-neon laser beam that scans the surface in a lateral raster motion. The scan pattern has sufficient overlap to ensure that the entire surface is covered. A clean surface has a predictable angle of light reflection, whereas a contaminated surface will scatter the laser beam. The integrating light collector amplifies the scattered light in the photomultiplier tube. The signals are then analyzed. A wide variety of defects can be detected, as indicated in Fig. 5.6. Point defects, area defects, and line defects are shown; they

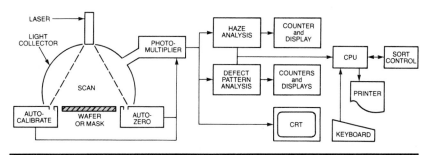

Figure 5.5 Operating principle of Surfscan. (*Courtesy of Tencor Instruments.*)

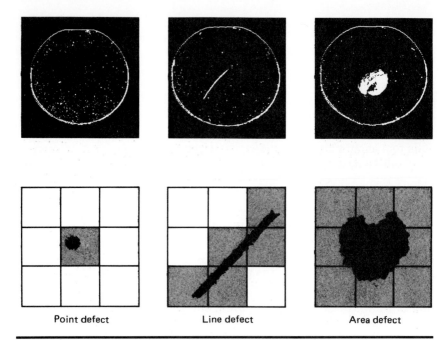

| Point defect | Line defect | Area defect |

Figure 5.6   Defect categorization of Surfscan. (*Courtesy of Tencor Instruments.*)

can arise from epitaxial spikes, scratches, cracks, particles, pits, protrusions, fingerprints, haze, and slips. The detection level of the systems is down to 0.1 μm.

Cleaning Methods

Preoxidation cleaning

After wafer-polishing operations, oxidation to grow the initial silicon dioxide layer takes place. Since silicon particles or other contamination will interfere with the growth of a uniform oxide layer, thorough cleaning is important. Particles may be held onto the surface only by van der Waals forces; if so, they can be removed easily by placing wafers in water. The high surface tension of a water bath with a wetting agent added will normally dislodge van der Waals–bonded particulates. The bonding forces of submicron-sized particulates can be high, depending upon ionization of wafer surfaces, clean-room humidity, and other room conditions. Figure 5.7 shows potential-energy curves for three types of bonds, along with diagrams of the manner in which the materials connect. The forces are interatomic and account for much of the behavior of particulates and other small contaminants on wafer surfaces. Silicon particles, however, may have oxidized on the

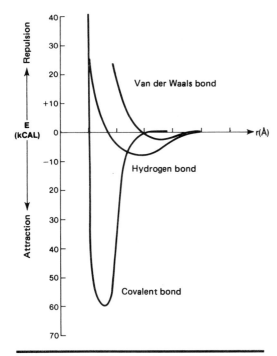

Figure 5.7   Potential-energy curves for different types of interatomic forces. (*Courtesy of Shipley Company.*)

wafer, thereby making a chemical bond along with the already-present physical bond. The use of mechanical nylon brush wafer scrubbers is recommended.[3] The brushes must be changed periodically to prevent loading and redeposition of particles onto other wafers.

Wafer chemical cleaning procedure

A chemical cleaning process for silicon wafers prior to oxidation is as follows:[5]

1. Load wafers (in Teflon boat) into a solution made up of 1 part hydrogen peroxide, 2 parts deionized water, and 1 part ammonium hydroxide. Note: A Teflon beaker must be used. Immerse for 20 min at 120 to 150°F. Avoid boiling the solution.

2. Rinse in deionized water for 3 to 5 min.

3. Immerse wafers into a solution made up of 1 part hydrogen peroxide, 2 parts deionized water, and 1 part hydrochloric acid. Note: A Teflon beaker must be used. Immerse wafers for 20 min. Caution: Use extreme care, because this mixture is exothermic.

4. Place wafer carrier in a rinse-dryer and rinse for 10 min followed by force-drying. This may be done manually as well by rinsing in deionized water and blow drying with filtered dry nitrogen.

This cleaning process is also used prior to either diffusion or metallization.

In addition to simple immersion, ultrasonics may be added to provide increased particle separation energy. Ultrasonics, using shortwave sound energy, works by creating a cavitation effect across the surface; as the cavitation energy encounters a particle, its force dislodges the particle. Thus removal of particles can be accomplished without physical scrubbing action, which could cause scratches or other surface damage. Placing the recommended cleaning solutions in an ultrasonic cleaning unit will provide the combined benefit of chemical and "sonic" cleaning. Ultrasonics can be used for cleaning all surfaces too delicate to mechanically abrade; the appropriate cleaning solution is simply placed in an ultrasonic generator and the generator is run for approximately 10 min.

After the wafer has been properly cleaned by one of the methods mentioned, it can be oxidized. However, oxidation should not be delayed more than 30 min after cleaning.

Pre-photoresist cleaning

Wafers coming out of oxidation can follow one of several paths. The preferred one would be loading cassettes directly onto the photoresist coater. Ideally, this would take place within 2 hr of oxidation, before moisture is absorbed by the wafer surface, before the wafers are exposed to an environment where particulates can be introduced, and before *any* form of film or other contamination can reach the virgin oxide surface. Coating directly from oxidation after a cool-down is not always possible because of changes in wafer production flow.

Wafers that must be held for several hours after oxidation should be completely enclosed in a plastic or glass holder to keep particulates off their surfaces. Such wafers can be loaded into cassettes from oxidation under a laminar flow hood and placed in a protective container. They should then be stored in a "nitrogen dry box," a plastic housing through which a slow flow of nitrogen is directed to keep the wafers dry and clean. Following an indefinite delay, wafers can be placed directly onto the coater *if* the coating equipment has a prebake fixture or track to heat or flash-dry the wafers before coating. If not, the wafers should be given a 10-min precoat bake at 100 to 200°C. This will promote better resist adhesion, which is particularly important on highly doped wafer surfaces. The bake is not always required on undoped silicon dioxide if the wafers are kept in dry nitrogen and if a primer is used. Higher bakes induce wafer distortion. The temperature at the surface should be high, but for a short period.

A third and less desirable path wafers may follow from oxidation is to be placed in a clean-room environment unprotected for several hours or longer. Since they will be contaminated by particulates, they must first be cleaned either mechanically in a wafer scrubber or by immersion in a sulfuric acid–hydrogen peroxide solution.

Following the chemical cleaning, wafers are rinsed and prebaked at 100 to 125°C, either on an in-line heated track or similar in-line bake device or in a microfiltered convection oven. Wafers are then ready for photoresist coating. Figure 5.8 outlines various paths wafers should follow between oxidation and photoresist coating.

Dry removal of organic contamination

Wet chemistry approaches tend to leave behind residues of their own. One way to avoid this problem and also circumvent chemical disposal problems is to burn the material off in an oxidation tube at 900 to 1200°C, an ideal reducing ambient. Lower temperatures may be necessary for this step to avoid changing junction depths or the use of low-temperature ashing in an oxygen plasma. Oxygen plasma removal is widely used for this purpose. It is frequently a standard process step (descum) after the developing step, during which veiling typically occurs. An organic layer burned off in a low-temperature ashing step leaves the surface clean and free of residues.

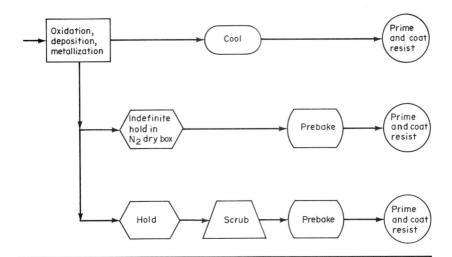

Figure 5.8 Wafer pretreatment cycles.

A slightly less desirable way to remove organic films is to expose the wafers to the dry-etch plasma gas normally used to etch the underlying surfaces. The gas mixture can be adjusted to favor resist attack and minimize, for example, silicon etching in a CF_4-O_2 plasma environment. The result will be removal of the resist film as well as a slight etching of the underlying semiconductor layer. Plasma cleaning may well become the best production tool, since it avoids toxic and environmentally unsound chemical disposal, can be processed in superclean, enclosed vacuum environments, and can be very highly rate-controlled. Thus, organic contaminants of any type are rendered volatile in any of the above oxidizing or reducing environments. The only method of removing all traces of organic contamination (negative resists) is 650°C heat treatment (ash) in air. The ammonia–hydrogen peroxide solution is the only reliable chemical resist film stripper, used as a clean-up after gross resist removal with Caros acid.

Inorganic ion removal

Inorganic ion removal is more difficult. It involves desorption of sodium and other ions that have been physically absorbed in the wafer surface. Oxidation serves to trap metal ions, especially porous deposited oxides. Removal of sodium ions with 5 N hydrochloric acid, using a 2-min rinse cycle, removes down to 0.0014 monoionic layer. That is still more than the 1×10^{-4} needed to invert the surface of a 1 Ω/cm silicon device. Desorption of fluoride ions is accomplished best with hot water rinses, and 3 min is usually adequate. Chloride ions are desorbed with a 2-min rinse in room-temperature water. Chloride ions seem easiest to remove, presumably because only weak van der Waals forces hold them to the silicon surface.

Inorganic atoms such as gold and copper are desorbed from a wafer surface most effectively in heated acidic hydrogen peroxide. Iron contaminants are readily removed by treatment for 60 sec in a 30% solution of hydrochloric acid (HCl). Figure 5.9 shows a cross-sectional schematic of a typical wafer-cleaning system.

Analytical techniques used to detect the ions and inorganic atoms are radiotracer testing, spark source mass spectrometry, and megaelectronvolt ion backscattering. In summary, metal ions are best removed with hydrochloric acid–hydrogen peroxide–water baths. Metal atom contamination is best removed with hydrochloric acid–hydrogen peroxide or sulfuric acid–hydrogen peroxide. A comparison of the immersion chemical cleaning process with the centrifugal process is given in Table 5.1.

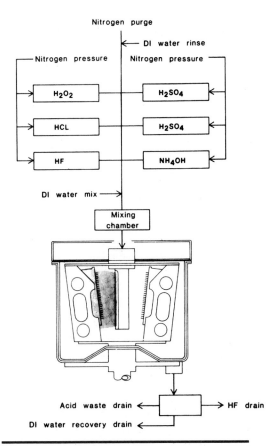

Figure 5.9 Wafer-cleaning system. (*Courtesy of FSI Corporation.*)

Ultrasonic wafer cleaning

Ultrasonics, like megasonics, will not pose the potential problem of either scratching or etching the wafer in order to remove surface soils. The natural uniformity of a wave of energy transmitted through a solution and against solid particulates on a surface is unique to ultrasonic cleaning. Uniformity of the cleaning helps to ensure reliability, a major concern when large numbers of VLSI slices are processed per hour. Ultrasonics, like megasonics, greatly reduces the problems of waste-treating large volumes of spent effluents. Unlike megasonics, ultrasonics relies on the cavitation energy of bubbles generated by sonic waves, energy that dislodges particles when bubbles collapse.

In some versions of megasonic cleaning, the wafers are stationary. Enhancing the cleaning efficiency by providing for wafer movement is

TABLE 5.1   Recommended Cleaning Procedures using Immersion or Centrifugal Spray Processing

Step	Immersion	Centrifugal spray
1	Removal of photoresist or other heavy organic residue 2–10 min of H_2SO_4 + H_2O_2 in bath	Same 3–4 min of H_2SO_4 + H_2O_2 (4:1) at combined flow of 650–700 cm^3/min
1A	Overflow rinse	Alternating line rinse and purge while rinsing wafers
2	Removal of residual photoresist film or light organic contamination 20 min of $NH_4OH:H_2O_2:H_2O$ (1:1:5–1:2:7) at 75–85°C	Same 3 min of $NH_4OH:H_2O_2:H_2O$ (fresh) (1:15) at a combined flow rate of 700 cm^3/min
2A	Overflow rinse	Alternating line rinse and purge while rinsing wafers
3	Removal of thin layer of silicon dioxide 60 sec in dilute HF at room temperature	Same 60 sec in dilute HF at room temperature (user's choice)
3A	Overflow rinse	Alternating line rinse and purge while rinsing wafers
4	Removal of inorganic (atomic and ionic) contaminants 20 min of $HCl:H_2O_2:H_2O$ (1:1:5–1:2:7) at 75–85°C	Same 3 min of $HCl:H_2O_2:H_2O$ (fresh) (1:1:5) at a combined flow rate of 700 cm^3/min
5	Overflow rinse	Alternating line rinse and purge while rinsing wafers
6	Withdraw to rinser-dryer for final rinse and spin dry	Spin dry

recommended. Also, adding transducers or generating sonic energy from several directions is useful in adding to the uniformity of particulate removal. Several energy sources, positioned at different angles, will provide wave energy to strike particles at their many different "angles of rest" on the etched wafer topography. Overlapping wave action, for VLSI and ULSI devices is almost a requirement because of the increasing aspect ratio of surface topography. Many patterns are now etched 2 to 4 μm deep with openings of only 0.5 μm. Thus, a 4:1 aspect ratio would not be uncommon, and it is in just this type of structure that a small particle (say, a small piece of resist) could be really stuck. Brushes might never dislodge such a defect.

Ultrasonic wafer cleaning can be broken into three distinct phases: a wash cycle, a rinse cycle, and a dry cycle. The wash cycle is responsible for getting particles, soils, or films off the wafer surface and into suspension in the cleaning solution; the rinse step is designed to simply carry away the suspended particles and soils; and the dry cycle is purely a water removal process. In the wash phase, a solution containing either a detergent or a nonionic surfactant is ultrasonically energized in a vessel that has rounded corners and no weld marks or exposed threaded fittings. In a study by Glenn Evans of Keithley Instruments, Cleveland, Ohio, it was determined that the concentration of the wetting agent cannot exceed 0.1% by volume. This is to keep the viscosity or fluidity low so as to not impede the effectiveness of the ultrasonic energy or reduce the chemical activity of the solution.

The transmission of ultrasonic energy can be further impeded if the cleaning solution has bubbles in it before cleaning commences. Thus, the solution should be degassed. Degassing a solution before ultrasonic cleaning is done by turning on the ultrasonic energy and using it to break up the bubbles. This typically takes about 15 to 30 min at approximately 65°C. When gas bubbles are highly concentrated, the temperature is elevated to approximately 90°C and the time is extended.

Ultrasonic solution temperatures for the wash cycle should be 60 to 70°C for aqueous baths. Temperature is important in this type of cleaning because it is directly related to the efficiency of ultrasonic propagation. Figure 5.10 illustrates this point; relative cavitation weight erosion is plotted against solution temperature. The data are

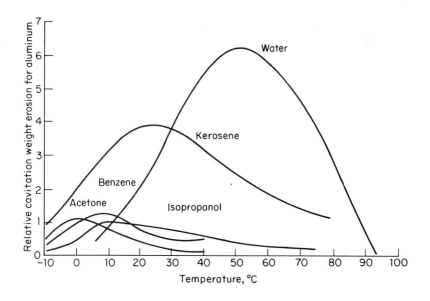

Figure 5.10 Cavitation weight erosion versus solution temperature.

based on the use of aluminum in several different mediums. Note that water achieves maximum ultrasonic propagation efficiency at approximately 55°C. Efficiency also is a function of solution cleanliness in the wash cycle. All baths should be circulated continuously through a submicron nominally retentive filter at very slow (2 to 5% of total volume filtered per minute) flow rate.

Many of the soils that must be removed at various stages of wafer processing require two or more different cleaning solutions. Resist stripping, for example, could involve a hydrogen peroxide–sulfuric acid (piranha) immersion, perhaps followed by an ultrasonic wash cycle to remove any residual resist particles left behind as the resist was carried off the wafer in the piranha. The wash phase, using any combination of chemical and ultrasonics, should be followed by a rinse phase to carry off the debris or soil removed in the wash phase. Rinsing by the "quick dump" technique is very efficient, since it moves substrates through progressively cleaner cycles and thereby conserves expensive 15 to 18 MΩ/cm water. The final step at the end of the quick dump rinse is a spray-off prior to the rinse step and immediately after the chemical wash; a short "quench" rinse is recommended to remove most of the surface chemistry and prevent drag-in of the cleaning chemical into rinse waters.

Cascade overflow and counterflow rinsing also are very popular production-oriented techniques. Cascade or overflow rinses generally move substrates through several successive tanks of increasingly cleaner solution. The flow rate of the water should be regulated to change the entire volume two to four times per rinse cycle, which allows time to carry off the particulates still suspended in the residual cleaning solution on wafer surfaces. Rinse water temperature and filtration are important. Filtration should be carried out with nominally retentive cartridge filters that provide good flow rate yet are rated down to 0.1 to 0.2 μm. Some processes use final polishing filters that are absolute-retention membranes sandwiched with prefilters. Rinse water temperature is best kept well above ambient, anywhere from 35 to 90°C, because the action of the water molecules on the wafer surface increases with temperature. Added rinse water action can be provided by bubbling nitrogen through the rinse water.

Wafer drying

The main problem in drying wafers is removing the water *without* adding additional contamination. The point immediately after complete hot deionized ultrapure water rinsing is critical, since at this juncture the wafer surface should be perfectly clean and free of particles. In many cleaning processes, a solvent vapor is used as a follow-up to superpure rinsing; the reliance is on water displacement to get the surface dry. Solvent vapors may, however, contain many contaminants, just as clean rooms, despite highly filtered air on all incoming sources, contain millions of defect-producing particles. Similarly, after the fresh chemical vapor deposition of an oxide or the evaporation of a metal layer, practically any attempt to clean the surface will result in contamination. Thus, solvent vapor zone drying may pose particle contamination problems because the quality of the solvent atmosphere is extremely difficult to control.

Evaporative drying is one way to keep an already clean surface clean yet remove its water. Evaporation of water from VLSI wafer surfaces is accomplished by providing laminar flow submicron filtered heated nitrogen over the wafer surfaces.

These "gas" dryers are ideally suited for full in-line automated production, since a gas drying unit can be interfaced with track wafer-handling systems. One potential problem is the possibility of particulates being left on the surface when the water evaporates off. In essence, water with particles suspended in it can evaporate around a particle and leave the particle behind on the wafer. This suggests the third drying alternative: centrifugal spin drying. The spin dry is a rapid means of getting water off the surface in a batch process. The mechanical action of centrifugal spinning forces water off the surface, which makes sure that particles are not left behind, as can happen in evaporative drying.

Cleaning of reworked wafers

Although "reworks" can be wafers that must be reprocessed through any step that produced an undesirable result, they will be defined here as wafers with defective photoresist imaging. These are wafers that demonstrate poor adhesion after development or have been scratched or damaged in exposure. Whatever the cause, wafers are brought back from postdevelopment inspection for resist stripping and reimaging. Several stripping methods are used for both positive and negative resists. Regardless of the stripping method used, moisture condensation, or "fog," tests after stripping show the presence of a "memory," or film, left after stripping, including plasma stripping. Therefore, after using any of the commercially available resist strippers (Chap. 12), one of two poststrip cleaning steps is suggested (Fig. 5.11). The first is a dip in a solution of 10 parts deionized water to 1 part hydrofluoric acid, followed by a deionized (DI) water rinse to 10-MΩ resistivity (at 25°C) and a spin-dry in a dry-nitrogen ambient.

The second method, taken from a paper by D. A. Peters and C. A. Deckert,[6] involves immersing the wafers in a solution of 7 parts

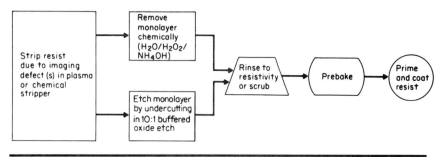

Figure 5.11 Wafer reworks: surface preparation.

deionized water, 3 parts 30% hydrogen peroxide (H_2O_2), and 3 parts 29% ammonium hydroxide (NH_4OH). Wafers should be immersed for 20 min at 80°C, followed by the same type of deionized water rinsing mentioned earlier for the post-HF etch step.

These two poststrip cleaning steps are designed to remove the monolayer of resist that usually remains after stripping. The monolayer is seldom detectable under optical microscopy; it is caused by a reaction between the photoresist and the silicon dioxide layer. Exclusion of either of these types of poststrip cleaning steps may cause subsequent resist adhesion failure or increased undercutting during etching.

An example of resist adhesion failure typically caused by poor wafer surface preparation is given in Fig. 5.12. Note that the resist line has completely lifted off the wafer and redeposited between two other resist lines. Because of the partially dissolved state of the resist during development, the lifted resist has "fused" with the other two images

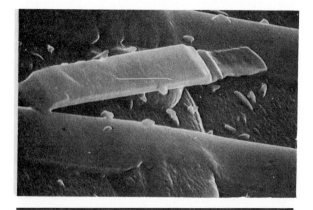

Figure 5.12 Positive-resist lifting during development caused by poor surface preparation; SEM at 30,000 × . (*Courtesy of Mann Labs.*)

and will not rinse off freely. This wafer will be stripped, recleaned, and reimaged. In addition, note the many small chunks of resist in the scanning electron micrograph. They are probably caused by wafer edge handling, wherein the brittleness of the resist has resulted in shattering, leaving many small pieces of resist scattered across parts of the wafer.

Metal cleaning

Aluminum layers may be cleaned in the same way oxide layers were—by scrubbing to remove particulates followed by a bake to remove moisture before photoresist coating. Reworked aluminized wafers should be stripped and then cleaned in the 7:3:3 solution of deionized water, hydrogen peroxide, and ammonium hydroxide, followed by a bake before photoresist coating. In some cases, baking in air will allow a very thin layer of aluminum oxide to form, and this oxide will provide a better surface for photoresist adhesion.

Wafer scrubbing and cleaning equipment

Wafer scrubbing and/or cleaning is used before resist coating, oxidation, diffusion, metallization, and vacuum, chemical, or vapor deposition.

Wafer scrubbing is a well-established technique for particulate removal from wafer surfaces. Since particles fall onto surfaces in clean rooms, scrub equipment should be designed to minimize exposure to air or provide reduced transport distances. Wafer breakage is another source of contamination. Equipment design should be concentrated on minimal wafer movement, and load arms should be used to reduce relative wafer motion. Construction materials in scrubbers also can be a source of contamination; anodized aluminum and stainless steel are relatively clean surfaces. Since wafer scrubbing seeks to remove all particles before resist application, wafers must be isolated from the environment after cleaning.

A typical cleaning-scrubbing process is microprocessor-controlled with light-emitting diode (LED) diagnostics in the event of vacuum failure, wafer out of position, or other abnormal events. The possible combinations for wafer surface cleaning include the mixing of mechanical brush scrubbing with high-pressure jet cleaning. A process could run as follows:

1. Brush scrub (detergent and water)
2. High-pressure jet scrub (detergent and water)
3. Rinse

4. Spin dry

5. Nitrogen blowoff

Wafers can be treated or scrubbed first with a 3-mil tufted nylon brush, pony belly hair, or microcloth surface. In order to keep particles from building up on the scrubber surface, continuous rinsing with water occurs when the brush is in the ideal mode. The microprocessor control allows spin speeds to be varied so that low speeds are used for rinsing and high speeds for final drying. Ideal scrubbing action provides enough wetness to the bristles or other scrubbing surface that a liquid meniscus forms between the brush and wafer surface. Thus, the scrubbing material does not actually touch the wafer; it only transfers its motion to the cleaning solution which in turn chemically scrubs the surface. In the high-pressure jet mode, a 4000-psi stream or solution, directed at a 60° angle, forces particles off the surface by overcoming the surface adhesion energy of particulates stuck on the wafer.

Scrubbing principles.   Water scrubbing, using either high-pressure jets or rotating brushes, is really not a physical-mechanical process in the sense we generally think of it; rather, it works by a transfer of forceful energy with a brush serving as a mechanical initiator. Scrubbing is very effective for all types of particle removal (organic and inorganic) as well as for removal of organic films, assuming the proper type of detergent is used. Earlier we mentioned the meniscus that forms between the bristle or nap and the wafer. This is a critical aspect of scrubbing that is highly dependent upon the brush construction material and the wetting properties of the wafer.

Wafer scrubber materials are typically polypropylene, helically wound nylon, tufted nylon, mohair, camel hair, or a polymeric polishing nap. All of these materials are hydrophilic and, when wetted, skid across a usually hydrophobic surface like a hydroplane. The wafer-cleaning medium (water and detergent) clinging to its brush carrier then sweeps the wafer surface and insulates a stiff brush material from an even harder silicon oxide or metal surface. Wafers are most effectively cleaned when their surfaces are hydrophobic so that the cleaning solution is actually repelled from the wafer surface and carries particles in its path.

Two basic types of brushes are used for wafer scrubbing: cup and cylindrical. The cup style, shown in Fig. 5.13, has an area of contact with the wafer that is large compared to that of the cylindrical type. The continuous stroking action of the bristles carries particles to the edge of the wafer and gets them away from the brush. Note that the bristles on the "cup" scrubber are parallel to the wafer's rotational

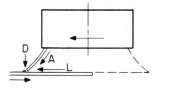

Figure 5.13   Mechanism of cup-type scrubber.

axis. The removal of particles is especially effective because the brush rotates in the direction opposite to that of the wafer, and pressure exerted by the fluid medium on a particle is lateral, not perpendicular. Further, the offset axes of the wafer and cleaning brush serve to move particles off the wafer surface. The cup-type brush covers a large area of the wafer; it offers scrubbing action in several different directions. That is important on previously etched surfaces on which particles get caught in crevices and can be swept out from only one angle.

The other major brush type is the cylindrical one shown in Fig. 5.14. It moves on an axis perpendicular to the wafer surface, and wafer and brush movement are clockwise. Automatic commercial scrubbers use brushes of this type, perhaps because the brushes always show fresh surfaces to the wafers. That reduces the problem to particle redistribution via the brush, which does happen with cup-type brushes. Occasionally, however, particles are lifted off the wafer and moved back to other spots if they do not separate from the brush as the brush lifts up and off the wafer. Although the cylindrical brush does not clean in as many angular directions as the cup brush, chuck rotation can be changed to overcome that limitation. Also, the cylindrical brush does not form "cowlick" patterns in its center as the cup style does from constant swirling at its axis.

Industry experts argue well for both types of scrubbers, and both types have given good performance in production. The cup and cylindrical types need to be positioned carefully onto the wafer surface and exert equal pressure in all directions. Unequal pressure or varying pressure will cause process variances in cleaning that may be translated into imaging variability. Some brush scrubber manufacturers use several pneumatic actuators to avoid this problem. Some applications, in which pressure may damage the wafer, require chemical cleaning only or low-pressure jet spray cleaning.

Jet spray cleaning.   Jet spray cleaning is really brushless cleaning with pressures from 250 to 6000 psi, and the method can be used by itself or in conjunction with the brush scrubbers. Whereas brush scrubbers excel in cleaning hydrophobic wafer surfaces because of the

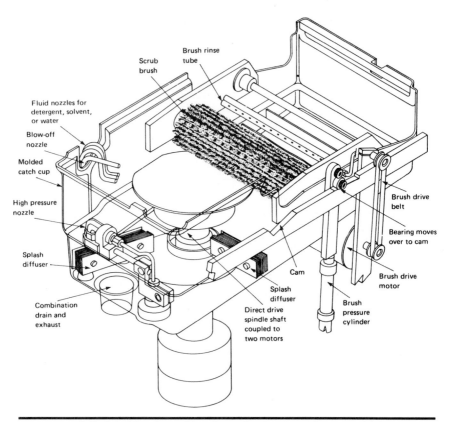

Figure 5.14 Cylindrical brush scrubber for wafer cleaning. (*Courtesy of Silicon Valley Group.*)

repellency effect and hydrophilic brushes, the jet spray creates a microscopic cleaning action that is very effective in reaching into submicron-size crevices, where brushes with mil-size bristles simply cannot reach, and dislodging debris. Also, the jet spray is useful for all delicate surfaces the scratching of which is a potential problem.

Jet spray cleaning requires a close wafer-to-nozzle distance in order to maximize cleaning fluid momentum. Second, the nozzle, to provide maximum effectiveness, should be movable across the entire area of a wafer and be positionable at multiple angles. Pressure should be constant because changes will be detrimental; among other things, they will cause poor positioning of a brush on a wafer surface. Microprocessor control is usually the answer for all types of wafer-cleaning methods; it provides the necessary real time in process control of solution pressure (or temperature, normality, agitation, etc.) to ensure high reliability.

Static electricity problems

The only two notable drawbacks of high-pressure spray cleaning are potential damage to highly delicate high-aspect ratio structures and static electricity generation. Actually, static electricity is a problem with all types of scrubbing equipment; it arises from the use of high resistivity. In jet spray cleaning, the high pressure generates static electricity if the solution passes through a long enough distance before it reaches the wafer.

Static electricity in any microelectronic process can cause severe particle attraction problems and even an explosion or fire hazard. Its elimination is necessary to remove a potential source of defects in the cleaning process: electric charges or energy forces that tend to keep particles "stuck" on the wafer surface. Elimination usually involves deionizing the air or surfaces on which electric charge forces aggregate. Static electricity in chemical solutions can be minimized or completely eliminated in one of several ways, as follows:

1. Substitute a nonpolar molecule–based cleaner or a proprietary nonaqueous cleaner. Generic materials such as methanol and isopropanol work well but must be used at low pressure and low temperature to remove the explosion hazard.

2. Static buildup in solutions is often caused by streams of chemicals being too long. Reduce the nozzle-to-wafer distance to less than approximately ½ in. Generally, reduced spray distance will eliminate most static electricity.

3. Additives to aqueous detergents and cleaners are effective static eliminators. For example, CO_2 canisters in deionized water systems are recommended.

Wafer-cleaning equipment

Wafer-cleaning equipment is either in-line or batch type, and many alternatives are available. Plasma cleaning has become a popular method because it eliminates the need for more hazardous wet chemicals. This same system may be used to treat the wafer to promote wetting of photoresists or etchants.

Other types of in-line wafer-scrubbing systems are available; they include those which provide high-pressure deionized water scrubbing only, thereby eliminating the possibility of wafer recontamination from overloaded brushes. The pressure range obtainable with these systems is 2500 to 6000 lb/in^2 or higher coming from a nozzle enclosed in an oscillating head.

The wafer is spun while the oscillating spray is directed onto the wafer surface. This is followed by a nitrogen purge to promote rapid drying. The systems are also programmable and contain most of the production-oriented wafer-processing steps. After any of these scrubbing operations, wafers are loaded onto an in-line dehydration baking system.

Batch-type systems are also used for cleaning, rinsing, and drying prior to photoresist coating. The systems can be programmed, and wafers are loaded in cassettes before each cycle. The disadvantage of batch systems is the inability to be fitted as part of an automatic wafer-processing line. It is offset by the fact that wafer progress must be interrupted between most major operations for inspection, so that completely hands-off automatic processing is seldom achieved. As a higher level of confidence in handling and processing equipment is achieved, more highly automated processes will be used.

Wafer Priming

Adhesion factors

Priming is a pre-resist coating of some material that will increase the adhesion of the resist so as to reduce the amount of lateral etching or undercutting. The adhesion of photoresists and electron resists to wafer and mask surfaces is a factor in the amount of intermolecular bonding between the resist and the surface onto which it is applied. In addition to this fundamental adhesive force, which is chemical in nature (molecular and atomic bonding), there are "external" forces and factors that influence the ability of resists to withstand lateral etching. The external forces include

1. Moisture content of the wafer and mask surfaces
2. Wetting characteristics of the resist on surfaces
3. Type of primer used and its method of application
4. Delays in the wafer-imaging process
5. Resist chemistry (solvent and resin types)
6. Wafer or mask surface smoothness
7. Stress forces in the resist coating
8. Contamination or surface defects that occur at the interface of the resist and wafer and mask surfaces
9. Resist-imaging factors (level of softbake, exposure, developing, and postbaking)

Each of these parameters needs to be characterized so that the amount of adhesion will remain constant despite fluctuations in process times or temperatures.

Surface moisture.   A hydrated surface has been shown to reduce resist adhesion. Figure 5.15 shows a schematic representation of absorbed molecular water on the surface of silicon dioxide. Primers such as HMDS react with wafer oxide surfaces to tie up molecular water. The additional use of a prebake, which increases the liquid contact angle of the surface, helps promote adhesion. The presence of water on the oxide surface allows the wet etchants to penetrate easily between the resist and the wafer or mask surface and cause considerable undercutting.

Resist wetting properties.   Ideally, the wafer surface should have a zero contact angle for spreading, but positive contact angles are required for increasing adhesion. Consequently, these factors work at cross-

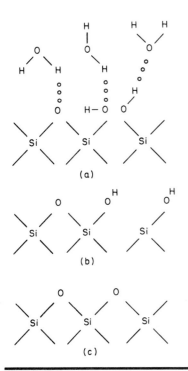

Figure 5.15 Thermal oxide surface conditions: (*a*) absorbed molecular water; (*b*) silanol groups (no absorbed molecular water); (*c*) siloxane (no absorbed molecular water).

purposes. Satisfactory coatings can be and are obtained, however, with the positive contact angles usually present in treated (primed and baked) oxide surfaces. In summary, the surfaces must be completely wetted with both resist *and* primer to provide good adhesion. For further discussion on wetting, see Chap. 6.

Primer type, concentration, and method of application.   There are several types of primers and various ways of applying them (see next section), and this must be considered in the overall picture of resist adhesion parameters. A considerable range of adhesion values can be obtained by varying these factors alone. Primers are generally silanes like HMDS and TCPS diluted in xylene.

Imaging process delays.   The chemical and physical forces that impact resist adhesion vary as a function of time. For example, humidity absorbed by wafers stored temporarily before coating or etching may change the amount of etchant undercutting. High surface energy states caused by high-temperature prebaking or postbaking will dissipate in time and change the interfacial resist-oxide adhesion. Delays between prebaking (or oxidation) and coating or postbaking and etching are particularly critical to resist adhesion and should be minimized. A process should call out the outer limits of these delays for a given resist system. Controlling critical dimensions at submicron levels requires specifying all process steps and variations that impact dimensional change.

Resist chemistry.   The resist resins and solvents and their physical and chemical interactions with semiconductor surfaces affect the level of resist adhesion. For example, the degree of molecular coiling in solvents plays a role in resist and substrate bonding. In some cases, the resist-resin system is a primary determinant of adhesion and can override even the need for a primer. AZ-111 is a good example of a positive resist whose adhesion to silicon dioxide precludes the use of a primer. All resists should be tested for this singular property, as well as other properties specified in the quality control section of Chap. 13. Resist adhesion varies considerably according to resist chemistry.

Surface smoothness.   The smoother and more physically uniform a surface, the less the mechanical, or "tooth," adhesion. Crystalline materials with variously shaped faces on their surfaces will allow for a considerable increase in resist adhesion. In applications where adhesion is particularly difficult, a mild microetch of the surface is recommended to set up a rougher surface that will then allow mechanical bonding in addition to the chemical or molecular bonding. Ideal microetches are either dilute versions of standard etches or crystal- and grain-highlighting solutions (see Chap. 11 and Table 12.2 on Remover 1112 as a microetching solution).

Stress forces. Stresses are introduced in the resist coating during spin coating. Even though softbaking may relieve some of them, others may remain to disrupt resist adhesion. In some cases, stress forces are exaggerated by processing. See Chap. 6 for information on reducing stress in resist layers.

Surface defects and contamination. Pits or raised areas (epitaxial spikes) on a surface may give rise to prime- or resist-coating anomalies by causing either incomplete coating or variations in the thickness of the coating. Primer coating thickness, for example, will cause variations in undercutting and subsequent line width changes. Surface irregularities, including dust particles, also can alter the stress properties in the coating and thereby introduce adhesion variability. Surface mapping for detection of irregularities and defects is an important quality control test. Lithographic yield is directly impacted by surface irregularities.

Imaging factors affecting resist adhesion and etch resistance. The degrees of softbaking and hardbaking have significant effects on resist adhesion, but exposure and developing effects participate as well. Increased levels of softbaking and hardbaking generally improve resistance to the etch environment and lateral etching. These factors are described in detail in Chaps. 7 and 10.

Primer types

Many solutions have been used to promote the adhesion of resists to wafer and related semiconductor surfaces. A list of adhesion promoters used mainly on silicon dioxide but also on other oxides, metals, and glasses follows:

1. Hexamethyldisilazane (HMDS)

2. Trichlorophenylsilane (TCPS)

3. Bistrimethylsilylacetimide (BSA)

4. Monazoline C

5. Trichlorobenzene

6. Xylene

HMDS is by far the most commonly used adhesion promoter; it is described in U.S. Patent No. 3,549,368 by R. H. Collins and F. T. Deverse (December 22, 1970). It is preferred for a number of reasons. The adhesion promoters containing chlorine are undesirable because of corrosion problems. In addition, their toxicity and overall bonding effectiveness are less than those of HMDS. Some hardbaking is nearly always required after treatment with all but HMDS, and the length of time over which an adhesion promoter remains effective on a wafer is several weeks with HMDS and only several hours with the others. Hardbaking is often used as a drying step, and postexposure baking or deep-uv flooding is used in place of high-temperature baking. This reduces wafer warpage and simplifies resist removal. Pinhole reduction also has proved to be best with HMDS, and the amount of undercutting is reduced.

The combined use of HMDS adhesion promoter and anisotropic dry etching solves the main problem with adhesion to oxides and metals, which is the loss of valuable IC "real estate" because of undercutting. Often, space between the source and drain on a device is added to compensate for the loss of area resulting from undercutting. The main idea is to keep the doped areas narrow and deep rather than shallow and wide, all of which allows increased device density.

The mechanism whereby HMDS bonds resist to the oxide surface is shown in Fig. 5.16. HMDS ties up the molecular water on a hydroxylated silicon dioxide surface with a portion of the complex molecule and bonds to the ends of the resist molecule with the other. Thus a twin-ended molecular bonding mechanism makes HMDS a surface-linking adhesion promoter.

Note the bed of CH_3 groups and other moieties left on the outside; their relative inertness and hydrophobicity are what keep the etchant from penetrating along the oxide-resist interface.

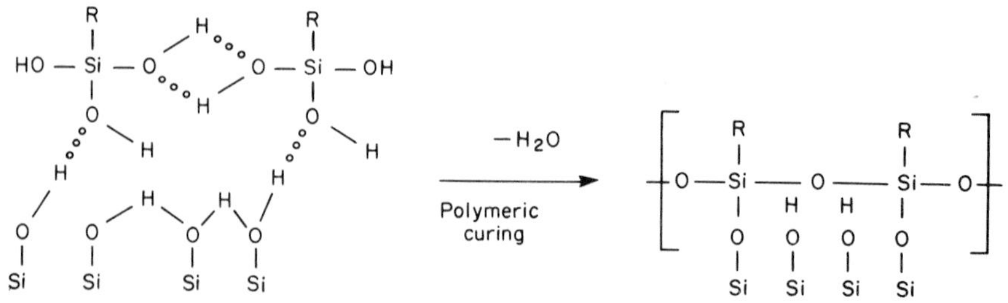

Figure 5.16   HMDS bonding mechanism with silicon dioxide.[3]

Primer application

The application of adhesion promoters to wafer surfaces is done by one of three coating techniques: (1) spin coating heated or ambient, (2) puddle or spray coating, and (3) vapor coating. The choice of one of these techniques will depend on the needs of the application and the production environment. The pros and cons of these methods are presented in subsequent paragraphs.

Puddle dispense on dip priming.   Immersion of wafers in a solution of adhesion promoter has the advantage of allowing a relatively thick layer of adhesion promoter to be built up. Dip priming is a batch process that allows a large number (cassettes) of wafers to be treated simultaneously; puddle dispense is performed in-line on track equipment. Immersion or puddle dispense is recommended for surfaces which are extremely difficult to bond resist to, such as quartz or very highly doped glass. The thicker layer of adhesion promoter offers increased bonding sites for the resist and substrate surfaces; the adhesion promoter actually acts as a conversion coating, whereby the entire wafer surface to be coated with resist is chemically converted.

Puddle-dispense priming offers another advantage in that the solution can be heated to accelerate the action of the promoter and increase the layer thickness. In addition, the concentration of the solution can be varied to increase or decrease the action of the adhesion promoter. Concentrations of HMDS from 10 to 100%, diluted with xylene, are used. Contact or process time can vary with the solution concentration, temperature of the primer, and degree of surface bonding difficulty.

Spin priming.   Spin priming is very popular because it fits into the already established wafer-processing equipment, as does puddle-dispense priming. By using the existing spin-coating heads and modules, HMDS or other adhesion promoters are simply pumped into the system and directed through the coating head to the wafer. Since most coating equipment has the possibility of multiple coating nozzles, no additional equipment or process area is required. Spin priming can be used in one of two modes: (1) pressure dispense, whereby a stream of solution is pumped onto the wafer and spun off, or (2) spray, whereby the nozzle directs a spray onto the wafer either before or during the spin cycle. In pressure-dispense priming, enough solution is pumped onto the wafer to create a puddle that covers the surface. That is followed by a 3000- to 6000-rpm spin for approximately 30 sec or until optical color change or fringing stops.

In spray priming, the nozzle partly or completely aspirates the solution into a finely divided spray usually just before the spin cycle commences. A membrane filter should be used to block particles from reaching the wafer surface. Of the two types of spin priming, the pressure-dispense technique offers the advantage of providing a given adhesion level in shorter time than by spraying, thereby permitting shorter cycles or increased wafer throughput. Spray priming also reduces reactivity due to evaporative cooling of the solution during the spray, although heated spray nozzles are used to override the cooling effect. Pressure-dispense priming is more economical than spray priming because less solution is used.

In spin priming, the concentrations are the same as for dipping. A 100% solution is preferred because it eliminates the possibility of incorrect mixing and the bonding reaction takes place faster than with diluted solutions. By the pressure-dispense technique, 100% HMDS is applied to cover the wafer and is left for 5 to 15 sec before spinning, depending on the level of adhesion needed. In spray priming, a 5- to 20-sec spray is typically required to provide an equivalent level of adhesion. In some applications, the layer of HMDS with 100% concentration is too thick, making complete residue-free resist removal difficult. Therefore, 50% dilutions of HMDS with xylene are recommended. Trichlorotrifluoroethane (Freon) is often the recommended diluent for use in spin priming because of its higher vapor pressure compared with xylene. The thinner for the resist system used is also a suitable diluent.

Spin priming, by the spray or pressure-dispense technique, offers a practical in-line method that does not seriously change wafer throughput or process flow. An added advantage of in-line wafer priming before resist coating is the surface-scrubbing action provided by the solution, which can remove dust or other particulate contamination on the wafer before it is encapsulated in the resist film. Spin bowls should be well exhausted and drained to prevent chemical mixing and redeposition of waste material.

Vapor priming.   Vapor priming is the use of adhesion promoter vapors to provide bonding of the resist to semiconductor surfaces. A schematic cross section of a vapor-priming system is shown in Fig. 5.17. Nitrogen is bled into the system at a rate of 3 to 5 L/min; it is directed into the HMDS chamber, and the pressure in the chamber causes vaporization of the solution. The vapors are directed into the wafer or substrate treatment chamber, where a 10-min dwell time provides an excellent level of adhesion of resist to silicon dioxide and similar surfaces, including aluminum oxide, polysilicon, quartz, silicon nitride, and silicides.

The advantages of vapor priming are that a very large number of wafers can be placed in a treatment chamber and the line process need not be interrupted when repriming or when special runs are made.

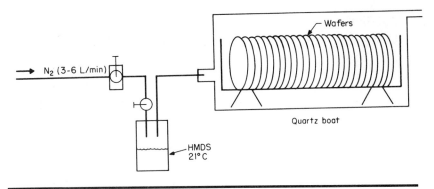

Figure 5.17   Vapor-priming apparatus.

The fact that only vapors touch the wafer eliminates potential contamination by solid particles, and priming uniformity is good. Since vapor priming is outside the wafer-processing equipment line, dwell time is not critical in determining wafer throughput, and several minutes of wafer exposure to the vapors ensures a thoroughly primed surface. The amount of HMDS used per wafer is reduced by vapor priming, which provides economic and environmental benefits. Other typical problems of spin priming also are eliminated, including:

1. Spin coater drain clogging caused by HMDS-resist reactions
2. Exhaust flow rate changes resulting in resist thickness or HMDS thickness changes
3. Resist surfaces contacted by HMDS splashback or vapor and changing resist exposure parameters

In Fig. 5.17 the reactor tube can be made of glass, quartz, or similar compatible material. The reservoir size for HMDS is 800 cm^3. After vapor priming, wafers should be placed in an absolutely clean, laminar flow clean air module and allowed to sit for 5 min. The clean air module dwell is designed to allow any condensed HMDS to evaporate from the wafer surface before the resist is applied. The maximum delay before resist coating is 60 min, and if delays exceeding this time occur, the wafers should be placed back in the priming chamber. The exhaust from the vapor-priming equipment should be directed to an exhaust hood or appropriate area to be kept away from the process environment.

REFERENCES

1. "Filtering AZ Photo Resists," Technical Data Sheet, Shipley Company, Newton, Mass., 1980.

2. M. L. Long and S. Harrell, "Detection of Wafer Contamination," Texas Instruments, Inc., February 1970.

3. K. L. Mittal, "Factors Affecting Adhesion of Lithographic Materials," *ACS Symposium Series,* No. 95, 1979.

4. J. A. Amick, "Cleanliness and the Cleaning of Silicon Wafers," *Solid State Tech.,* November 1976, p. 47.

5. C. A. Deckert, "Modes of Photoresist Silicon Dioxide Adhesion Failure," RCA Laboratories, David Sarnoff Research Center, Princeton, N.J., 1979.

6. D. A. Peters and C. A. Deckert, "Removal of Photoresist Film Residues from Wafer Surfaces," *J. Electrochem. Soc.,* May 1979, pp. 883–886.

7. G. Evans, Keithley Instruments, Cleveland, Ohio.

8. Pieter Burggraaf, "Wafer Cleaning Systems," *Semiconductor Intl.,* July 1981.

6

Photoresist Coating

Coating Mechanisms

Spin coating

High-speed centrifugal whirling of silicon wafers has been the standard method for applying resist coatings in IC manufacturing for many years. "Spin coating," as it is more commonly called, will probably be used for this purpose for a long time, or as long as the IC industry uses flat circular wafers, which lend themselves well to this coating technique. Spin coating has remained basically the same since it was first used for applying thin uniform resist films on wafer surfaces. The coaters have, however, become more sophisticated in many respects. Spin coaters offer multiple-head cassette-to-cassette wafer handling, a far cry from the original home-built single-head spinners that were manually loaded and unloaded. The acceleration, coating speed, and many other parameters are computer-programmable to a wide range of settings. Spin coating technology is well entrenched and is apparently capable of handling the wafer-coating needs of IC manufacturers.

The spin coating dynamics are shown in Fig. 6.1. The basic param-

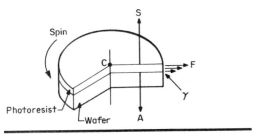

Figure 6.1   Dynamics of spin coating.[1]

eters at work are the centrifugal force F caused by the velocity of the spinning wafer (whose first derivative is the acceleration of the spin coater), the adhesion of the photoresist to the substrate, the force S of the solvents being driven from the photoresist, and the surface tension at the edge of the wafer. These are simply the mechanical forces that exist, and they only partially determine the quality and quantity of photoresist film remaining at the completion of the coating cycle.

Spin coating fluid dynamics.   The actual spinning process is a study in fluid dynamics. The process begins after resist has been dispensed onto the wafer surface and ideally has reached equilibrium. This "equilibrium" stage is portrayed in Fig. 6.2, the first of a series taken with a high-speed 16-mm motion-picture camera. The second stage, also occurring within the first few hundredths of a second, is "wave formation." Note the crest of photoresist forming after only a few rotations of the wafer. This stage is shown in Fig. 6.3. The next significant change in the behavior of the photoresist fluid occurs in the "corona" stage, in which a real crownlike structure forms. This occurs after approximately 30 rotations, and it signifies a running-out of the bulk of photoresist fluid that supported the integrity of the wave. A corona is shown in Fig. 6.4.

The next major change takes place when most of the excess resist has been carried off and the waves and corona no longer exist. At that point, centrifugal force acts on the remaining resist by scrubbing it off the wafer, generally in the shape of a spiral. The "spiral" stage creates thousands of resist droplets as the fluid reaches the wafer edge and is thrown off. The droplets, traveling at high speeds, are a possible source of splashback or redeposition onto the wafer. This once-

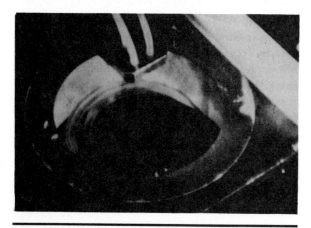

Figure 6.2   Spin coating equilibrium stage.[2]

Figure 6.3 Spin coating wave stage.[2]

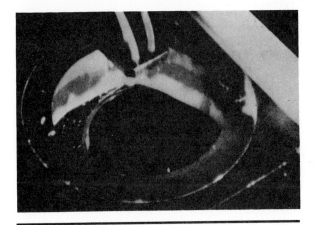

Figure 6.4 Spin coating corona stage.[2]

common problem has been solved by providing plastic splashguards that catch the droplets and the resist thrown off from the waves and carry them downward along a negative-sloping surface. Larger spin bowls accomplish the same result. Resist droplets redepositing on the still-fluid film will cause a streak or visual nonuniformity. A solid dust particle will cause a similar coating defect. Figure 6.5 shows the "spiral" stage.

The duration of these various stages is largely a function of the acceleration rate of the spinner. The higher the acceleration rate, the more rapidly these changes occur and the more uniform the final resist coating is.

The dispensing in this test was done by "static flood," wherein the

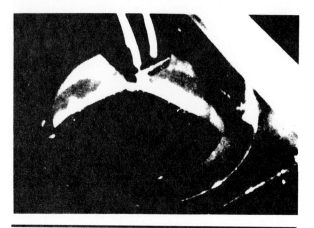

Figure 6.5   Spin coating spiral stage.[2]

resist volume is on the wafer surface *before* spinning commences. "Dynamic coating," wherein the photoresist is dispensed during wafer spinning, may result in uncoated sections and less resist coating uniformity unless the acceleration and rotation are set very low, which will allow time for the resist fluid to equilibrate.

Spin coating parameter calculation.   Several formulas have been developed to predetermine photoresist coating parameters on the basis of some known parameters. One such formula is

$$T = \frac{KP^2}{\sqrt{W}}$$

where T is the resist thickness, in angstroms, K is the whirler or spin-coater constant, P is the percentage of solids in the photoresist, and W is the whirler or spin coater rpm/1000.

Assuming a required 11,500-Å coating thickness, the resist solids content is 16.8%, and the spin coater constant is 87, the calculation is as follows:

$$\sqrt{W} = \frac{87 \times 16.8 \times 16.8}{11,500} = 2.14$$

$$W = 4.6 \quad \text{and} \quad \text{rpm} = 4600$$

Therefore, a 4600-rpm spin speed would yield an 11,500-Å-thick coating. The spin coater constant would always be 100 if the same coater were used. If the formula is applied to several spin coaters, their individual performance differences must be tested. The testing is done by plotting resist thickness versus spin speed for any number of spin coaters.

Another formula is

$$t = \frac{kS^2}{(\text{rpm})^{\frac{1}{2}}}$$

where S is the percentage of solids in the resist, t is thickness after spin coating, and k is the constant of proportionality, which varies with resist viscosity. Thus the solids content squared is directly proportional to the thickness and inversely proportional to the square root of the spin speed.

The photoresist thickness determination can also be made on the basis of weight fraction. Note that a 2-μm film can be obtained by any of the following combinations:

1. 1350J at 3000 rpm
2. 1350J of approximately 95% at 2000 rpm (95 parts 1350J to 5 parts thinner)
3. 1375 of approximately 90% at 5000 rpm (90 parts 1375 to 10 parts thinner)
4. 1375 of approximately 83% at 3000 rpm, equivalent to 1350J

Some spin speed versus thickness curves for typical resists are given in Fig. 6.6. All IC fabrication lines should, as a matter of course, monitor spin on resist thickness periodically, despite quality control checks on resist lot viscosity and percent of solids. The number of variables at work (see next section of this chapter and Chap. 13) necessitate an in-line functional test for something as critical as resist thickness. All imaging parameters are tied directly to the photoresist thickness including the final size of etched geometries, which determines chip pass-fail parameters.

Spin Coating Parameters

Resist thickness control is vital to maintaining overall control of an IC fabrication process. As IC geometries get smaller, more attention is given to the parameters surrounding control of resist coatings. Since the parameters are generally interdependent, they will be discussed as a group.

Viscosity and thickness determination

Viscosity is related to solids content, and solids content is typically preselected to coincide with spin speeds in the range of 3000 to 6000

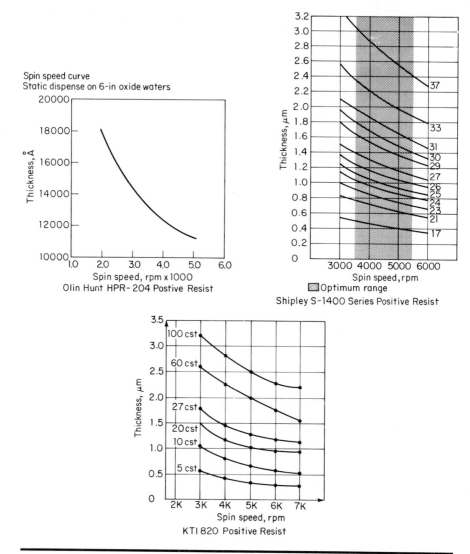

Figure 6.6 Spin speed versus resist thickness curves.

rpm. However, often there is a viscosity range that is best for optimal coating properties.

Edge bead

The selection of a particular viscosity can be for reasons of uniformity alone, including the effect of edge bead. Several parameters are important in minimizing resist edge bead, for example:

1. Edge bead width goes from 40+ mils to about 3 mils within 9 sec and stabilizes after 9 sec of spin time.

2. Edge bead height begins to stabilize after 9 sec of spin time, going from 30+ mils to less than 15 mils. Resist viscosities tested were 4.5, 17.0, and 3000 centistokes (cSt).

3. Edge bead formation can be minimized by providing a negative wafer taper, beginning ⅛ in from the wafer edge and moving outward.

A paper by M. W. Chan[8] provides a complete analysis of edge bead parameters. A typical photoresist edge bead is shown in Fig. 6.7.

A special resist viscosity may be chosen for reasons other than uniformity. For example, lower-viscosity resists flow more rapidly on a wafer surface and permit shorter spread-time cycles. A low viscosity may also be chosen for better wetting properties on devices with deep trenches or similar topography, especially when thicker versions of the same resist cause "skip coating" or nonwetting of small valley areas. A good rule for choosing viscosity is to use as thick a coating as resolution permits. This provides maximum etch protection. Thus the minimum viscosity should be dictated by minimum coating thickness requirements for pinhole-free etching. The maximum viscosity is

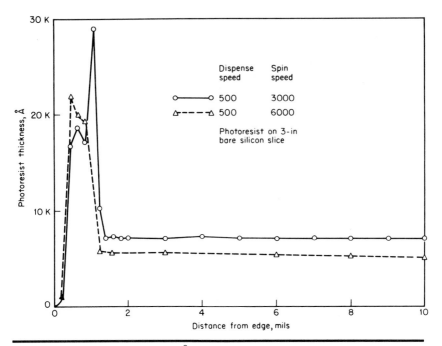

Figure 6.7 Resist edge bead profile.[8]

somewhere above the point at which pinholing is eliminated (as caused by inadequate thickness) and nonuniformity of the coating becomes undesirable.

Since almost all resists have varying molecular weight and rheologic properties for a given viscosity, individual resist testing is necessary to optimize the viscosity for a given process.

Nozzle position

H. Denk[6] showed that the nozzle tip should be close to the wafer surface to prevent splashing effects. A distance of $\frac{7}{32}$ in was found to be much better than $\frac{1}{2}$ in, probably because less fluid turbulence resulted. Optimum nozzle distance also is a function of orifice diameter and resist viscosity.

Wafer chuck

The chuck diameter should be large enough to consistently hold wafers and prevent breakage. Small chucks with large wafers cause several problems, including wafer breakage, "dishing" from excessive vacuum, and noncentered wafers. The wafer chuck should be large enough with respect to the wafer to incorporate multiple vacuum ports to spread the vacuum force across the wafer. This prevents dishing, which is illustrated in Fig. 6.8. The chuck also should be at ambient temperature, since temperature variations on the wafer surface will cause coating nonuniformity. Wafer chuck vacuum should be at a minimum to prevent dishing, yet safely above the point at which wafers leave the chuck occasionally from centrifugal forces. The wafer chuck also should be completely level, because tilted chucks can cause backside coating during the spread cycle. The wafers also must be level as well as centered when loaded onto the chuck for priming and/or coating. Chucks should have a low particle-generating surface, such as polished aluminum oxide or sealed, anodized aluminum. A

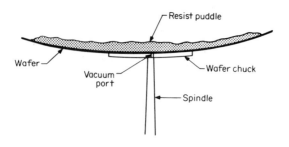

Figure 6.8 Wafer spin coating cross section showing dish effect.[6]

superflat, optically polished chuck surface will greatly reduce the need for vacuum clamping and will produce very flat wafers during coating.

Coating motor

Spin coater motors are subject to variations and should be strobe-checked frequently and calibrated. As they wear, checking becomes more critical. In production, wafer motors heat up and change speed as a result. Several studies have proved that coating motor warm-up with the equivalent to 15 to 25 spin cycles helps to stabilize the motor and therefore provide increased coating thickness reliability. Motor performance in production also is a function of spin speed. Speeds in excess of 6000 to 7000 rpm are avoided for that reason, as well as because of increased wafer breakage and noncentered wafer problems.

Dispense-and-spread cycle

The resist volume dispense varies with wafer diameter, but a general guideline is to dispense a volume of resist that, after the spread cycle and immediately before acceleration, covers all but the last ⅛ in of the wafer edge. This volume is generally about 2.5 to 3.0 mL for a 3-in-diameter wafer, 4.0 to 5.0 mL for a 4-in-diameter wafer, and 10 mL for 8- to 10-in-diameter wafers. Longer dispense times will provide improved coating uniformity, and up to 5 sec of dispense time for 3- to 4-in wafers is suggested. The increased time ties directly to the spread cycle, which should be long enough to permit a uniform radial spreading of the photoresist to within ⅛ in of the wafer edge. Figure 6.9 shows two spread patterns. The photoresist should reach equilibrium and be essentially static before spinning commences. In production,

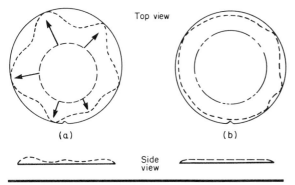

Figure 6.9 Resist dispense and spread. (*a*) Undesirable (dynamic). (*b*) Ideal (static equilibrium).[6]

particularly when large substrates (masks or wafers) are involved, a low-speed spread cycle is used. Although this increases wafer- or mask-coating throughput, it decreases coating uniformity. The static equilibrium puddle of resist is optimum for best coating uniformity, unaided by low-rpm spread cycles that create waves in the fluid that are never fully removed in high-acceleration spinning.

Spin and ramp velocities. High-spin ramping (time from 0 rpm to desired rpm) provides better coating uniformity than low-spin ramping over a range of ramp accelerations of between 500 and 40,000 rpm/sec. The faster the resist is removed from the wafer surface, the more uniform the coating. This shows the same results as low initial spin speeds compared with immediate acceleration to the preset velocity. In all cases, the more time allowed for the resist to evaporate while spinning, the greater the drying and setting-up tendencies, all of which makes the resist less fluid as centrifugal force attempts to "scrub" it off. Spin ramps within the range just mentioned do not have any appreciable effect on edge bead; they affect only the center–to–wafer edge coating thickness profile.

A spin ramp of 8000 to 20,000 rpm/sec is suggested to provide maximum coating uniformity without causing spin motor wear problems or excessive wafer breakage.

Spin velocity or speed should be based on several factors: desired thickness and uniformity of photoresist, spin motor wear, and wafer breakage. Most wafers should be spin-coated in the 3000- to 7000-rpm range to balance these factors. The high end of the range (6000 to 7000 rpm) is desirable in applications in which higher uniformity (± 50 Å) is required. When greater coating nonuniformity is tolerable, the 4000- to 6000-rpm range is preferred for the other reasons mentioned (minimal wafer breakage and motor wear).

Most photoresists have a spin speed versus thickness curve with a flattish "tail" toward the high-rpm end. It is desirable to be as far along the flatter portion of this tail as possible so that slight variations in spin velocity will not result in significant thickness changes. As velocity is reduced, spin-coated thickness becomes increasingly sensitive to velocity change. A typical spin time and speed versus thickness curve is shown in Fig. 6.10.

Spin time, bowl exhaust, and striations. Most resists begin to achieve thickness equilibrium after about 8 to 9 sec of spin time at greater than 5000 rpm. Additional spin time up to 20 to 25 sec does add slightly to coating uniformity, and therefore most wafers should be spun for 15 to 20 sec minimum, depending on their diameter. Larger

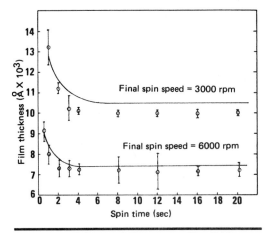

Figure 6.10   Spin speed time versus film thickness.

wafers will require more spin time to achieve the same coating uniformity as a smaller-diameter wafer for an equivalent spin velocity.

Venting of solvent vapors is required from an operator health and safety standpoint. The actual spin bowl, however, should be vented only minimally so that a solvent atmosphere is present. Excessive venting or exhaust of any spin bowl can cause coating nonuniformities in the form of radial striations (Fig. 6.11). Although some positive photoresists are formulated to prevent these striations, many resists are not, and in exaggerated cases, striations will occur even with striation-free resists.

Striations are only small undulations in the resist film, but they are large enough to cause variations in line geometries after developing or etching. A quick method of detecting striations is to look for a radial "sunburst" pattern on the mask blank surface just after coating. If a pattern cannot be detected, give the plate a short exposure (blanket) equaling about 20% of the time normally used. Then develop the plate for several short intervals and check in between for the occurrence of the striated pattern. Coating striations can be eliminated by simply spinning for an extremely short time (3 to 5 sec), by providing a solvent atmosphere in the spin bowl, and by greatly reducing the air movement (exhaust) in the spin bowl area. Figure 6.11 shows radially striated resist films.

Other spin coating parameters that affect resist film negatively include an excessive spread cycle that allows *too much* drying to occur before the actual spinning begins. A second area is excessive exhaust to remove solvent fumes, and often a mild downdraft to simply create a slow downward directional movement is sufficient to keep fumes

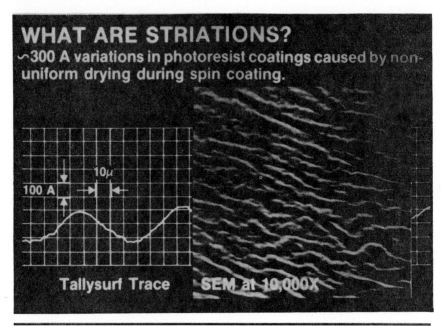

Figure 6.11   Radial striations.[3]

contained in the spin bowl. Finally, the throat of the spin bowl, with various-size openings should be optimized to reduce venturi effect. Venturi effect is strong downward air flow through the spin bowl caused by too narrow an opening at the top. Striations are typically 400 Å high and 40 μm from valley to peak. The exhaust flow should be constant in the coating bowl, because solvent atmospheres will affect coating and drying parameters.

Other spin coating influences.   The number of possible factors at work in the resist spinning process is surprising. Some of the factors to consider are shown in Fig. 6.11. The initial dispense involves, as a variable, possible overdrip from the dispense tube. The wafer diameter and chuck diameter should be nearly equal, and temperature in the bowl and downdraft exhaust influence coating quality. The air quality is very important, because particulates entering the spin bowl have a good chance of being lodged in dried resist films. Just prior to coating, a short squirt of primer, normally used for adhesion promotion, is especially beneficial in "scrubbing" particles off the wafer surface. After the spread cycle, resist skinning can set in as the solvents begin evaporating. Some processes provide a solvent atmosphere in the spin bowl to reduce skinning and striations, both of which are caused by rapid drying. It is important to dispense highly filtered resist immediately after priming or after a solvent scrub–spin clean, then allow a short spread cycle so the liquid equalizes and relaxes on the wafer, follow with an acceleration that uniformly spreads the fluid without allowing it to overtake itself, and finally remove the excess. Right after coating and *before* sending the wafer on the softbake, a short air dry is helpful in allowing the resist to even out on its own. For wafers with high steps, short air drying keeps more resist on corners, from which it tends to pull away in the softbake. Spinning too long before the air dry can set the film up too fast and reduce coating uniformity. Figure 6.12 shows a schematic of the coating sequence.

The solvents in the resist naturally play a major role in resist coating characteristics. The solvent system used in the 1350J and 1450J resists and evaporation rates for each are as follows:

Component	Evaporation rate (relative to butyl acetate at 100)
2-Ethoxyethyl acetate 2	21
Butyl acetate	100
Xylene	~50

The ratios of these solvents to each other in the resist formulation is critical as well, and the solvent ratio and even the composition may

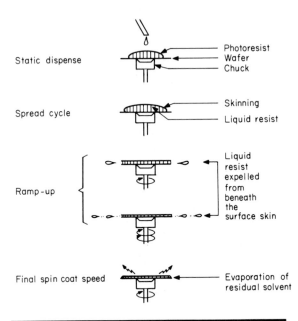

Figure 6.12 Physical factors in spin coating. (*Courtesy of R. Martin, Shipley Company.*)

need to be changed to produce spray coatings that are uniform. In many cases, it is possible to substitute low- or high-boiling solvents to change coating and drying properties in various types of coating equipment.

Radial resist dispense.   One technique used to increase the coating uniformity of resists is radial dispense application. This is accomplished by a special radius (1 to 2 in) while the resist is being dispensed. Figure 6.13 shows a resist solution being applied and the pattern made by the motion of the moving dispense head. This technique is extremely important to future device fabrication, and the MTI system really anticipated the need for better coating control. Small variations in resist thickness can result in not so small changes in the exposure, development, and pattern dimension control area. Customers who have used this system in production report excellent thickness uniformity. The thinnest coating that is safe to use in terms of pinhole and etch resistance has been reported to be in the 6000- to 7000-Å range. The advent of dry processing, in which etchants regularly remove over 1000 Å of the unexposed resist, pushes the need for thicker coatings up fur-

Figure 6.13   Radial dispense technique for resist coating. (*Courtesy of MTI.*)

ther. Thus, resolution improvement requires thinner coating, but the various process steps that erode the protective layer require additional thickness.

Coating Measurements

Stylus techniques

Photoresist coatings can be measured with a stylus-type instrument. The instrument is destructive and relies on the mechanical force of a riding stylus to trace the topography of the resist surface. The stylus leaves a slight scratch on the photoresist film, and this raises the question of its ability to measure absolute height, since it probably sinks partially into the film itself. The stylus technique is useful for tracing the shape of a photoresist or IC surface and mapping its topography in very small sections or across the entire wafer surface.

Optical measurement

Photoresist coatings are checked for thickness by optical techniques. (See Chap. 4.) The techniques include the ellipsometer, infrared laser scanning systems, image-sharing eyepieces on optical microscopes, SEM, and light section microscopy. Ellipsometer systems with direct copy printout offer a nondestructive production-oriented approach that does not rely on operator interpretation as the microscope approaches do. Many studies have shown that, despite extreme magnification and color or image phase separation, operators measure differently. Also, non-operator-dependent optical systems are faster and do not delay the process, which allows much greater wafer throughput.

A spectrophotometer technique used widely is a computerized automatic film-thickness-measuring system. In this system, reflected light from the specimen surface is focused through a monocular viewer by the objective lens onto a mirror located at the underside of the spectrophotometer. The light from the image passes through a hole in the mirror and is then measured. The computer program is based on an average refractive index, and the wavelength range is automatically scanned under computer control.

SEM measurement

The scanning electron microscope is used to measure photoresist coatings. This approach is used only when extreme accuracy is required and a small area is to be checked. The SEM can be used to calibrate or cross-check measurements taken with another system. It offers accuracy in the 50-Å range, and it is especially useful for cross-sectional analysis and resist sidewall angle measurement.

Beta ray measurement

Beta ray backscatter techniques are used to measure particularly thick (greater than 2 μm) coatings of photoresist. Isotopes of carbon and promethium provide a source of beta rays, which are directed through and scatter off the substrate. The low atomic number of photoresist requires that a carbon source be used, and the Microderm is a commonly used beta-ray-based measuring system. This system may be used to measure any number of metal film thicknesses, including metal deposits. Because of its nondestructive nature, it is particularly advantageous for measuring layers of gold or other metals.

Spin Coating Equipment

Track versus nontrack

There are many types of commercially available spin coating systems, both single head (masks) and multiple head (wafers). Most systems provide cassette-to-cassette loading and unloading and are "parameter programmable." The basic differences in spin coating equipment are not in the coating heads, but in how wafers are conveyed *from* coating through softbake. One type is a nontrack processor that coats and bakes on the same spinner and uses a microwave source for heating. Nontrack or batch systems provide priming, coating, softbaking, hardbaking, and even dehydration baking and uv flooding. Thus, a nontrack coating system is differentiated from a track system in that it breaks the work flow up into distinct functions connected by cassette unload and load mechanisms. This has the advantage of more specific control by process step but is less flow-oriented and more operation-intensive.

The in-line system shown (Fig. 6.14) is a track processor. It takes up more floor space but is oriented toward high volume and automated processing, including the integration of exposure, developing, and etching in a single line. Track and nontrack systems complement each other; each offers distinct features suitable for a particular process.

Cycle and equipment times

A typical coating program[7] for wafer coating would run as shown in Table 6.1. Acceleration plays an important role in coating uniformity, and it is essential to preset and control it on coating equipment.

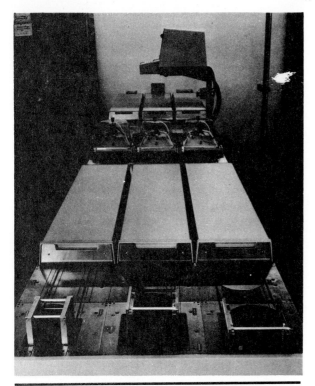

Figure 6.14 Wafer-coating system. (*Courtesy of GCA.*)

TABLE 6.1   A Typical Coating Program

Step	Sample time
Wafer clean (scrub)	6 sec
Spin dry	6 sec
Dehydration bake	6 sec
Spin cool	7 sec
HMDS dispense	2 sec
Spin	3 sec
Resist dispense	3 sec
Resist spread (static)	2–3 sec
Resist spin	15–20 sec

Equipment features

A key parameter in wafer processing is throughput. Coating equipment manufacturers typically supply this information on their data sheets. Wafer throughput in the cycle shown in Table 6.1 can vary as much as 50 to 200 wafers/hr, depending on the system.

Programming features are another important consideration in selecting a system. Minimizing operator interface is important, as it is in measuring coating thickness, and systems that offer minimal operator involvement provide processes with greater and higher-quality throughput. The tradeoff, however, is control. In order to establish a hands-off wafer process system, a very high degree of reliability and built-in process control *must* be provided. Without this control, a significant amount of expensive scrap will be produced.

Other programming features are wafer sort, search, and send-and-receive capability. Keeping the computer program simple and easily changeable is important, since the wafer process parameters frequently change as new materials and techniques are introduced. Service is always cited as the single most important need for anyone using a large in-line system, but equipment that is well engineered will require only minimal service. Many options for wafer-coating equipment exist—shop carefully.

Since wafer sizes change regularly, consideration for multiple-size handling is important. Most systems will process wafers of several diameters.

Other important considerations are, for example, the coating head mechanism itself. A common problem is dripping or plugging of solvent or resist in nozzles. Some systems offer retractable nozzles; others incorporate a "suckback" mechanism. Either can have its own set of advantages or disadvantages, but this aspect of coating equipment should be explored.

Under "Spin Coating Parameters," the need for wafer centering during spin coating was mentioned. This is another equipment-related parameter to explore, and some systems have wafer-centering guides to provide this capability.

Contamination is a key problem in lithography, particularly at the resist application stage. Selecting a piece of coating equipment includes determination of particles added per pass. This is the number of particles contributed in a single operating step. The amount of surface area exposed and the number of times wafers are handled are major factors in contamination, not to mention wafer contact with various parts of the coating system. Contamination specs for equipment should be so established as to be compatible with the level of lithography supported by a process.

Spray coating

Numerous types of spray equipment are in use throughout the industry. Most have enclosed spray chambers, although some have open spray chambers. All the units have Class 100 laminar flow air directed into the spray chambers to ensure maximum cleanliness. Some are characterized by a vertical downdraft of laminar flow air; others have horizontal laminar flow air.

Loading stations can be placed on one side or both sides of the coating chamber, and a transfer mechanism can be used to move parts in and out of the chamber. There are a number of options for loading and unloading substrates. Some versions have a conveyorized continuous processor that permits a high degree of both production output and part consistency.

In addition, a number of variables can be used to produce various types of spray coatings. These variables will be explained briefly and then the optimum methods for spray coating will be discussed. The variables are as follows:

1. Dilution of the resist. Low-viscosity resist is required for spraying, in the 4- to 20-centipoise (cP) range, for 2- to 5-μm coatings (dry).

2. Traveling speed of the spray gun. The spray gun traverses the indexing table as it sprays. Speeds from 5 to 20 in/min can be achieved. Generally speaking, a higher speed will produce better results, and a gun speed of 15 in/min is typical.

3. Table index. As the spray gun sweeps back and forth from front to back in the spray chamber, the table on which the parts to be coated are placed indexes forward at various intervals. It is possible to index the table between ½ and 2 in at a time. A small index distance is desirable, and a ½-in index is suggested.

4. Nozzle size. Several different nozzle sizes can be used with spray guns. Typically, they range from 0.013 to 0.030 in. To reproduce coatings in the 2-μm or less range, the 0.030-in nozzle is suggested. To reproduce coatings in the 2- to 5-μm range, the 0.015-in nozzle is suggested, and to reproduce coatings that are 5 μm or greater in thickness, the 0.020-in nozzle is recommended.

5. Gun pressure. Basically, two types of gun pressure can be used: high or low. When low pressure is used there is less atomization of the photoresist. The mechanism for applying photoresist in this type of equipment relies on the use of a superheated hydrocarbon vapor as the carrier. In the spray nozzle, the photoresist is atomized into a conical spray pattern that can be adjusted by varying the nozzle pressure, nozzle size, or type of spreader used in the nozzle. Low pressure is recommended for spraying photoresist, and high pressure is used primarily for materials that are more difficult to atomize.

6. Spreader. The spreader is simply an orifice in the spray nozzle that shapes and controls the spray pattern. Spreaders are supplied

with various types of orifice sizes, ranging from 20 to 70 mils, and with differently shaped openings. Also, there may be more than one orifice. The more elongated the spray pattern, the dryer the surface; the rounder the spray pattern, the wetter the surface. A wider spreader will result in less thickness and more overspray. The spreader is a good adjustment for wetness or dryness of the sprayed coating.

7. Height of the gun. The recommended distance between the gun and the parts to be coated is about 6.5 in. A 1-in change in either direction will result in approximately a 15% change in photoresist thickness.

8. Type of supply. There are two ways in which the photoresist can be supplied to the gun: cup supply or supply direct from the photoresist quart bottle. The cup supply is simply a small cup located 8 to 10 in above the gun. The "quart attachment" permits standard quarts of photoresist to be plugged directly into the end of the supply tube. A vent tube is provided on the side of the supply tube, and this permits air to enter as the solution level drops. This mechanism automatically offsets the differential pressure that would be created as a function of solution level. A 1-in drop in the solution level will result in approximately a 2.5% change in photoresist thickness. The cup supply is ideal for prototype runs involving small quantities of a given photoresist; in prototype runs it is important to use good flushing. Extreme uniformity can be produced with spray coating by creating a series of closely spaced overlapping coatings. The flow properties of the resist should allow a leveling to take place before resist baking. For example, if you were using a ½-in index and the spray pattern were approximately 6 in across, then any given spot in a finished coating would have received 12 separate applications of photoresist.

These parameters are intended as guidelines for any spray system using a heated vapor carrier. They also may apply to smaller, manual spray guns or other types of spray equipment. In summary, spray coating equipment does an excellent job of applying photoresists. Coatings from 1 to 10 μm can be applied with excellent uniformity.

Roller coating

Roller transfer has been used to apply a "backside" photoresist coating and to coat liquid crystal device (LCD) and thin-film substrates. In backside coating, photoresist is applied to the entire wafer surface before scribing. The scribe line areas are imaged and the resist is removed. The wafer is then laser- or diamond-scribed, and the active device areas, which are covered by a layer of resist, are protected from silicon dust contamination.

A positive photoresist is well suited for this or *any* application in which a large area must be protected by a resist image. Negative resist, to provide the image protection, must be exposed and polymerized. Any dust or opaque spots (chrome or other particles) on the photomask will leave a portion of the negative-resist film unpolymerized. A pinhole will result after developing, and that will, in this case, leave the substrate exposed to silicon dust contamination.

The mechanism for applying photoresist by roller transfer is shown in Fig. 6.15. Several possible variations on the mechanism shown below are used, but the basic concept is the same.

A threaded rubber coating roller is used for transferring resist to delicate substrates. By varying the configuration of the groove, the amount of liquid resist applied can be controlled. Thread angles of 70 to 90° are used, and the thread-per-inch density varies from 100 to 50. The standard thread-per-inch types used are 48, 64, and 96. The other thickness-determining factor is solids content of the resist. The combinations of various roller-thread configurations and resist viscosities permit a wide range of coating thicknesses. Rolled coating equipment often includes an in-line infrared drying system (Fig. 6.16). The roller coater equipment shown here is from Gyrex Corporation in Santa

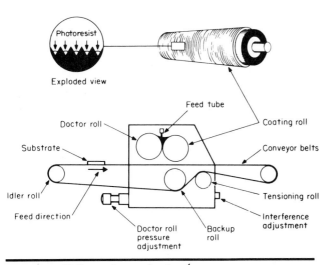

Figure 6.15 Roller coater mechanism.[4]

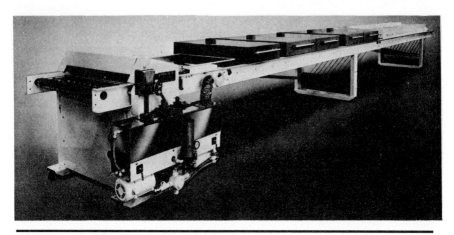

Figure 6.16   Roller coater and drying system.[4]

Barbara, California. There are also other suppliers of roller coater equipment.

REFERENCES

1. G. F. Damon, "The Effect of Whirler Acceleration on the Properties of the Photoresist Film," *Kodak Seminar Proceedings,* 1965, p. 34.
2. MOS Technology, "High Speed Motion Picture of Wafer Being Coated," Valley Forge, Pa. 1976.
3. Technical Data Sheet (D1300) on Positive Resist, Shipley Co., Newton, Mass., 1980.
4. Technical Data Sheet on the Model 630/730 Coating and Drying System, Gyrex Products, Santa Barbara, Calif.
5. P. O'Hagan and W. J. Daughton, "An Analysis of the Thickness Variance of Spin on Photoresist," *Kodak Interface Conf. Proc.,* 1977, p. 95.
6. H. H. Denk, "Optimization of Photoresist Processing with the IBM 7840 Film Thickness Analyzer," *Kodak Interface Microelectronics Seminar Proc.,* 1976, p. 28.
7. Technical Data Sheets on "Wafertrac" Coating System, GCA Corp., Sunnyvale, Calif.
8. M. W. Chan, "Another Look at Edge Bead," *Kodak Interface Seminar Proc.,* 1975, p. 16.
9. T. Stoudt, Azoplate Corp., Murray Hill, N.J., private communication.
10. W. Daughton and F. Givens, "An Investigation of the Thickness Variation of Spun-on Thin Films Commonly Associated with the Semiconductor Industry," *J. Electrochem. Soc.,* vol. 129, no. 1, January 1982.

Reference 9 was a source of background information for this chapter.

Softbake

Softbaking Parameters

Softbaking is the process step wherein almost all the solvents are removed from the resist coating, thereby rendering the coating photosensitive. It plays a very critical role in photoimaging because a number of subsequent parameters are affected by the process, including adhesion, exposure, development, geometry control, and solvent content.

Adhesion

Adhesion of the photoresist is partially a function of softbaking. Incomplete softbaking results in poor resist-oxide bonding and subsequent lifting of resist images in the developer. Extended softbaking is used to achieve improved adhesion to highly doped phosphorus oxides, quartz, and chrome masks. Overbaking will cause film stress and brittleness and can reduce adhesion.

Exposure

As mentioned earlier, resist coatings cannot be imaged until their solvents are removed. They become photosensitive, or imagable, only after softbaking. Oversoftbaking will degrade the sensitivity of resists by either reducing the developer solubility or actually destroying a portion of the sensitizer. Undersoftbaking will prevent light from reaching the sensitizer; in the case of a negative resist, incomplete cross-linking is the result in exposed areas. Positive resists also are incompletely exposed if considerable solvent remains in the coating. However, "apparent" increased sensitivity results because of increased resist dissolution rate in exposed areas.

Development

Positive novolak-type photo resists that are undersoftbaked are readily attacked by the developer in both exposed and unexposed ar-

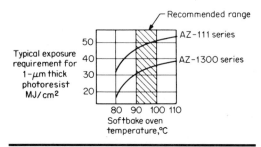

Figure 7.1 Softbake temperature versus exposure requirements.

eas. This causes apparent increased photosensitivity and can cause problems because the image area also is attacked and therefore will provide less etching resistance. Negative resists have a similar problem in that incomplete polymerization in exposed areas results in developer attack, "scumming" in image areas, and subsequent poor etch resistance. Oversoftbaking positive and negative resists will usually result in increased exposure times, but it will also increase exposure and development latitude.

Line control

Reduced softbaking makes line control difficult, because the image is more sensitive to exposure and developing steps. That is because the dissolution rates of both exposed and unexposed areas increase substantially with increased solvent retention. Increased softbaking reduces the response of the resist in exposure and development and provides increased line control capability. Here the inverse is true because dissolution rates are lower in all cases. In addition, the differential solubility is closer to optimum when most of the solvent has been removed from the resist film.

Since softbaking is integrally tied to the other process steps, it is necessary to run characterization tests on all these operations before selecting a final sequence. Figure 7.1 shows the relationship between softbake and exposure for AZ-1300 and AZ-111 series positive photoresists.* Note that, as the bake temperature increases, the curve flattens out. The best process control is afforded at higher temperatures, because changes in softbake temperature or time, inevitable in any process, are increasingly less significant to changes in exposure

---

*Process conditions: softbake time, 20 min; exposure source, high-pressure mercury vapor lamp; development time (immersion), 60 sec at 21°C; developer makeup, 1 part AZ-351 to 3 parts H_2O and 2 parts AZ-311 to 3 parts H_2O.

time. At softbake temperatures below 80°C, exposure time is very sensitive to changes in bake time or temperature. Note also that exposure time is significantly less at lower softbake temperatures, a condition caused by the increased rate of developer attack on all areas of the exposed coating. *Many control and yield problems are a result of attempts to increase exposure throughput by reducing softbake temperature.*

Another parameter impacted by softbake is absorbance. Absorbance changes at different wavelengths for various bake times. This type of analysis may be useful in optimizing a software process.

The relationship of softbake to line size is another parameter to be investigated. The line width stability of the photoresist image is tied directly to softbake temperature *and* the effect that various temperatures have on exposure and development. Since measurements are made after development, the result is a cumulative one, reflecting the total effects of softbake, exposure, and development. In general, a longer or higher-temperature softbake provides better line control, but this trades off with exposure time.

The change in solubility rates in the developer as a function of softbake (time and temperature) is well known. Figure 7.2 shows this relationship for four different softbake temperatures. Minimizing the dissolution rate of the positive unexposed resist measured is very im-

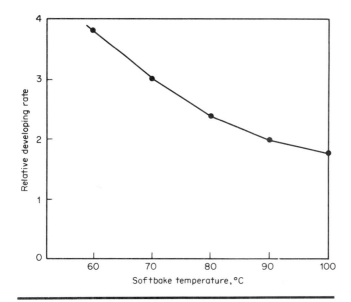

Figure 7.2 Development rate versus softbake temperature.

portant in preserving the pinhole resistance and integrity of the resist image. Increased resistance to the developer in negative resists is a function of complete polymerization *as well as* complete softbaking. In positive resists, the softbake alone determines this parameter. Increasing the softbake temperature to a maximum (below thermal flow, and just above the steep portion of the softbake versus exposure dose curve) is recommended because it provides high development and line control latitude.

Solvent content

The goal of softbaking is solvent removal. Figure 7.3 shows the relationship between residual solvent and baking temperature (30 min for three major types of positive resist). Note that at 90°C, where most softbaking is done, solvent content is still high, and it is not until coating temperatures exceed 140°C that weight loss drops off considerably. Some of the loss is water.

Another interesting change during softbaking is thickness reduction. Figure 7.4 shows the changes in resist thickness over various softbake temperatures. In the 50 to 90°C range, the thickness is relatively stable, and only at what we normally consider postbake temperatures (greater than 100°C) does the resist undergo additional shrinkage. This relationship is important to remember, because checks of resist thickness on different spinners should standardize the softbake to eliminate this variable.

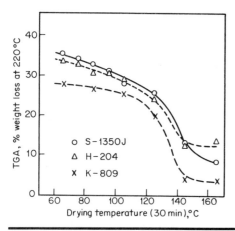

Figure 7.3 Solvent content versus softbake temperature.[2]

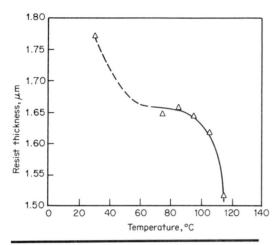

Figure 7.4 Resist thickness versus softbake temperature.[2]

Optimizing softbake temperature

Several analytical techniques are used to determine just how long and at what temperature to bake coated resist films. Since the level of softbake has such noticeable leverage in determining the functional sensitivity of the resist, it is important to be able to quantify it and preestablish optimum parameters. Thermogravimetric analysis (TGA) uses the mechanism of volatilization, or polymer weight loss, to measure solvent loss and thermal decomposition. A sample is first resist-coated and placed immediately, without softbaking, into a thermal environment and baked at approximately 70 to over 180°C, possibly up to the carbonization temperature or the temperature at which the resist becomes disordered graphite. A sample profile from a TGA test is shown in Fig. 7.5. One aspect of TGA to observe is the gap between the point at which solvent loss is complete and that at which decomposition begins and damage to the sensitivity of the resist can occur.

Softbaking Techniques

Convection

Convection baking is used in prototype IC fabrication processes. (Figure 7.11 shows the mechanism of solvent removal for convection and other softbaking methods.) Convection softbaking is performed in an oven cavity, although convection-drying effects occur even during

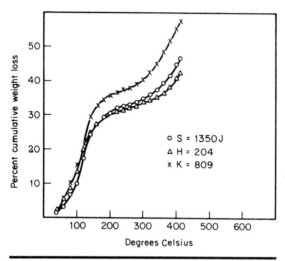

Figure 7.5 Thermogravimetric analysis (TGA) profile for softbake optimization tests.

other types of drying, such as infrared, in which a convection air current will aid in the baking process.

Ovens have many aspects that make them undesirable for IC production environments. One problem is temperature variation in the oven cavity, which may cause nonuniform softbaking. Another source of nonuniform temperature is poor recovery time when wafers are moved in and out of the oven. Different ovens will recover differently and give varying softbake results. Loading and unloading tests, as under actual conditions, should be simulated to calculate heat loss effects. A typical temperature profile for a convection oven is shown in Fig. 7.6.

Contamination is another problem with softbaking in convection ovens. Since good air movement (and exhaust of solvents) is necessary, particles often enter the cavity and are deposited on partially dried resist coatings.

A key advantage of convection oven softbaking is the repeatability achieved once a process is established and the oven is properly characterized. Ovens are typically very stable and do not fluctuate with age as other sources do. Nor are they variable with surface types as infrared and microwave heating sources are. Metals, oxides, and wafer films of all types do not introduce major variables in softbake cycles with convection ovens, because wafers are heated as a mass and not from the surface.

Equipment types. A good softbaking convection oven is clean (submicron filtered), offers rapid recovery, has extremely uniform

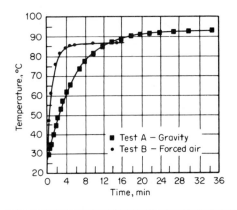

Figure **7.6** Convection oven temperature profile.[3]

heat distribution in the oven cavity, and exhausts solvents efficiently. Figure 7.7 is a diagram of an oven that is used in IC production. The uniformity of the oven cavity needs to be ±0.5°C at 40 to 204°C. Typical air temperature uniformity should be ±1°C at 100°C and ±2°C at 200°C. Drift, over a 24-hr period, is ±0.75°C at 100°C and ±0.50°C at 200°C. These figures can be achieved mainly because this oven

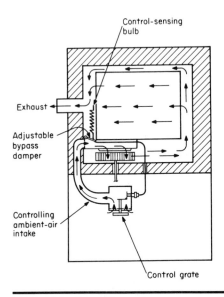

Figure **7.7** Convection oven cross section.[4] (*Courtesy of Blue M Corporation.*)

operates without a heating element. Air turnover up to 45 ft³/min is recommended for 90°C softbakes.

Infrared

Infrared radiation includes all wavelengths of radiant energy from long-wavelength visible red light (approximately 8000 Å) to short radio waves. A breakdown of infrared sources by wavelength shows that resists absorb 85 to 90% of the long-wavelength (75- to 76-μm) infrared sources. Long-wavelength *IR* sources are very efficient in resist curing. The "far" infrared (farthest from the visible spectrum) has the lowest frequency and comes from the lowest-temperature sources.

The combination of cleanliness, rapid solvent removal, temperature uniformity, good recovery time, and safety is essential for an efficient softbake in IC fabrication. Infrared softbaking is widely used to cure or desolvate resist-coated wafers and masks. One of the benefits of infrared baking is rapid drying that can be accomplished in-line without interrupting the process flow. Infrared softbake times are fractions of convection bake times, and the drying is done "inside out." Solvent removal from the resist-wafer interface toward the surface provides more thorough drying and prevents the formation of "skins," bubbles, wrinkles, and pinholes, which occur more frequently with other baking methods that remove solvents from the surface of the resist initially. Only longwave (6- to 20-μm) infrared radiation can penetrate to the resist-oxide or resist-chrome interface first and then be reflected back through the coating to the surface. Because reflectance of the far-infrared wavelength plays a key role in solvent removal, substrates with varying reflectance will require varying bake times. The thermal gradient is such that the coolest part of the coating is the surface, and the evaporation actually cools the internal part of the coating. The greatest heat occurs at the driest part of the coating where evaporation is complete and the solvent removal mechanism is nearly complete. This "inside out" mechanism for solvent removal is desirable, since it does not allow the entrapment of solvents that will bubble up or cause large voids in an image area. Shorter-wavelength infrared sources are less efficient for photoresist softbaking because they offer less penetration and generally less uniformity than planar or panel long-wavelength infrared sources.

The theory of heat radiation is explained by the Stephan-Boltzmann law for heat radiated from a black body. This law may be useful for theoretical modeling of energy transfer from various sources to substrates, but simpler and more practical methods are recommended. For example, the wavelength (for any source of radiant heat) versus source temperature can be plotted, and so can be wavelength versus relative energy intensity. [Both the temperature and the wavelength are radiated from an ideal (black-body) source.] The heat rise occurs when the infrared radiation is absorbed by the organic molecules in the photoresist coatings, thereby inducing rotation and vibration of atoms or groups of atoms. These molecular motions cause heating of the absorbing mass. Since infrared radiation is readily absorbed by many materials, the solvents in the photoresist do not need to be preselected for optimal baking reactions.

Surface temperature relates to wafer temperature. The actual surface temperature of both the wafer and the source is dependent on many variables, including the reflectivity of the substrate, heating element materials, and the design of the heating source.

A considerable amount of energy is lost between the source and the substrate. Since silicon dioxide is a good absorber, infrared softbaking of photoresist is very efficient. On aluminum, which reflects much of the energy, baking times will be different. The absorption of infrared energy by organic materials occurs more easily at lower temperatures, generally with source settings of 500°F and actual surface temperatures of about 90°C, which are the same as the convection oven temperatures commonly used. Overall, relatively low source temperatures are recommended for curing photoresist coatings. A low-temperature, long-wavelength source, like a radiant glass panel, is appropriate for most applications. The color or dye in the resist also plays a role in energy absorption, and dye can be used to adjust not only image contrast but also exposure and infrared drying properties.

In all cases, infrared softbaking sources should be checked periodically for consistent output, usually with an infrared pyrometer. Solvent vapors will coat onto the infrared lamps or glass panels and will reduce the effectiveness of the output. These lamps and reflectors should be cleaned regularly to remove the "solvent clouding." This is where a panel heater with no reflector is preferred, since the panel is too hot to condense the vapor. Reflectors are desirable, however, because they can prevent expensive loss of heat and radiant energy. They should be placed around the sides of a panel or other type of heater. Cleanliness of the softbaking area must be maintained, and positive-pressure laminar airflow units should be placed in the softbaking area.

In typical wafer production areas, cassette-to-cassette softbaking units based on infrared sources are used. An in-line system is shown in Fig. 7.8. Such units are programmable to vary the cycle for the different types of surfaces used, since each will need a different amount of energy for a given resist thickness.

The same type of softbaking mechanism is achieved with the flat-track system. A temperature range of 30 to 200°C $\pm$ 1°C is available

Figure 7.8 In-line programmable coating and softbake system. (*Courtesy of Silicon Valley Group.*)

in independently controlled infrared zones. The heating element, drive system, and insulation are sealed to minimize wafer contamination. The softbake oven can be used in-line with coating or separately as a stand-alone unit. A heated nitrogen curtain is often used to provide baking uniformity, and each oven uses a single infrared panel to avoid nonuniform "cold spots."

Conduction

Resist softbaking by conduction is a very effective method. Some of the earliest softbaking techniques for ICs involved the use of hotplates; coated wafers were placed on laboratory hotplates that had thick metal tops. Heat was readily conducted through the wafers and forced the solvents out from the oxide-photoresist interface. The "inside-out" curing technique (similar to infrared) is efficient from an energy conservation point of view and leaves the photoresist free of surface skins and solvent entrapments. This technique generally results in a uniformly cured coating. The oxide-photoresist interface, where softbaking begins with conduction techniques, is a very important part of the cross-sectional structure. Incomplete curing at this interface will cause excessive resist breakdown and undercutting during etching. Providing complete solvent removal at this interface promotes better photoresist adhesion and better etch resistance later.

Conduction softbaking has been successfully adapted to in-line wafer-handling systems. Such systems are completely programmable to provide a wide variety of conditions, including an air spin dry *before* softbaking to help reduce resist viscosity and increase step coverage. Thus when the coated wafer reaches the heated conveyor, its partially dried condition permits less viscosity drop and reduced flow of resist from steps. Several other process variables can be programmed. Some units *individually* sense wafer differences and vary the cycle accordingly. This type of process adjustment from wafer to wafer can be accomplished at any of the stages in the lithography sequence.

An example of this type of in-line conduction baking is an in-line hotplate system that extends the advantage of the laboratory hotplate to production processing. A Teflon-coated aluminum plate provides the heat source for softbaking, which typically requires 30 sec. During baking, a cover encloses the wafer and hotplate, providing a low-pressure atmosphere, and the solvents are vented through a 0.2-μm filter at the end of the cycle. Digital temperature control, along with a film heater compensator (for edge losses), provides uniform and controlled softbaking.

Microwave

Microwave softbaking is another production method for desolvating photoresist coatings, as well as an energy source for rapid dehydration baking prior to the application of an adhesion promoter. Although microwave energy is not a widely used energy source, its potential for semiconductor fabrication is considerable. In fact, more development and refinement of this technology will increase its application.

Microwaves are a low-energy, high-wavelength form of radiation, and microwaves are readily absorbed by certain polar groups and molecules in photoresist coatings. The radiation is converted to heat, and the amount of heat produced is a function of the absorptivity and dielectric loss factor of the absorbing species. In general, molecules, although electrically neutral, have electrons that cause the molecule to have positive and negative (polar) ends. A "polar molecule" is one that exhibits an electric dipole because of the asymmetrical distribution of electrons in its atoms or a particular spatial arrangement of these atoms. When photoresists are placed in a changing electric field, the polar molecules align with the direction of the field. When the electric field is removed, the molecules' electrons return, by elastic recovery, to their original positions. In a microwave field, the energy is turned off and on several million times each second, which causes the same rate of electron movement in the polar photoresist molecules. The molecular friction that results from this movement is what generates the heat, an effect called "dipole rotation." Since the heat is generated only within the material in the electromagnetic field, the area around the material remains at room temperature.

Another way in which microwave energy generates heat is through the mechanism of "ionic conduction." This happens mainly in liquid materials when free ions are flipped back and forth in the oscillating field. As the ions collide with nonionized molecules, heat is generated. Some materials do not block microwave energy and are therefore said to be "transparent" to microwaves. The materials that do respond are the dielectric or semiconductor ones including wafers and most plastics. Metals, however, do not respond to microwaves unless they are formed into thin strips or coils. Teflon, polypropylene, and polyethylene are notable plastics that do not respond to microwave energy. Thus, fluids can be simultaneously heated and transported through tubing made of one of these materials without heating the tube.

The source of microwave energy is usually either a klystron or a magnetron; both are shown in Fig. 7.9. In a klystron tube, a strong electron beam is focused by oxide magnetic fields through a series of resonant cavities. Electron bunching causes the energy to move into the last cavity. The magnetron operates in a similar way, except that the electron beam is formed into a spiral shape by magnetism and the resonant cavities lie outside the spiraling beam. The magnetron is typically less expensive initially and more efficient; the klystron lasts longer and is capable of generating more power.

The basic equation for electronic heating is

$$p = ke^2f$$

where p is the heating power generated in the material, k is the dielectric factor of the heated material, e is the strength of the electric field, and f is the frequency of the field. From this formula, we can see that higher frequencies have greater potential for power generation, given the same electric field.

Microwave energy is transmitted from the klystron or magnetron

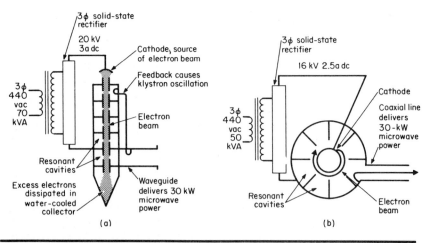

Figure 7.9   Cross sections of (a) klystron and (b) magnetron.[6]

tube to the work by a waveguide. "Waveguides" are variously shaped metal structures that have the strongest energy fields in their centers.

Automatic microwave baking of wafers with and without photoresist has become a production tool, and microwave ovens have been forgotten for this application. An example of the in-line microwave system, included as part of a clean, coat, and bake unit, is shown in Fig. 7.10. The microwave energy provided in this system eliminates hot spots by creating a parabolic field and rotating the wafer on the spin chuck directly under the microwave source. The wafers are given a microwave dehydration bake, which takes approximately 12 sec. After a spin cool, prime, and resist coating, softbaking occurs with the same unit, typically taking about 10 sec. Key advantages of this single-unit design are reduced wafer handling and less physical machine space, which offer the possibilities of reduced contamination and a more efficient production area. The system is microprocessor-controlled, and a variety of process variations can be programmed.

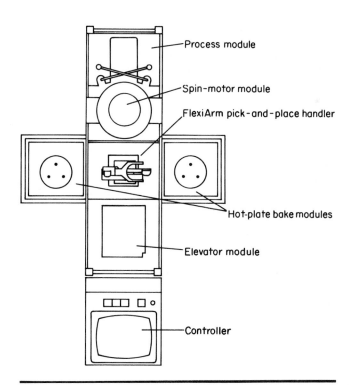

Figure 7.10 Coat and bake system with hotplate units, coater, and computer control. (*Courtesy of Machine Technology, Inc.*)

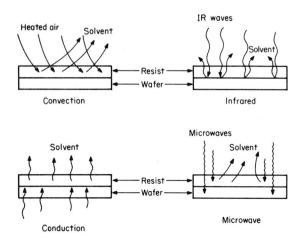

Figure 7.11   Convection, infrared, microwave, and conduction solvent removal mechanisms.

Vacuum softbaking

Vacuum softbaking is a highly efficient way to desolvate resists. Vacuum baking avoids the problem of resist surface "skinning," and tests on chemical resistance (etching and implant) have shown vacuum-baked samples to have better resistance than infrared- or convection-baked samples. This is probably due to more gradual and complete solvent removal. One drawback of vacuum baking is its inability to be put in-line. However, batch vacuum baking will provide substantial wafer throughput. Figure 7.11 shows various solvent removal mechanisms.

REFERENCES

1. Technical Manual, Shipley Company, Newton, Mass., November 1980.
2. J. M. Kolyer, F. Z. Custode, and R. L. Ruddell, "Thermal Properties of Positive Photoresist and Their Relationship to VLSI Proceedings," *Kodak Interface Seminar Proc.,* 1979, p. 150.
3. D. J. Elliott, "Positive Photoresists as Ion Implantation Masks," Society of Photooptical Instrumentation Engineers, vol. 174, April 1979, p. 153.
4. Technical Data Sheet, Bulletin No. 721, Blue M Engineering Co., Blue Island, Ill., 1980.
5. R. W. Jeffery, "Advantages of Process Heating with Infrared," *Circuits Mfg.,* October 1969, p. 7.
6. G. M. W. Badger, "Microwave Heating," *Machine Design,* October 1970, p. 88.
7. Data Sheet on Model 6330 Hot Plate Wafer Oven, Machine Technology Inc., Sunnyvale, Calif., 1980.
8. T. Batchelder and J. Platt, "Bake Effects in Positive Photoresist," *Solid State Tech.,* August 1983, p. 211.

8

Exposure

The exposure (ion beam, electron beam, laser beam, x-ray, and filtered mercury sources) of resist materials for IC fabrication has been the subject of many papers, studies, and symposia, and for good reason. The parallel shrinking of IC geometries and reduction of chip costs have placed significant process-related pressure on the exposure step of the imaging process. Exposure is the *only* step in the imaging sequence during which wafers must be individually processed; all other steps allow for batch, conveyorized, or in-line processing. This creates a throughput problem relative to the other steps, and as it turns out, throughput usually trades off with resolution. For several reasons, exposure throughput can always be increased when resolution is decreased. The task of resist suppliers and exposure equipment suppliers alike has been one of providing an exposure process in which the photoresists and equipment provide maximum resolution *and* maximum throughput, and balancing many parameters is necessary to achieve that end.

In this chapter we will discuss the exposure mechanisms of photoresists and nonoptical resists, as well as the various process factors that affect exposure. The optimization of the exposure step in wafer imaging will be discussed in detail, and the types and capabilities of equipment will be reviewed. Finally we will discuss trends in the exposure area, including resists and equipment to meet future needs.

Exposure Reactions

Negative optical resists

The exposure parameters for any resist material are in large part determined by the chemistry of the resist. All the most commonly used

---

Reference 11 was a source of background information for this chapter.

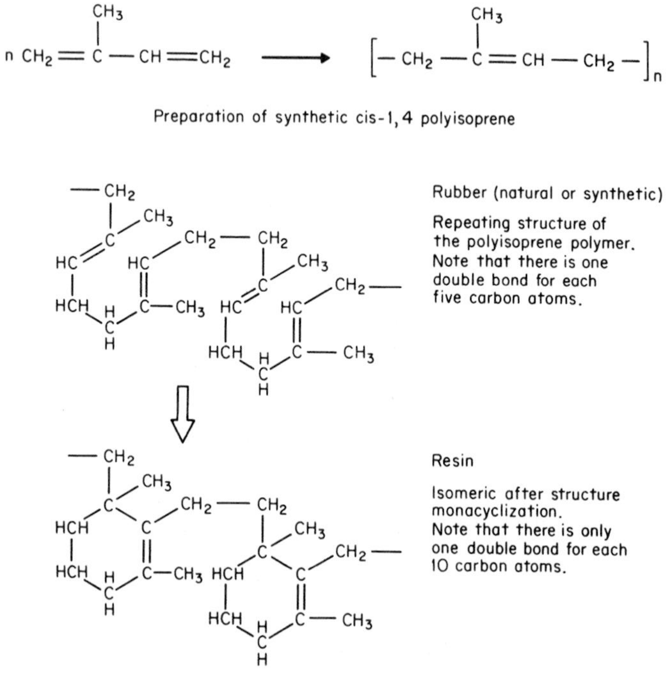

Preparation of synthetic cis-1,4 polyisoprene

Rubber (natural or synthetic)

Repeating structure of the polyisoprene polymer. Note that there is one double bond for each five carbon atoms.

Resin

Isomeric after structure monocyclization. Note that there is only one double bond for each 10 carbon atoms.

Figure 8.1 Polyisoprene resin formation.[1]

negative photoresists in IC fabrication have similar molecular structures and exposure mechanisms. The resists are formulated with two basic components: a photosensitive cross-linking material that gives the resist its exposure characteristics and a nonphotosensitive synthetic rubber compound that provides the etch resistance, adhesion, and film-forming properties. The rubber component, a cyclized cis-poly (isoprene), is shown in Fig. 8.1. This structure is formed by reacting the isoprene structure with the poly (cis-1,4-isoprene) and then with the cyclized state, following the sequence in Fig. 8.1. The molecular weight of the final cyclized polymer is in the range of 60 to 150,000.

The photoactive compound in these negative resists is a bisaryldiazide, which, when exposed to light, cross-links the rubber material and makes it insoluble in the solvent developer, typically a mixture of xylene isomers. The structure of the photoactive compound is shown in Fig. 8.2. A good chemical explanation of the reactions is given by Thompson and Kerwin.[3]

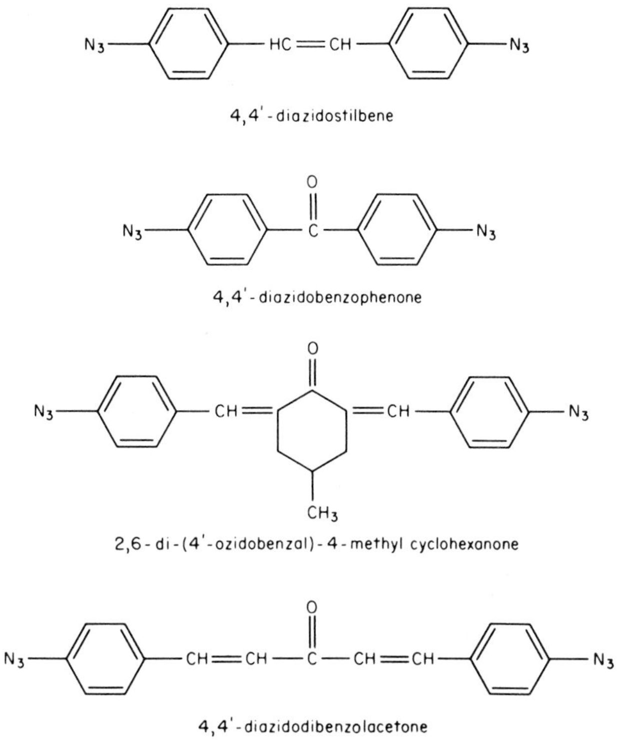

4,4'-diazidostilbene

4,4'-diazidobenzophenone

2,6-di-(4'-ozidobenzal)-4-methyl cyclohexanone

4,4'-diazidodibenzolacetone

Figure 8.2 Structures of bis-azide sensitizers for cyclized rubber resists.[1]

Positive optical resists

Positive photoresist such as Shipley 1350J and Hunt HPR-204 are composed of three basic ingredients: a photoactive compound (PAC), resin(s), and the carrier or solvent system. The photoactive compound gives the resist its developer resistance and special absorption properties. The structure of the photosensitive compound is given in Fig. 8.3. This material, an o-napthaquinonediazide, is often called an "inhibitor" because it greatly reduces the rate of developer attack in the unexposed film. Most unexposed positive-resist films are very developer-resistant; they show attack rates of only 10 to 20 Å/sec. If the resin is cast alone, without the sensitizer or "inhibitor," the developer dissolves the resin at a rate of 150 Å/sec. The resin used in these positive resists is a low-molecular-weight novolak resin, whose structure is given in Fig. 8.4. The solvent systems for these resists are mixtures of n-butyl acetate, xylene, and cellosolve acetate. The ratio of sensitizer (PAC) to resin is usually about 1:1. The

Napthoquinone diazides

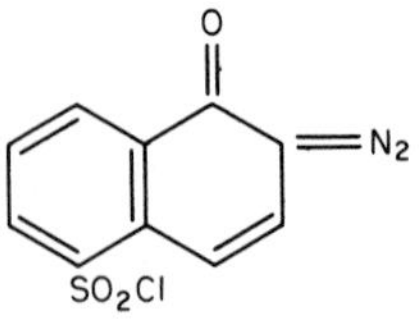

Napthoquinone-1,2-diazide-5-sulfochloride

Benzoquinone diazides

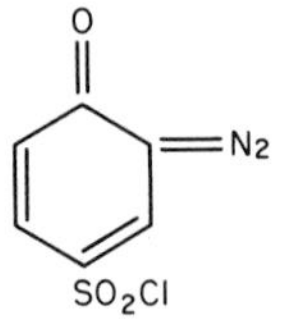

Benzoquinone 1,2-
diazide-4-sulfochloride

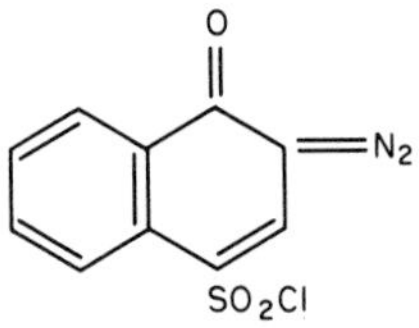

Napthoquinone-1,2-diazide-4-sulfochloride

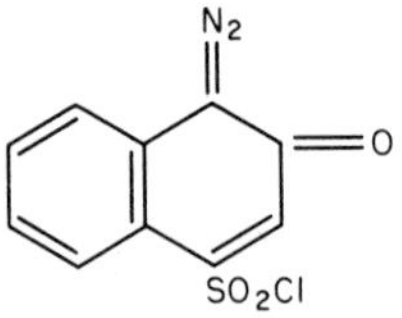

Napthoquinone-2,1-diazide-4-sulfochloride

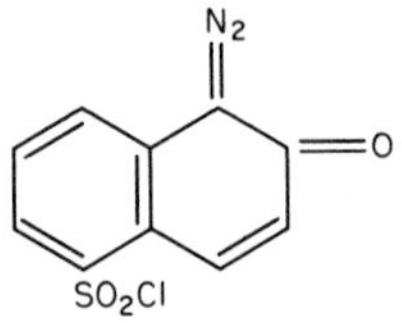

Napthoquinone-2,1-diazide-5-sulfochloride

Figure 8.3 Positive-resist light-sensitive elements.[1]

coated and dried film is exposed, and exposed areas are rendered highly
soluble so that the dissolution rate in the developer increases to 1000 to
2000 Å/sec, or about 100 times faster than the developer dissolution rate
of the unexposed resist areas. This differential solubility is the primary
means of image formation in positive resists. The actual photolytic reaction is shown in Fig. 8.5, and in the reaction, photochemical decomposition of the diazide causes rearrangement and hydrolysis into a carboxylic
acid. The exposed film is now base-soluble and readily dissolves in the
aqueous alkaline developing solution.

Several exposure mechanisms have been proposed for the diazo positive resists, and some degree of controversy exists as to what actually
happens, mainly because of the short lifetime and complexity of intermediate reactions. The intermediate reactions are rapid and difficult

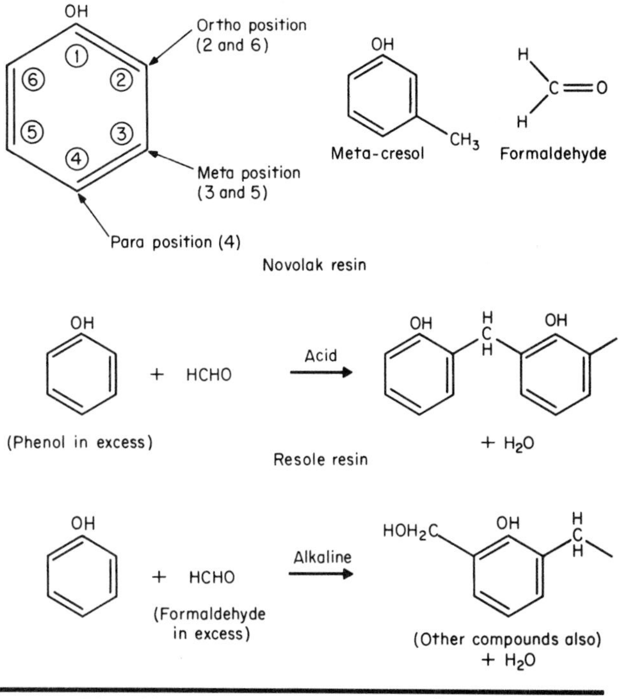

Figure 8.4   Phenolformaldehyde resin reactions.[1]

to isolate, and several possible paths to the same end product are possible, as shown in Fig. 8.5. The PAC shown in the figure is representative of any compound of the o-naphthaquinonediazide moiety in which R is compatible with the diazo group. Similar studies, in which 1- and 1-o-naphthaquinonediazide are connected to a mono- or trihydroxybenzophenone, have shown similar photochemistry. Findings indicate that many diazo sensitizers that are embedded in phenolic resins have similar photolytic reaction mechanisms. Figure 8.5 also shows how AZ-type positive resists can be used as negative resists, i.e., by exposure through a mask in a vacuum. In the vacuum, exposed areas are cross-linked (when two or more diazo groups exist in PAC), and subsequent blanket exposure in air with *no mask* causes dissolution of the previously *unexposed* areas in the developer and hence a negative-resist image. The recommended developer for this "reversal" process would be the standard aqueous alkaline developer.

Several other aspects of resist chemistry are worth mentioning because they are noticeable and may cause concern. For example, many positive diazo resists are very light amber in color when manufactured

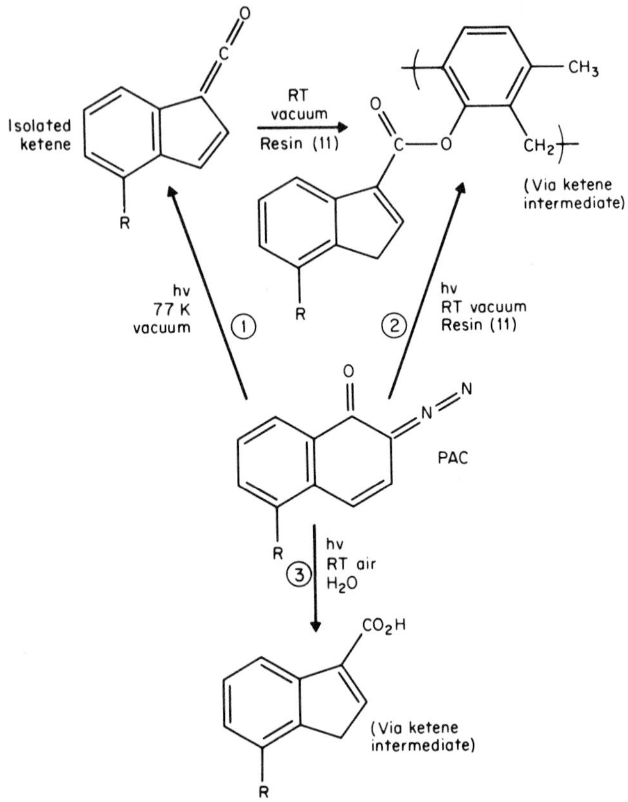

Figure 8.5 UV-induced decomposition pathways for *o*-naphthaquinonediazides.[2] (*Copyright 1979 by IBM.*)

but darken with age. This is a function of the amount of oxidation that takes place in manufacturing. Small production batches will have less chance to oxidize than a larger mix, and sometimes differences in color from batch to batch because of oxidation differences are detectable. This discoloration of the resin is largely a cosmetic effect and has not been tied to change in the functional properties of the resist. Resist suppliers have strict controls on incoming resins, and any that are too highly oxidized *before* resist manufacturing are discarded. The key parameters that affect resist performance from a quality control standpoint are covered in Chap. 13.

Positive resist, as it ages, produces a diazo dye (in the liquid state). This also darkens (dark decay) the material further and is a result of decomposition of the sensitizer. In fact, sensitizer decomposition of the PAC is identical for exposure to light, exposure to heat, and the dark reaction, or "dark decay" (chemical decomposition). In the process of chemical decomposition, the carboxylic acid groups, which give these positive resists their developer solubility, are not formed. Instead, the intermediate products join with nondecomposed PAC and form the dyes. In concentrations as low as 1 ppm, these dyes may produce noticeable color changes ranging from red to orange to yellow. However, the amount of decomposed sensitizer produced by these storage and aging reactions is small and will not noticeably affect the functional properties of the photoresist. If decomposition exceeds 5 to 10% of the total PAC content, functional differences will be noticeable. Notably, the loss of sensitizer (inhibitor) will cause an increase in the developer dissolution rate, and cause an apparent increase in the photosensitivity of the resist. However, under normal storage conditions, the amount of decomposition in 1 year at room temperature is typically less than 5%. See Chap. 13 for recommendations on storage conditions. Many optical positive resists are in the same chemical family as the type 1300 novolak resist and will behave similarly in imaging. Each, however, will require its own characterization and optimization.

x-Ray, e-beam, and ion beam resists

Poly (methyl methacrylate), a thermally degrading polymer, has been used as an electron beam resist for some time. Many other types of positive and negative electron-sensitive resists have since been investigated (see Chap. 3). PMMA is a positive working material with a sensitivity of 5×10^{-5} C/cm^2 and is an extremely high resolution resist. A typical e-beam resist is poly (glycidyl methacrylate-coethyl acrylate), shown in Fig. 8.6. Sensitivity to this resist is less than 10^{-5} C/cm^2, but etch resistance to both plasma and wet etchants is reportedly good.

The exposure distribution for a typical electron beam is produced when the beam enters the resist layer: The electrons lose energy as they are diffused through absorption by the film; collisions (elastic and inelastic) occur; and resultant scattering takes place. While this happens, the secondary and backscattered electrons that are produced represent separate energy sources, including heat x-rays. All these energy sources cause chemical reactions in the resist. The energy scattering is similar for photosensitive resists, since light is reflected back from oxide and metal films and produces secondary exposure. Infrared radiation also is present and plays a role in the total energy exchange that takes place during exposure. Figure 8.7 shows the dosage profile of an electron beam in a 2-μm resist film on a silicon substrate. As with positive and negative optical resists, exposure is measured either by degree of solubility achieved after exposure to the electron beam (positive resist) or by degree of cross-linking created for a given

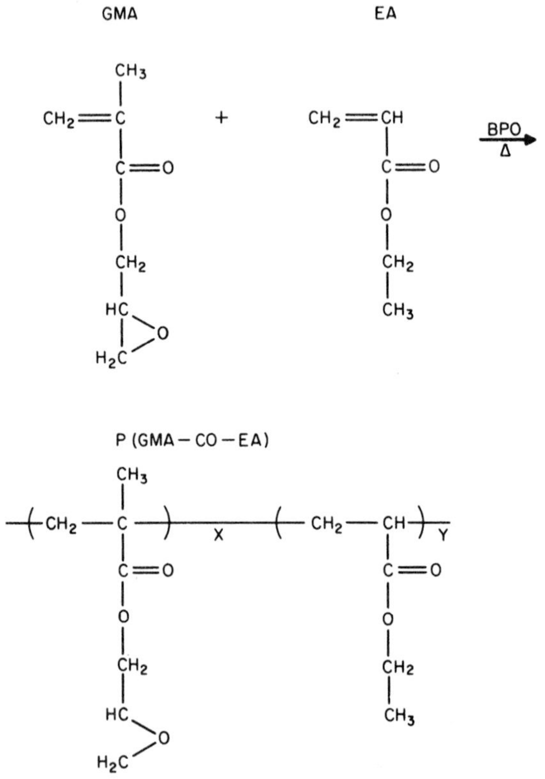

Figure 8.6 Poly (glycidyl methacrylate-coethyl acrylate) structure.[3]

dose of electron energy (negative resist). All radiation-sensitive recording materials have a rate of desired reaction (developing rate) for a given exposure to the energy source. This is loosely defined as the "contrast" of the material, which is discussed in detail in "Exposure Optimization."

Resist Sensitivity

Optical resists

Although there is a strong interdependency among the various photoresist imaging steps, exposure is the most important. The exposure step is the first point at which the CAD-generated IC patterns reach the wafer surface. The latent resist image is the geometry that is tightly held in the IC process, through developing, etching and ion implantation, until it is "finalized" as an electrically functioning struc-

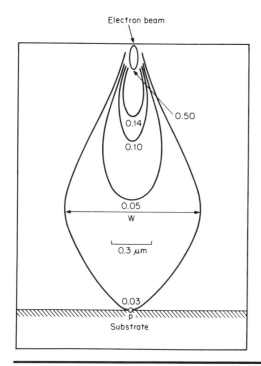

Figure 8.7 Electron dose profiles, Monte Carlo technique, 2.0-μm resist film, silicon substrate.[3]

ture in silicon. Thus, considerable emphasis is placed on generating the right image in the exposure step. Properly determining the sensitivity for a resist and running the exposure process efficiently requires a complete understanding of not only the resist exposure chemistry but also parameters that affect it and how to control them. In addition to sensitivity determination, process optimization (latitude studies) and process control are major components of resist exposure as an imaging step.

Spectral absorbance. The sensitivity of resists to the exposure radiation is expressed in many ways, the initial one usually being the "spectral response curve." This is a measure of the response of the resist to the light spectrum used for exposure, usually a mercury vapor source, as shown in Fig. 8.8. The spectral response curve gives a general indication of how effective any given light source may be in exposing a resist. If the resist absorbs strongly in an area in which the light source has strong emission lines, relatively long or short exposure times may result. Reabsorption (measured with an International Light 411T Photometer and a SC-100B Sensor or similar instrument)

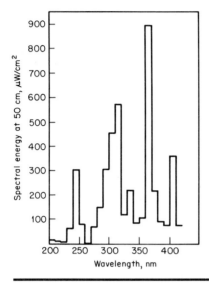

Figure 8.8 Mercury arc lamp spectrum. (*Courtesy of Canon.*)

that is not coincident with a strong emission line in the lamp may result in long exposure times. Spectral absorption profiles for a commonly used positive photoresist (Shipley S-1400) is given in Fig. 8.9. Spectral absorbance curves do not, however, give quantitative information on exposure times. They do indicate the wavelengths and the intensity of light transmitted through a resist layer. More critical is the functional chemical change in the resist as a result of the exposure. The inverse of transmission is absorbance; curves of absorbance versus wavelength are shown in Fig. 8.9.

Actinic absorbance.   Actinic absorbance is another way of evaluating resist sensitivity, and it has been used to test positive-resist systems. "Actinic absorbance" is defined as the difference between the absorbance of the unexposed and the exposed resist, shown in Fig. 8.10 for AZ-1300 series positive resists. These data were generated with a 200-W mercury vapor source. Actinic absorbance indicates the amount of absorbed energy by wavelength across a spectrum, but it does not guarantee that the absorption will translate directly into a linear developer solubility relationship. Indeed, the absorption at a given wavelength may not carry with it a functional group that participates in the developer solubility or decomposition reaction. Actinic absorbance is, however, more indicative of resist sensitivity than spectral absorbance.

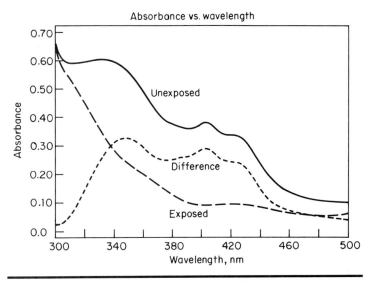

Figure 8.9 Spectral absorbance profiles for Shipley S-1400 positive photoresist.

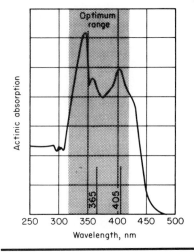

Figure 8.10 Actinic absorbance of AZ-1350J photoresist.[4]

Photochemical speed. "Photochemical speed," defined as the reciprocal of the exposure required for complete reaction of the sensitizer, has been used to compare photoresist exposure sensitivities. Photochemical speed is tested by exposing the resist and continuously sensing the decreasing absorbance at the exciting wavelength until all the sensi-

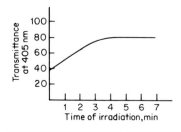

Figure 8.11 Increase in transmittance at 105 nm with time (positive photoresist).[5]

tizer is reacted and a constant absorbance value is achieved with continued exposure. The photochemical speed is then given as a value directly proportional to the time required for any resist to reach the constant value. A plot of transmittance versus exposure time is given in Fig. 8.11, and the 405-nm light was chosen for exposure because most positive photoresists have an absorbance peak near 405 nm. A complete description of this test is given by Loprest and Fitzgerald.[5] This is essentially a monochrometer study, and the response of a photoresist to a specific wavelength is measured. The apparatus for determining photochemical speed is shown in Fig. 8.12, and results of the tests are given in Table 8.1. The monochrometer was a Bausch and Lomb 33-8608 unit.

Another method for measuring photoresist sensitivity, the spectrophotometric, is described by Ilten and Patel.[8] The authors found high correlation between a theoretical model and experiments to derive quantum yield figures on the disappearance of the photoactive compound as a function of experimentally derived halftimes. The quantum yield figures are given in Table 8.2. This work employed the three primary wavelengths used in commercially available exposure aligners, which allowed direct use of the data in production processes.

Reciprocity. Negative resists typically suffer from reciprocity law failure, whereas positive resists do not. Exposure (absorbance) and re-

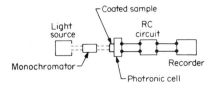

Figure 8.12 Apparatus for continuous measurement of absorption of exposed resist coatings.[5]

TABLE 8.1   Photochemical Speeds of Resists

Material	Exposure energy,* J/cm²	Speed, 1/exposure energy	Standard deviation, %
AZ-111	0.126	7.94	4
AZ-1350	0.113	8.85	6

*Energy required for complete decomposition of photosensitive material (time at which constant transmittance is reached multiplied by incident-light intensity).

sist thickness, within the range used in lithography have a linear relationship that supports the Lambert-Beer law. This is important in processes in which high-intensity pulses from zenon or mercury-zenon exposure sources are used and extremely high intensities are achieved in very short times. This type of exposure is used in photomasking applications, in which evolved nitrogen from the exposure process would force very flat chrome-on-chrome or chrome-on glass substrates

TABLE 8.2   Primary Quantum Yield of AZ-1350 as a Function of Intensity

Percentage T of neutral-density filter	$t_{1/2}$	$a \times 10$	$\phi (\lambda)$	Average ϕ
Bandpass filter 3650 ± 100 Å				
100	34	0.666	0.15	
41.7	73	0.278	0.16	0.15 ± 0.02
33.5	98	0.223	0.15	
12.1	310	0.081	0.13	
Bandpass filter 4047 ± 100 Å				
100	32	0.518	0.20	
42.6	53	0.221	0.28	0.23 ± 0.04
32.4	84	0.168	0.23	
11.8	277	0.061	0.19	
Bandpass filter 4358 ± 100 Å				
100	32	0.727	0.14	0.16 ± 0.02
42.2	63	0.307	0.17	
31.2	84	0.227	0.17	
11.5	272	0.083	0.15	
No bandpass filter				
		$I_0(\text{erg/cm}^2/\text{sec}) \times 10^{-3}$		
100	2.1	9		0.2
50	4.2	4.5		0.2
32	6.9	3		0.2
10	19.8	1		0.2

apart and destroy the contact. The short bursts of high-intensity energy permit enough time between bursts for the nitrogen to dissipate. Also, special channels may be etched to carry the gas to the edges of the plates.

High-intensity exposure is used in step-and-repeat wafer aligners. Again reciprocity between exposure time and intensity is needed, especially on aligners that integrate energy intensity and time. In negative resists, the failure of reciprocity occurs at the high-intensity, short-exposure-duration end of the spectrum, where insufficient time for photoconductance within the high-molecular-weight polymer will retard the time to a nonlinear point.

Diffraction. Diffraction is a problem to all lithographers. It occurs close to the edges of microstructures on masks and reticles, where it is manifested by optical fringes in the shadow area of the structure. Diffraction of light occurs in all modes of image transfer (projection, proximity, and contact printing), and the intensity distribution of the scattered light is calculable according to diffraction theory referenced several times in the literature. Figure 8.13 shows typical patterns of diffraction.

Calculating the actual edge is important in deriving intensity. The edge of the resist is defined by the edges of the diffraction pattern, where incident exposure energy matches the threshold energy level of

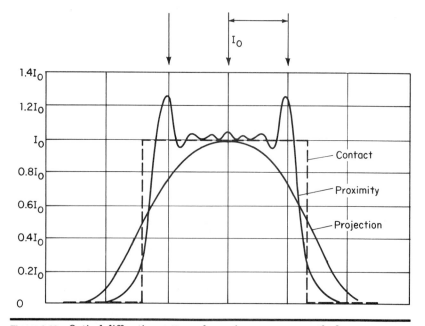

Figure 8.13 Optical diffraction patterns by various exposure methods.

the resist; that is where sufficient exposure exists to produce an image that results in a structure developing cleanly to the resist-substrate interface. Diffraction essentially steals useful exposure energy and scatters it away from the functional image area.

Diffraction patterns and their respective intensity are calculated according to the type of imaging. In projection printing, the diffraction is classified far-field or Fraunhofer ring diffraction. This is commonly seen as fringes on nonflat wafers where run-out or defocus exceeds 1.5 to 2.0 μm.

In proximity printing, Fresnel patterns or near-field diffraction occurs. The region of near-field diffraction is calculated with the formula W^2/λ: The reticle or mask feature width W is squared and divided by the exposing wavelength. The advantage of multiple-wavelength exposure is mixing wavelengths and thereby providing enough intensity gradients. In contact printing, the distance between wafer and mask is less than a wavelength, so the only effects produced must be in the region of geometric shadowing provided by the mask.

Modulation transfer function. In lithography, the resolution capability of resist systems is frequently limited by the aerial image contrast of the exposing radiation. Contrast in the image is intensity maxima divided by intensity minima, I_{max}/I_{min}. The degree of contrast, in turn, is determined by many factors in the optical system, including energy coherence, wavelength, and f number of the lens. The standard method of calculating the image transfer capability of the exposure tool is to measure the modulation transfer function according to

$$M_{mask} = \frac{I_{max} - I_{min}}{I_{max} + I_{min}}$$

This calculates the modulation of the mask as a function of V. The intensity levels are measured as the aerial image exits from the mask measured at the resist plane, and the MTF of the exposure aligner is

$$\mathrm{MTF}_V = \frac{M_{image(V)}}{M_{mask(V)}}$$

MTF is then plotted against spatial frequency, as shown in Fig. 8.14. This measures the true image transfer capability of the aligner. The figure shows various coherence values S, which vary from zero (totally coherent) to infinity (completely incoherent illumination). The normalized spatial frequency shown on the abscissa is expressed as $v\lambda f$, and $\frac{1}{2}f$ equals the lens NA.

Although a completely incoherent optical system produces images with twice the resolution size of a coherent system (see $S = \infty$ versus

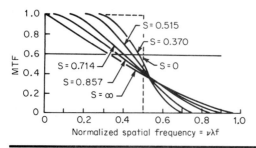

Figure 8.14 MTF of an ideal imaging system as a function of illumination coherence.

$S = 0$ curves), the incoherent system loses contrast rapidly with increasing resolution (or spatial frequency). It has been empirically determined that a 0.6 to 0.4 MTF value is needed to deliver sufficient image contrast in commonly used resists. New resists that perform well at low contrast and low MTF values are being developed. At 0.6 to 0.4 MTF, the optical system is best optimized for partially coherent light. This forms the best compromise between complete coherence (with unacceptable image diffraction rings) and complete incoherence with unacceptable contrast to produce the needed resolution in current resists.

Contrast in optical lithography. Image contrast, defined as the difference in density or the optical gradient in an aerial image, is a highly critical parameter in optical lithography. The contrast between dark and light areas from a mask determines the edge resolution and finite pattern resolution of microimages. Contrast has become an increasing concern as pattern dimensions reach below the 1-μm level. The response of a photo-, electron, or other resist will determine the reaction with the incident signal of contrasting energy shapes. The important part of contrast relates directly to how the line widths are affected by the contrast in the exposure image.

Figure 8.15 shows a model of a pattern element on a mask and the resulting types of signals of high and low contrast that can result. The corresponding change in line width for a given change in contrast is illustrated. Note that, for a given change in the exposure energy going from E_1 to E_2, the line width changes from L_1 to L_2, all with the high-contrast example. However, with a low-contrast optical system, the same change in the amount of exposure energy as in the high-contrast example will result in a much greater change in the resist pattern width, from L_1 to L_2. Thus, the contrast performance of an optical system is highly critical to good, economical production of advanced integrated circuits. The need for this contrast is predicated on the ex-

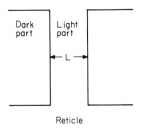

Reticle

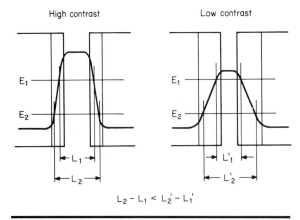

$$L_2 - L_1 < L_2' - L_1'$$

Figure 8.15   Contrast model.

pected real fluctuations in exposure energy. A high-contrast optical system will therefore deliver better exposure latitude than a low-contrast optical system.

Working at the submicron level brings with it many optical effects that were of little consequence at 2-μm-range geometries. The sensitivity of an advanced lithography process to reflectance, scattering, image contrast, and other optical parameters is many times greater at submicron levels than at approximately 2-μm levels or greater. More discipline and attention to process control is required for submicron processes. The overall subject of contrast is best defined by the modulation transfer function of the optical system.

Characteristic curves. The most useful method for comparing photoresist sensitivity is the "characteristic curve," which is derived by plotting the developed photoresist thickness as a percent of the spin-coated thickness and as a function of the exposure dosage. The characteristic curves for positive and photoresists are given in Fig. 8.16. The contrast, or gamma, is evident in the shape of the curves.

For a given exposure level, the change in intensity produces a

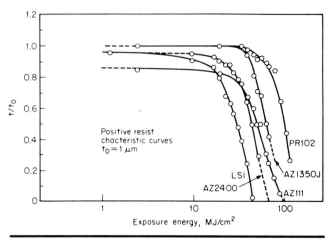

Figure 8.16 Characteristic curves: positive resists.[3]

greater change in resist geometry size in negative than in positive resists, even when corrected for differences in photosensitivity. This contrast behavior is analogous to "gamma," a term usually applied to photographic films. Gamma is derived as the tangent of the angle formed by the straight-line portion of the D log E curve when extended to the log E axis. It is a measure of the ability of a photosensitive emulsion or coating to form images (hard) over a wide range of film densities. For example, it would be very desirable in photomechanics to have a photosensitive coating that would expose only through the transparent portion of the negative; if the opaque area were not completely opaque, the ideal photosensitive material would not expose. This higher contrast in positive-resist systems is important in forming high-resolution structures, because the resist can more accurately reproduce the incoming "image" energy from the exposure aligner and mask and not respond to diffuse light that arises when the exposure is projected through air or reflections of low intensity from the wafer.

In projection printing, positive resists have this better mask reproduction capability because of their higher contrast compared with negative resists. The resulting developed image is more closely a 1:1 reproduction of the mask geometry. This better fidelity in positive resists is particularly important in such applications as aluminum masking, where the reproduction of closely spaced lines of equal width separated by spaces of equal width is important. Even though computerized mask compensation programs are used to precorrect for the expected response of the photoresist, the latitude that exists in this type of situation is considerably less with a negative photoresist.

Higher exposure contrast in positive resists also generally provides straighter resist sidewalls. The straighter sidewall means good image fidelity with respect to mask dimensions and more accurate geometry reproduction after etching. Resist users try to compensate for sloping resist sidewalls by reducing the resist thickness, something also done to improve image resolution. However, thinner coatings, particularly below 7500 Å, add to the defect and pinhole counts and ultimately reduce device yield.

The characteristic curve replicates the behavior of the resist in production, and it is a useful analytical tool in selecting and characterizing resist materials.

Image reversal.   In some applications it may be useful to reverse the image of a positive resist and make it act as a negative resist. For a discussion of this technique, see Chap. 3.

Nonoptical resists

The exposure parameters for e-beam resists are seriously impacted by the backscatter of electrons from the substrate (Fig. 8.17), leaving a sloped-wall resist image and detracting from the resolution. Nevertheless, characteristic curves for electron resists are used in the same way as those for optical resists to show energy versus normalized resist thickness. The contrast is the slope of the linear portion of the plot of thickness remaining (cross-linked resist) versus log electron dose carried out to 100% film remaining. The general exposure problem with electron beam resists is that the slow resists have good process

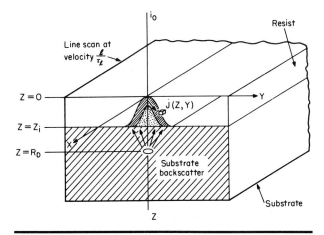

Figure 8.17   Three-dimensional electron scattering model of resist on a substrate.[3]

(etch) resistance, whereas the high-speed resists are sensitive to wet and dry etching.

Optical resists for electron beam exposure. Modifications of the well-characterized and -understood positive photoresists (diazoquinone sensitizer/novolak resin) are candidates for e-beam, as well as deep-uv, lithography.

Optical and e-beam exposures have been compared in positive resists based on sensitizers of the diazoquinone type. A bifunctional sensitizer was mixed with a cresol-formaldehyde novolak resin (Alnoval 429K, average molecular weight = 5000).[13] The solvent system was 2-methoxyethyl acetate, and the solution was spin-coated. An interesting aspect of this experiment was the comparison of fractional sensitizer destruction under the optical and e-beam exposure sources. The initial higher solubility rate of the resist under e-beam exposure and the change in direction of the plot after saturation led to negative-resist image behavior. This corroborates earlier studies on high electron dosages with positive working resists to cause negative-resist imaging capability. The solubility of the unexposed resist, in this case, is greater than that of the highly e-beam-exposed area, and a "strong" normality positive-resist developer is used to remove the now more soluble unexposed resist.

Exposure Optimization

The most difficult singular task in IC fabrication is the optimization of photoresist exposure, particularly because it involves (1) achieving proper resist thickness per specific surface, (2) selecting softbake that, in itself, has restrictions bearing on other parameters, (3) calculating the line width requirements and tolerances on those geometries with respect to the mask and exposure system, (4) defining the developer strength, agitation, and temperature parameters with respect to all the preceding, (5) balancing all the preceding with respect to the hardbake conditions, and (6) fixing the etching conditions. The etching conditions act as a "chicken" step because some of the etch parameters, such as etchant type and technique, must be established before the preceding conditions are set. In short, the exposure step acts as the first link in a chain reaction that ends with the etched structure. Optimization of photoresist exposure requires a disciplined ordering of process priorities with a thorough understanding of the functional interrelationship. See Chap. 7 for more information on exposure versus other imaging steps.

Exposure versus resist thickness

The first step in the optimization process is ensuring a known and re-producible photoresist thickness. This step assumes that the user has already run the types of tests described under "Exposure Reactions" and "Resist Sensitivity" or at least can predict, on the basis of avail-able data, the expected response characteristics of the resist in ques-tion. Photoresist thickness must be consistent, because the softbake, exposure, developing, hardbake, and etching steps are calculated or modeled on the basis of a specific thickness. Methods for ensuring re-sist thickness uniformity from wafer to wafer, cassette to cassette, and day to day are outlined in Chap. 6.

The most serious problem associated with a change in or loss of photoresist thickness control is a change in line width, and a typical relationship expressing this potential problem is plotted in Fig. 8.18. Although the line width change was only about ¼ μm overall for a 20% variation in photoresist thickness, this change, as we shall see, is on top of many other small variations in the process that will ulti-mately produce an unacceptable deviation in process specifications. This can happen much more easily if all the possible variables fall in the same direction. For example, while an increase in thickness of 20% causes the one-quarter variation in line size (wider line), a pro-portionately stronger increase in light intensity (happening on the same lot of wafers) would have a cancellation effect on the added thickness. However, if thickness is reduced and intensity of exposure

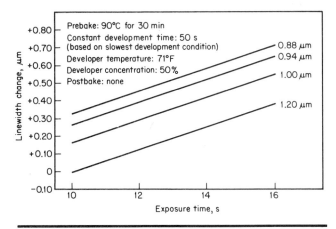

Figure 8.18 Resist thickness and exposure time relationship ver-sus line width change.[7]

is increased simultaneously, the net effect is perhaps a ½-μm increase in line width instead of zero change. A convenient way to precalculate the effects of these variations is to use the root-sum-square formula, as follows:

$$\text{Total deviation} = \pm \sqrt{x^2 + y^2 + z^2}$$

$$= \pm \sqrt{\left(\begin{array}{c}\text{aligner}\\ \text{variation}\\ \pm 10\%\end{array}\right)^2 + \left(\begin{array}{c}\text{resist}\\ \text{variation}\\ \pm 7\%\end{array}\right)^2 + \left(\begin{array}{c}\text{developer}\\ \text{variation}\\ \pm 15\%\end{array}\right)^2}$$

$$= \pm \sqrt{10^2 + 7^2 + 15^2}$$

$$= 19.34\%$$

Small variations can accumulate into significant changes, and included here are only three of the six previously mentioned process variables that impact on IC imaging. This formula is called the "theorem for the standard deviation of the sum of the individual variables." Thus each process variable shown in the equation is treated separately and may either aggravate or cancel the effects of the other variables. With movement from thickness variation versus exposure to the other five key imaging variables tied to exposure, a typical total variation is accumulated.

Exposure versus softbake

Exposure variation as a function of softbake is a critical variable requiring optimization in a photoresist process. Both positive and negative photoresists are insensitive to exposure in the liquid phase (as a solution or as a freshly cast film). The degree of solvent removal and the temperature and time of the softbake directly impact on the exposure parameters. Negative optical resists that are not completely softbaked will not completely cross-link when exposed, and the developer will attack and cause scumming in those areas. The unexposed negative-resist areas will be washed away by the developer, as they normally would be. Positive optical resists left only partially softbaked may not be completely reacted (sensitized) when exposed, but they will appear to be *more* photosensitive because the developer will aggressively dissolve away any positive-resist film that retains a relatively high solvent content. The unexposed positive-resist areas also will be attacked by the developer at a much higher rate than if the coating were properly baked. In short, the degree of softbake can have drastic effects on exposure parameters. Since the softbake temperature has a more direct effect than bake time on resist sensitivity, anyone installing a new resist process should plot softbake tempera-

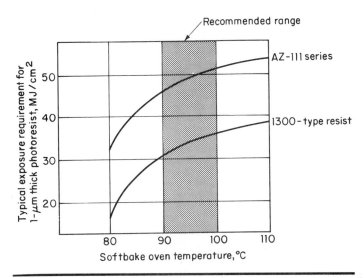

Figure 8.19   Softbake temperature versus typical exposure requirement.

ture versus exposure time for a given typical softbake cycle and a given typical development cycle (usually 60 sec of immersion or 15 to 30 sec in spray). This plot is given in Fig. 8.19 for two positive optical resists (AZ-111 and AZ-1370).* The shaded area of the graph represents a recommended operating range, since softbake temperatures above 100°C will begin to reduce the apparent sensitivity of most optical resists.

Resist processes may be optimized for throughput, resolution, or any combination of those parameters. As an example, throughput optimization calls for undersoftbaking, underexposure, and standard development and resolution optimization calls for oversoftbaking, standard exposure, and highly diluted developer processing. Most processes are established to derive maximum throughput for the minimum required resolution and critical dimension control.

Standing waves

The role of standing waves in resist imaging is important mainly because IC geometries are small enough that standing waves may act as a barrier to extending optical lithography.

Standing waves have been known about for some time, and Born

---

*Process conditions: softbake time, 20 min; exposure source, high-pressure mercury lamp; development time (immersion), 60 sec at 21°C: developer makeup, 1 part AZ-351 to 3 parts H_2O and 2 parts AZ-311 to 3 parts H_2O.

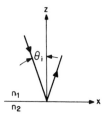

Figure 8.20 Coordinate system of exposing radiation incident on a wafer.[8]

and Wolf, in *The Principles of Optics* (2d ed., Pergamon Press, Oxford, 1965, p. 277) review the mechanisms of light wave interference and propagation. The coordinate system they use is shown in Fig. 8.20. The wafer is the xy plane, or the reflecting surface, and the xz plane is the plane of incident light. The mathematics to explain the occurrence and behavior is contained in a paper by Ilten and Patel.[8]

Simply put, "standing waves" are periodic variations in the incident (z axis) exposing energy intensity. These intensity variations are caused by interference between the incident and reflected light during wafer exposure. Figure 8.21 illustrates this in a coating of AZ-1370. The spacing of standing waves is calculated by deriving the optical

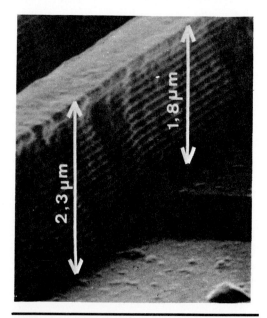

Figure 8.21 Micrograph of standing-wave patterns in photoresist. (*Courtesy of Paul Tigreat.*)

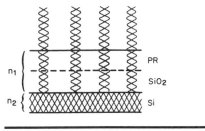

Figure   8.22 Standing-wave   principle   in photoresist.[8]

thickness that determines the frequency of the reflected waves. The point at which the incident waves cross the reflected waves causes an exposure minimum called an "antinode." The "nodes," or exposure maxima, are the outer limits of these two crossing wavelengths of energy (Fig. 8.22). The data in the figure were generated with monochromatic light (4047 Å) on steam-grown oxide, and some experiments connected with this evaluation were run at the 4358-Å wavelength; both wavelengths were used in conjunction with projection printing. Standing waves occur most dramatically in resist images that are projection-printed with monochromatic light. When exposure aligners that employ two or more wavelengths are used, the probability of multiple cancellation is increased (depending on optical thickness). However, standing waves have been observed from resist images that have been contact-, proximity-, and projection-printed with polychromatic light, with the full mercury spectrum.

Elimination of standing waves.   Since the occurrence of standing waves seems almost inevitable regardless of the type of resist, exposure equipment, or lamp used, process-induced methods are necessary to minimize the effect. Since the formation of standing waves depends on the proper frequency mixing (constructive and destructive interference) of incident and reflected light, and since the optical thickness determines the reflecting portion of this exposure reaction, we look first to methods of varying the optical thickness and thereby reducing the reflectivity.

The total optical thickness is the sum of the oxide and photoresist layer thicknesses, and reflectivity is maximized if total thickness is exactly an even-quarter multiple of the incident wavelength. Experimental proof of this theory is provided by data from Ilten and Patel.[8] Several wafers were tested with varying optical thicknesses. Where there is a one-quarter multiple of the 4047-Å exposing wavelength, reflection from the wafer is greater, exposure time for photoresist is lower, and haloing or standing waves occur. In those cases where standing waves were eliminated, exposure times were greater because of the reduction in reflected light.

Another technique to reduce standing waves is the use of a postexposure bake of 10 min at 100°C. Theoretically, this may cause a redistribution of the sensitizer from high- to low-concentration regions. Since the intensity distribution in a standing wave is considerable, with up to 9 times variation between maxima and minima in a distance of only 650 Å, the postexposure bake does have a significant effect on image line width after developing. Further, the short bake time and relatively low temperature does not allow reflow of the resist to occur. Some resists, having lower reflow temperatures, should be postexposure-baked at 90 to 95°C. The postexposure bake, by evening out the initial latent image intensity maxima and minima, will help to provide better line width control over a variety of optical thickness ranges.

Dyes can be applied to wafers as antireflection coatings to absorb the wavelength(s) used in exposing the resist. Nigrosin and methyl orange are suggested for the 4047-Å wavelength, and methanil yellow is suggested for all combinations of optical thickness as an antireflection absorbing layer.

The reasoning behind putting a dye in positive photoresist is to eliminate long-range, geometry-dependent reflections. When topographic features in a device are covered with a reflective film such as polysilicon or aluminum, the light from an exposure tool is often reflected off the film into portions of the resist which should not be exposed.

This turns out to be especially troublesome in the case of g-line steppers because all positive resists tend to be much more transparent at 436 nm. That means that reflected light is able to travel much further through the resist before it is attenuated. The long-range reflections are not to be confused (as they often are) with the classic case of line width control over a step.

Short exposure times and concentrated developers also are used to eliminate standing waves. In this approach, the stronger developer simply "chews" through the intensity gradient, which has already been reduced by the short exposure. This technique is not suitable for device designs in which geometries of less than 1.0 μm are employed, since the stronger developer makes line control more difficult and the postexposure bake could be used without sacrificing line control.

Exposure Aligners

Optical effects in resist exposure

The many different surfaces presented by the wafer in exposure provide a variety of light-scattering phenomenon. Figure 8.23 shows the various surfaces that interact with the incident exposure energy as

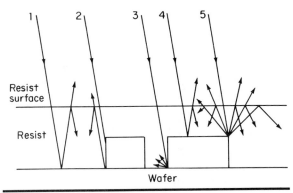

Figure 8.23 Surface reflections and light-scattering patterns encountered in microlithography.

well as the reflections, interferences, absorbances, and related optical phenomenon. The mask or reticle is the first of these surfaces encountered in projection lithography. Resist chemistry alone accounts for much of the absorption and changing of light quality (wavelength) as a function of refractive index. Thus, all of the optical parameters of the resist layer(s) and other chemical reactions (photolytic, etc.) occurring in real-time modes (while exposure is occurring) need to be part of the equation. Intensity changes and wavelength changes are occurring constantly in this model.

The variations in wavelength and intensity are especially critical in microstructure science because the pattern element size frequently represents a sizable percent of the excursion in intensity. For example, when exposure is at 436 nm, variations in wavelength behavior occur at ¼-wave multiples of the 436-nm exposure line, or in 109-nm increments. This dimension represents the allowable tolerance on geometries of 2 μm and will have increasing significance on smaller pattern elements.

Several commercial products that have addressed the problems of light or energy scattering in the optical lithography process can be summarized as follows:

1. Antireflective coatings spun onto resist and wafer surface

2. Oxides (black chrome) on masks and reticles

3. Dyes or absorbers added to resist

4. Specific optical thicknesses (resist-oxide thickness combinations) that avoid quarter-wavelength multiples of the exposing energy

5. Wavelength mixing to avoid nonuniform intensity distribution in resist layers

6. Chemical and physical collimators placed on surfaces or incorporated in the chemistry or structure of surfaces or films

7. Restructuring of VLSI device design and/or process to alter surface topography; for example, redesign to reduce step heights as in isoplaner technology.

8. Multiple layering of films (resist, oxide, other) to serve as diffusers, absorbers, or collimators of incident exposure energy

The use of the various antiscattering techniques becomes more important as the number of layers used in VLSI fabrication increases. The average number of masking levels has continued to increase with device complexity. Some bipolar devices have as many as 15 individual layers, whereas the average number of layers remains around 7, including MOS and bipolar technologies.

Surface reflections from the many different layers vary with the types and structures of the layers. The amounts of reflection from both aluminum and silicon vary as functions of resist thickness. The thicker coatings absorb considerably more of the reflected exposure energy, which shows a possible way to reduce the impact of reflections. Thick coatings for submicron lithography are used only when multilayer resist systems are used, because thin resist layers are more easily processed to provide resolution and pattern geometry control.

By interfering with the light, reflections from resist and substrate interfaces result in light coupling at regular intervals in the resist, which causes line width variations expressed as functions of all the geometrical parameters described previously.

Exposure aligner types

Three basic types of mask aligners are currently in use for fabrication of ICs: projection scanners, projection step-and-repeat machines (steppers), and contact/proximity aligners. Although there are many differences of opinion concerning the relative merits and deficiencies of each, all are currently being used for volume production, and all are likely to be used for some time to come.

With the scanning technique, the wafer is exposed in a succession of scans, but the effect is the same as if it received whole-wafer exposure; i.e., there is no opportunity to realign to local targets in midwafer.

The stepping mask aligners operate in one of two modes. The "first-generation" steppers use fixed steps, with no references to local targets in midwafer. Steppers of later design have the capability to re-reference periodically, typically at each field exposure step. The most sophisticated steppers can make adjustments to x, y, ϕ (rotation), z (focus), and two-dimensional xy plane tilt to attain the ultimate in overlay accuracy and resolution. Scanning aligners can, at most, readjust focus on the fly, although certain compensations can be made to minimize the necessity for midwafer adjustments. The benefits of stepping include $1\times$ to $10\times$ reduction of mask features, reducing defect and dust sensitivity in printing, resolution 25 to 40% better than scanning exposure at $1\times$, easier exposure of 8- to 10-in wafers, and high-resolution reticles to transfer the pattern.

The third type of mask aligner, the contact/proximity aligner, is considered the simplest and least expensive of the three. Here the resist on the surface of the wafer is either brought into direct contact with the mask and exposed (contact) or positioned within a few micrometers of the mask and exposed using collimated light (proximity). The mask in this case must contain the complete pattern (for all dies) at the actual size of the final features $(1\times)$.

Problems can occur because contact printers tend to suffer from short mask life that is due to wear in the contact process. In contact/proximity systems, image distortion that is produced when light is diffracted at the edges of features in the pattern may be another problem.

Hard contact printing. Hard contact printing was the original method of printing IC patterns on wafers. Emulsion masks, followed by iron oxide and then chromium masks, were brought into hard vacuum contact with resist-coated wafers. As circuit dimensions were reduced and the need for more efficient production (higher yields) increased, other methods of contact were found to reduce rejects associated with contact between the photoresist and the photomask image. Defects typically found included "gel slugs" from silver halide emulsion masks and particles of photoresist sticking to the mask after contact and redepositing on the next wafer exposed. Positive photoresist is especially brittle and shatters easily in contact printing. In addition to the debris from resist coatings and mask emulsions, particles from the environment become lodged between the photomask and the coated wafer, which caused poor contact in isolated areas and induced further damage to both the resist coating and the mask. Hard-surface masks are less forgiving in terms of tolerating particulate contamination, because emulsions can absorb both particles and even cushion the epitaxial spikes and other mechanical abrasions better. Hard-contact (high-vacuum-pressure) printing is now used mainly for research and development or for very low resolution applications.

Soft contact printing. Soft contact printing emerged because separation of mask and wafer offered the simplest way to solve the bulk of the problems. As contact between the photoresist and the mask is reduced, light begins to "stray" and resolution is more difficult to achieve. This means that a whole spectrum between hard contact (high vacuum) and projection (above 2 to 3 mils) printing could be evaluated and assessed for the tradeoff between resolution loss (via less contact and more mask separation) and contact-related defect density. The first step was soft contact printing, whereby the vacuum is reduced to prevent the extreme pressure between mask and resist. A dramatic reduction of defects was proved with soft contact, and resolution at acceptable levels close to those achieved with hard contact was still maintained.

Proximity printing.   Proximity printing is exposing resist by projection with only a narrow gap of one to several mils between mask and wafer. Positive photoresists work better than negative resists because of low oxygen sensitivity and lower MTF performance. Proximity printing, because of the relatively narrow gap between mask and wafer, posed the problem of mask contact with the wafers either by accident in loading or unloading or through calibration checks with still calibration plates. Thus projection printing, or exposure in which the mask and wafer are separated by more than several mils, became the next new exposure technology.

Projection printing.   Projection printing eliminated the mask-wafer contact problems but created new ones of its own: depth-of-focus and image contrast. The first 1:1 magnification projection aligners were of the refractive type by which a single wavelength of light would be passed through a lens system. Later, 1:1 systems utilized the full mercury spectrum along with mirrors to direct the light onto the wafer. In refractive projection aligners, the entire wafer is exposed at once with a cone of light from the aligner; the reflective projection system (Perkin-Elmer type) provides an aperture of intense and highly collimated light through which the wafer is exposed. A carriage scans the wafer under the slit and indexes it, as in ploughing a field (boustrophedon) until the entire wafer is covered. Figure 8.24 illustrates an early refractive lens (1:1) used in IC imaging.

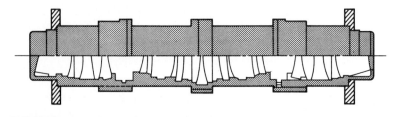

Figure 8.24   Projection printing (refractive) optical system. (*Courtesy of Perkin-Elmer Corp.*)

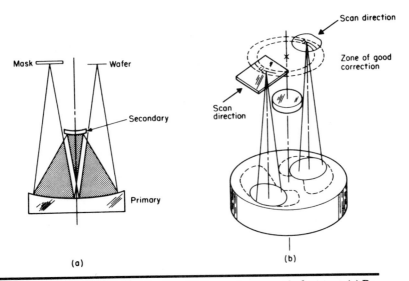

Figure 8.25 The Micralign all-reflecting, folding projection optical system. (*a*) Basic optical system consisting of two concentric spherical reflecting surfaces. (*b*) Zone of good correction. (*Courtesy of Perkin-Elmer Corp.*)

The cost of producing a multielement lens such as the one shown in Fig. 8.24 would be high, and the lens weight would be several hundred pounds. Providing temperature stability alone for such a lens would require a very expensive control chamber. Reflective-scanning projection printers have become widely used because of their optical simplicity, cost-effectiveness, resolution capability, and dimensional stability. The optical system shown in Fig. 8.25 is composed of two reflecting surfaces. The light passes through a point on the mask to the primary mirror, then to the secondary mirror, and then back to the primary, where it is finally imaged onto the wafer. The aperture slit width through which the light passes before it reaches the mask determines how much area is imaged in a single scan. The series of scans occurs with the wafer and mask held in a folded projection system (Fig. 8.26) so that they scan synchronously. The folded projection system is held by a scanning carriage (Fig. 8.27) that moves the entire structure back and forth in front of the stationary light source (Fig. 8.28). As shown in Fig. 8.26, the entire path of light and image is "bent" 5 times with mirrors until it reaches the photoresist coating. The lamp shown in Fig. 8.28 is a 1000-W capillary mercury arc having relatively high pressure (230 atm) and containing two tungsten electrodes spaced about 1 in apart. Air jets cool the bulb during exposure, and the heated air is removed by an exhaust fan. A fan is an important feature in all exposure aligners because heat can cause not only shifts in the spectral output of the lamp but also dimensional changes

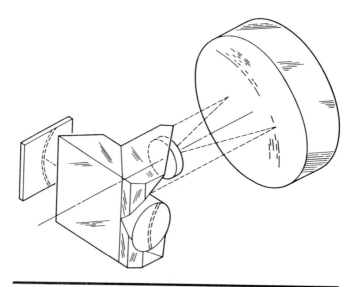

Figure 8.26 The Micralign all-reflecting folding projection system: three optical elements. (*Courtesy of Perkin-Elmer Corp.*)

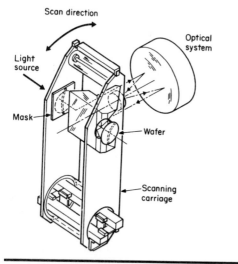

Figure 8.27 Projection and scanning system. (*Courtesy of Perkin-Elmer Corp.*)

in the mask, all of which distort the wafer and the dimensions of the images formed (latent and developed) in the photoresist.

The curved aperture slit referred to earlier allows for intensity and resolution adjustment because a wider slit would increase wafer throughput and reduce exposure time and resolution compared with a

Figure 8.28 Mercury capillary arc lamp for Micralign apparatus. (*Courtesy of Perkin-Elmer Corp.*)

narrower slit, which would provide the reverse, i.e., longer exposures and higher resolution. Slit width may vary from 1 to 4 mm.

A light-integrating system that is used automatically reduces the scan speed in proportion to a decrease in lamp output and increases scan speed when the lamp increases in intensity. Air-bearing wafer alignment and vacuum clamping of mask and wafer provide frictionless operation and viewing of the wafer with magnifications up to 250. Wafers can be loaded manually or with the standard cassette system, and two chucks are provided so that one can be loaded or unloaded while the other is holding a wafer in the exposing mode. The amount of magnification and distortion is theoretically near zero, because the light, imaged onto the slit and then onto the mask, simply reflects what the mask has on it and only mirror-transmits the image, thereby removing the most common source of distortions in optical systems—the lenses. There are mechanical adjustments to the folding mirror array that permit tilting of the mirrors.

Focus wedges and similar preexposure resolution tests should be run at the beginning of a shift to correct any changes and optimize for 1:1 exposure settings. The primary advantage of the scanning 1:1 optical reflection-projection printer is relatively high throughput at acceptable resolution levels. In order to benefit from the resolution advantages provided by this aligner, mask and wafer flatness combined should be a fraction of the total depth of focus of the aligner.

The Micralign 500 Series of projection mask aligners is a 1:1 magnification scanning system with submicron resolution and high throughput. The optical system shown in Fig. 8.29 provides exposing radiation from 2400 Å through the visible part of the spectrum. This wide range can be broken into select, discrete exposing wavelengths by use of a special family of filters. Special fabrication techniques are used to reduce light scattering in the uv portion of the spectrum, and the entire unit is sealed to eliminate particulate contamination. Figure 8.29 shows a diagram of the optical system of the Model 500.

Other capabilities of the system include automatic focusing and au-

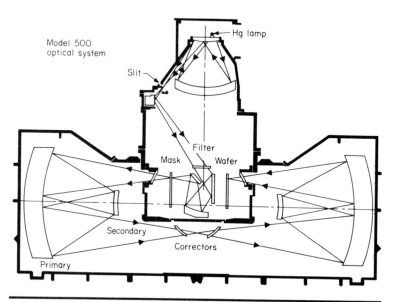

Model 500
optical system

Hg lamp

Slit

Filter

Mask   Wafer

Secondary

Correctors

Primary

Figure 8.29   Optical system of the Micralign 500. (*Courtesy of Perkin-Elmer Corp.*)

tomatic wafer flattening, which measures the contour of the wafers and optimizes it before patterning. An example of the resolution capable with the Micralign 500 Series is shown in Fig. 8.30. This particular shot is the deep-uv imaged RD2000N resist developed by Hitachi, showing pattern resolution of approximately 1 μm over 0.4-μm etched

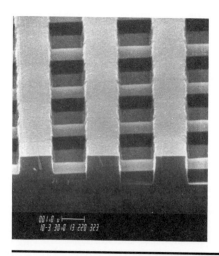

Figure 8.30   Resist image patterned with the Micralign 500. (*Courtesy of Perkin-Elmer Corp.*)

steps. The space between resist images is submicron, about 0.9 μm. One of the functional aspects of the 500 Series printer is that there are three wavelength regions: uv-4 range ($\cong$400 nm), uv-3 range ($\cong$300 nm) and uv-2 range ($\cong$260 nm).

Scanning reflective projection has a natural advantage over the refractive lens systems used for step-and-repeat projection in the area of optical distortion, which is theoretically zero on the reflective systems. Overlay accuracy is greatly influenced by optical system distortion, and the higher throughput 7× and 5× magnification steppers have more distortion than slower 10× systems. The 500 Series is manufactured to a surface accuracy of $\lambda/200$ (λ = 6238 Å). The total distortion in the 500 Series Micralign, shown along with all other specifications in Table 4.1, is ± 0.25 μm, 3-σ.

Machine-to-machine overlay. The overlay accuracy of a single machine is very critical to device performance, especially when all mask levels are exposed on a single machine. In production, however, multiple machines are used, with the result that mask levels are divided among several machines. Matching machine to machine is therefore as important as matching overlay in a single aligner. The alignment and distortion errors among machines are calculated by combination, using a root-sum-square calculation. In an experiment to measure overlay error between two Perkin-Elmer Micralign Model 500s, wafers were imaged for one mask level on the first machine, then taken to the second machine for patterning a second mask level. This experiment utilized a test pattern with optical vernier alignment targets that could be electrically probed for a resistance reading, a technique that has a reported precision of 0.01 μm. The accuracy and repeatability of this overlay measurement test is better than simple optical measurement techniques. Distribution of 80 of these probable patterns over the wafer surface is made to determine errors in x and y directions. The results of the measurements are shown in Fig. 8.31, which represents the summary of over 4000 individual measurements. The machine-to-machine overlay results are impressive: 99.7% of the errors are within 0.35 μm, and 95% are within 0.25 μm.

Step and repeat. The basic optical scheme for the GCA stepper system is outlined in Fig. 8.32, and the fiber bundle is shown in Fig. 8.33. These figures show a dramatic departure from the conventional illumination systems in the manner in which light is carried to the wafer. Step-and-repeat wafer imaging would not be practical with standard collection optics (Fig. 8.34) because of extensive exposure times. Note the small percentage of used output from the three-dimensional view in Fig. 8.35. An illuminator breakthrough[12] was accomplished by sur-

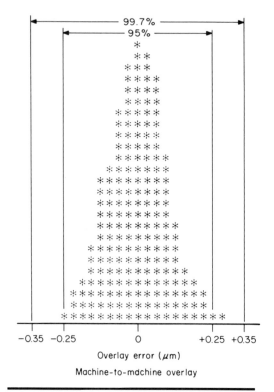

Figure 8.31 Image overlay results from Micralign 500.

rounding the mercury arc lamp with four sets of collection optics. In this way, the energy is transferred through the transfer optics and into the fiber bundle. Exiting from the fiber bundle through a secondary source, the light is sent through condenser lenses, where uniform illumination is achieved. The light source is a 350-W (high-pressure) mercury arc lamp with a 4 mW/cm^2 intensity of g-line radiation in the object area. The "inbound" ends of the fiber bundles are shaped elliptically to match the shape and size of the mercury arc image. This provides efficient use of space and gives the fiber bundle a diameter that is compatible with the size of the entrance pupil of the microobjective and its distance from the object plane on the output end.

This illuminator, collecting from four sides, is about 7 times more intense than conventional sources. This increase in intensity is provided by using a dichroic mirror in place of a heat filter and thereby increasing transmitted g-line energy from 80 to 95%. In addition, the fiber bundle is kept very short (previously longer) without sacrificing the rotary integration function that only fibers provide. The mathe-

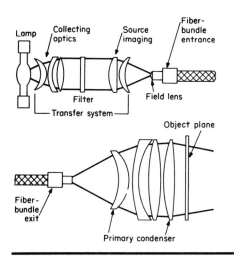

Figure 8.32 Schematic of general form of a Mann stepper illuminator. (*Courtesy of GCA Corp.*)

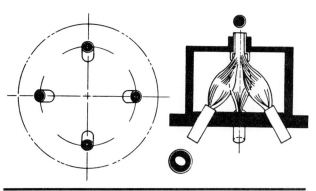

Figure 8.33 Four-branched fiber bundle for GCA stepper.[12] (*Courtesy of GCA Corp.*)

matical model describing the efficiency of this system correlates well with actual test runs on AZ-1350B at ½-μm thickness. The model predicted a 230-msec exposure requirement, and actual tests produced a result of 200 msec. AZ developer was used for 60 sec at 50% concentration.

The high resolution of a positive resist can be exploited with the capabilities inherent in a direct-step-on-wafer aligner at $1\times$, $5\times$, or $10\times$ magnifications of the reticle.

Wafer exposure by step-and-repeat imaging techniques is performed by a variety of optical systems. Since the introduction of the GCA sys-

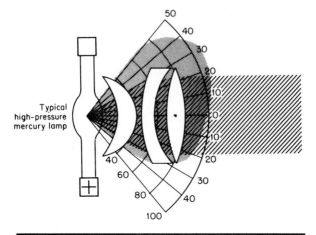

Figure 8.34 Collection optics of a typical illuminator showing relative intensity distribution with respect to the output of a mercury arc lamp.[12] (*Courtesy of GCA Corp.*)

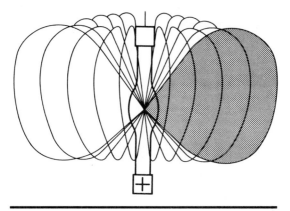

Figure 8.35 Three-dimensional view of the output of a typical mercury arc lamp.[12] (*Courtesy of GCA Corp.*)

tem previously discussed, close to a dozen different optical step-and-repeat exposure aligners have been manufactured. One example is the Censor/Perkin-Elmer wafer stepper.

The illuminator schematic of the Censor/Perkin-Elmer stepper is shown in Fig. 8.36. A mercury arc lamp is the source of *g*-line energy, which is collected in a dual-condenser system and passed through four cold mirrors, a filter, a lens, a shutter, and a light integrator. The light then goes through a uniformer to multiple condensing lens elements which form the rectangular field. The light intensity is such that exposure times for 1-μm-thick positive optical resist films are in the range of 50 to 200 ns. The light uniformity of the Censor system,

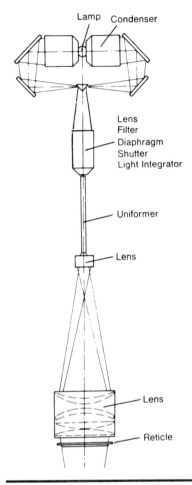

Lamp   Condenser

Lens
Filter
Diaphragm
Shutter
Light Integrator

Uniformer

Lens

Lens

Reticle

Figure 8.36 SRA-100 illuminator and optical system.

an important parameter for any submicron imaging system, is an impressive ±2%.

The high uniformity and good resolution are supplemented by a local or repetitive alignment strategy. This process is automatic to keep the stepping frequency high, and it is performed with respect to six parameters (x, y, ϕ, z_1, z_2, and z_3) and leveling. In Fig. 8.37, two alignment strategies are compared. Note that, in the upper half of the wafer, the die sites are exposed in equidistant steps. That causes a high percent of defective dies, since only one alignment is made, referenced to marks on the wafer, and exposure fields are determined with the aid of a laser interferometer. The problem with this approach is that

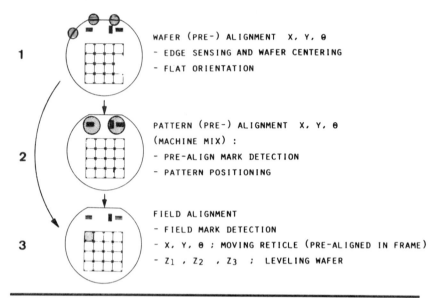

Figure 8.37 Alignment strategies of the SRA projection aligner. (*Courtesy of Perkin-Elmer Corp.*)

distortions from earlier processing, represented by the curved black lines, result in misalignment on all total-field exposure systems. Distortions arise from a variety of sources, including wafer dimension shifts from high-temperature excursions, magnification errors in the total-field exposure system (homogeneous and inhomogeneous), stepper-to-stepper leveling variations, and temperature changes in the stepper itself.

On the lower half of the wafer shown in Fig. 8.37, these problems are largely solved by the local die-by-die alignment, even though some of the same distortion errors, such as machine temperature changes, remain. However, when each individual die is sighted and printed, the distortion can occur constantly between exposures as long as the next mask patterns are imaged nicely over the previously printed, etched, and doped image. In fact, the entire pattern and the wafer and stepper can and do move, in differing degrees, throughout the IC fabrication process. Machine-to-machine errors also occur, but the moment the exposure occurs, freezing in place the site of the etch and doping step, previous and subsequent shifts, drifts, and dimensional changes have little consequence. Only distortions that occur within the exposure fields are not accommodated by the individual die alignments.

The reticle alignment marks are in the four corners outside the pattern area, small 400- by 600-μm rectangular areas. The wafer has its own set of marks, which are derived from the first layer reticle and are 10× conjugates that get imaged on the first wafer exposure. On the reticle they are the 400- by 600-μm marks reduced to 4- by 60-μm marks on the wafer. Subsequent reticles will have only the 400- by 600-μm rectangles or window marks, whose long sides are directed to the center of the image or optical axis. Since the lens is designed for perfect diffraction-limited imaging at 435 ± 3 nm and alignment is performed at 547 nm, some distortion will occur. Although the sagittal aberrations are minimized, the radial ones are more distorted, which is the reason for the alignment marks being placed in radial directions. When the alignment light is projected down onto the wafer, going through the alignment window marks, the wafer and reticle image is composed in a plane behind a semitransparent mirror. The entire image is scanned with the use of a rotating mirror across a slit in front of a photodiode. The result is a profile of the intensity of the image, and the position of perfect focus occurs when the intensity slope profile is at a maximum. The in-focus and out-of-focus images are shown in Fig. 8.38, along with an outline of the ideal and real image profile. Note the differentials between foot (bottom) and shoulder (sidewall) and top image as an image moves from ideal to an out-of-focus position. Also, the actual

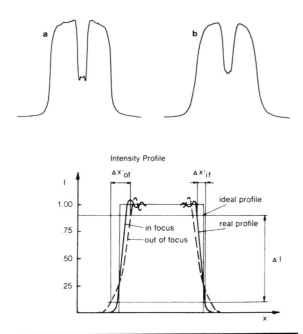

Figure 8.38 Optical images (in focus and out of focus) and image profiles (ideal and real) of the SRA projection stepping aligner. (*Courtesy of Perkin-Elmer Corp.*)

profiles from the chart recorder of the autofocus-autoleveling photodiode mechanism are shown next to the charted profiles. Note that a 10-μm out-of-focus image has greater sloping sidewalls, lower shoulders, and a more shallow central trench.

In some cases, alignment marks are etched off the wafer or lost by a similar process (deposition thermal growth of oxide). When this occurs, the software in the stepper will skip to the marks that are left or will refer to the correct coordinates in the foregoing field, using lost ϕ, x, and y if all three are lost. The other process changes, namely temperature, will not affect the individual die because total run-out is divided evenly by number of image fields across the wafer diameter. These "first order" temperature effects, with the Censor system, are barely measurable after individual dies are imaged by the step-after-step align and exposure. The production environment can therefore be less critically controlled.

Focusing and leveling are calculated by processing signals already put into the system. The actual focal position is calculated by the stepper computer by taking the coordinates of the previously shown intensity profile and figuring the slope of the curve. The precise focal point, at which a signal is sent to expose the resist, occurs when the second derivative of the slope is equal to zero. Leveling is provided by running four of the focus points just described. All of these focusing and leveling operations are performed directly through the lens, which tends to be more accurate than indirect distance methods that use reference points to simulate the wafer-reticle positions. The problem with these calculated reference points is that actual wafer and reticle surfaces move as a function of room temperature. The movement of parts in an optical system becomes more critical as the numerical aperture increases and depth of focus decreases.

Wafer throughput is the most severely criticized aspect of stepper exposure, especially systems that prealign each individual die to ensure good overlay performance and good yield. Real throughput is a function of many parameters and should really be measured by the total number of good dies. In all aligners, there are several components of throughput in the system, including machine loading and unloading time and internal wafer servomechanisms that transport the wafer between alignment and exposure intervals. The total machine time is added to align time and exposure time per wafer as the first major component of throughput. This must be calculated along with the total dies processed per unit time and then reduced by the number of dies lost that can be attributed to the exposures aligning process. Shortening the machine time and taking all measures to improve aligner device yield are the two areas to focus on, and they are summarized as follows:

1. *Higher resolution.* Use the maximum resolution pattern the system can sustain. Higher resolution makes possible the use of die reduction and more die per wafer, hence higher die yield. Higher-resolution resists, such as resists with very high contrast, can be used to extend even the maximum resolution of the stepper system signal by image "clipping," whereby a sloped signal is straightened by a contrasting resist.

2. *Reduction of machine time.* Wafer movement, alignment, and exposure times can be shortened by one of several strategies, such as more intense exposure lamps or collecting optics, faster resists, reducing exposure times and increasing development times, or more rapid servomechanisms to move wafers through the system. High-acceleration servomotors, for example, are employed in the Censor system to facilitate shorter machine times. Another feature of the system is the high scanning frequency used to locate alignment marks.

3. *Reduction of the total number of exposures.* The die format established the number of total exposures, and flexibility within the stepper system can optimize the fields used to image the wafer.

The Censor stepper system provides a variety of exposure fields within the 145-mm object circle that fits within the 6- by 6-in reticle area. The largest possible square is 102.5 mm on a side; the smallest rectangle dimensions are 64 by 130 mm. The smallest field that can be printed separately is 65 mm on a side, 10× reduced on the wafer.

To translate these exposure image fields into actual throughput, refer to the die format curve in Fig. 8.39, where throughput, in wafers per hour, is plotted against the square root of die area, in mils. One example of using this curve would be a 260-mil^2 die. With one die per exposure and a 1:1 aspect ratio, throughput would be 24 wafers/hr. In contrast, a 360-mil^2 die would yield 48 wafers (of 4-in diameter) per hour. All of the running parameters shown are number of dies per exposure. The fixed process overhead values for these data include an exposure time of 0.8 sec per field with 70 fields, a machine overhead time of 4 sec per wafer, and a 1470 resist layer 1 μm thick. The 0.8-sec time includes translation time, align and expose time, and simultaneous exposure of as many die fields as possible.

The die aspect ratio, which is really controlled by the device designer, can be modified according to the chart to provide an optimum fit with a given optical system's image format. This die fitting will add to the number of wafers per hour. Finally, added yield can be derived by pellicle protection of the reticle, a standard practice for any optical reduction or 1:1 projection printing system.

The Censor/Perkin-Elmer system represents one of several production wafer-imaging systems that anticipate the needs of VLSI device

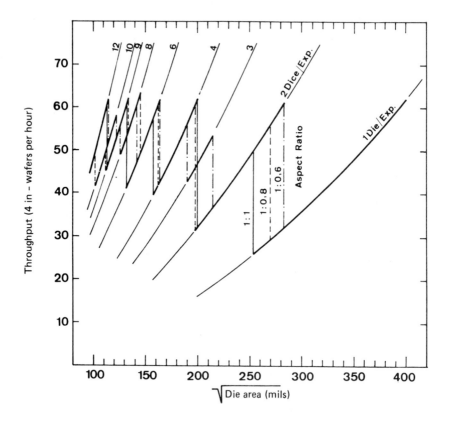

Typical throughput diagram for different die sizes and aspect ratios. Running parameter is number of dice per exposure. Assumed process values: 0.8 sec per exposure field; 70 fields, 4 sec overhead time per wafer

Figure 8.39 Die format curve used in optical stepping throughput calculations.

fabrication by addressing the region of patterning below 1 μm. The working resolution, or resolution achievable in a production mode, is about 0.7 to 0.8 μm, meaning that the maximum resolution is approaching 0.5 μm with optimized resist processing and good overall process control. The primary attributes that make submicron production resolution possible for this optical system are

1. Elimination of vibration frequencies that degrade imaging

2. Through-the-lens focusing and feedback sensing

3. Autoleveling at each die site to offset wafer distortion, increase overlay accuracy, and in general compensate for dimensional changes of all kinds (machine, reticle, wafer, pattern)

4. A 0.35 numerical aperture lens with sizable depth of focus (± 6 μm)

5. Registration accuracy of ± 0.1 μm for each exposure, a built-in way of accommodating the mix-and-match approach in which different aligners are mixed on the same set of mask levels for a device and must be registered or matched to each other as each successive layer is imaged and the pattern is transferred over preceding ones

Lenses for submicron lithography. A variety of 5x and 10y lenses for submicron imaging are used in optical wafer steppers; they are listed in Table 8.3. In small-field-size (14-mm) applications, it is possible to image 0.56-μm resolution at the 365-nm (i-line) wavelength. At the diffraction limit, using a 248-nm excimer laser, resolution of 0.33 μm is specified. Up to 70 wafers per hour (6 in diameter) can be exposed by using the wide-field g- and i-line lenses. The trend toward shorter wavelength is clearly heading toward 193-nm excimer laser imaging, at which point theoretical resolution reaches 0.33 μm. All these lenses are refractive and are limited to a single wavelength, a factor that restricts exposure throughput.

TABLE 8.3   Lenses for Submicron Microlithography

	Zeiss 10-78-34	Zeiss 10-78-46	Zeiss 10-78-52	Nikon 1010	Canon UL 1011	Zeiss 10-78-48	Laser-based system
Magnification	10×	5×	5×	10×	5×	10×	5×
Wavelength λ, nm	i-Line 365 nm	g-Line 436 nm	365 nm	365 nm	436 nm	365 nm	KrF Excimer 248 nm
Numerical aperture (NA)	0.315	0.38	0.32	0.35	0.43	0.42	0.2–0.38
Diameter of exposure field	11.6 mm	20 mm	23 mm	14.16 mm	21.2 mm	14.1 mm	14.5 mm
Diffraction limit of resolution $L = \lambda/2NA$	0.58 μm	0.57 μm	0.57 μm	0.52 μm	0.51 μm	0.43 μm	0.62–0.33 μm
"Pilot-line" resolution $L = 0.65\ \lambda/NA$	0.75 μm	0.75 μm	0.74 μm	0.68 μm	0.66 μm	0.56 μm	0.81–0.42 μm
"Production" resolution $L = 0.8\ \lambda/NA$	0.93 μm	0.92 μm	0.91 μm	0.83 μm	0.81 μm	0.70 μm	0.99 μm–0.52 μm
Rayleigh depth of focus $F = \lambda/2(NA)^2$	$\pm$ 1.84 μm	$\pm$ 1.51 μm	$\pm$ 1.78 μm	$\pm$ 1.49 μm	$\pm$ 1.18 μm	$\pm$ 1.04 μm	3.1 μm–0.86 μm

SOURCE: H. L. Stover, *SPIE*, vol. 633, 1986.

Alignment. Wafer step-and-repeat system performance depends on several subsystems in addition to the lens. The alignment system accuracy is perhaps second in importance to the lens, since it determines lithographic overlay performance. Through-the-lens, on-axis alignment at the exposure wavelength is the most inherently accurate type of alignment because it removes wavelength-dependent focus offsets and preexposure motion in the system, another potential source of error. Most alignment systems lose laser-based alignment *not* at the exposing wavelength.

Alignment strategy can be die-by-die (most throughput-burdened and most accurate), global marks only (least throughput-burdened, least accurate) or *n*th die alignment, whereby a selected number of strategically placed sites are aligned. This represents the best way to balance throughput and accuracy to meet the needs of a process. Figure 8.40 illustrates die-by-die versus *n*th die alignment sites and shows a typical alignment system.

Overlay. The overlay budget is composed of several elements that are compared to the lighographic approach in Table 8.4.

The reduction of registration error is the most likely way to reduce the overall overlay budget. The difference between 5× optical and x-ray is so small that optical technology will emphasize marginal im-

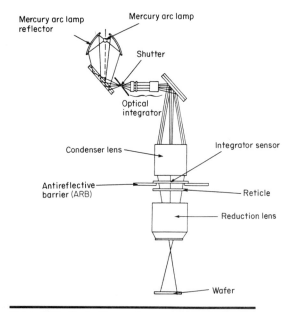

Figure 8.40 Projection system components of a Nikon wafer stepper. (*Courtesy of Nikon Precision Inc.*)

TABLE 8.4 Overlay by Lithographic Approach

	1× optical	5× optical	1× x-ray	E-beam direct write
Lens distortion	0.06	0.1	0	0
Registration error	0.1	0.1	0.1	0.05
Magnification error	0.15	0.01	0.1	0.01
In-plane distortion	0.05	0.05	0.05	0.05
Mask-to-mask errors	0.085	0.085	0.085	0.085
Root sum square	0.21	0.15	0.17	0.11

Source: B. J. Lin, IBM Corporation.

provements to keep it competitive with x-ray in terms of resolution. Registration can be reduced to 0.05 to 0.07 µm, which will permit optical lithography to reach a resolution level of 0.18 µm.

Optical step-and-repeat imaging at 53 magnification. The use of reticles at 5× final image size provides a good compromise between maximum resolution and maximum field size for device production. Optical stepping at 10× permits resolution improvement of 20 to 30% beyond 5× imaging, but the loss of effective field size is too great for economic IC production as measured in wafer exposure throughput. Optical stepping at 1× greatly increases wafer throughput but lowers the available resolution and greatly increases the difficulty of reticle manufacturing with submicron geometries compared to 5× imaging. The projection system components of a commonly used 5× stepper are shown in Fig. 8.41.

This system is typical of wafer steppers operating at g- and i-line wavelengths in that it extends the useful life of mercury arc lamp light source technology. The uv energy is collected to maximum efficiency and "homogenized" through the optics, where a condenser lens system collimates the light. Before reaching the reticle, the energy level is sensed to permit integration for reproducible exposure doses in resist films. This is a fundamental need to compensate for lamp output decay over time.

The reduction lenses used in 5× optical steppers offer a wide range of field sizes, numerical apertures, and corresponding resolution at the principal mercury lines of 436, 405, and 365 nm. Numerical apertures from 0.35 to 0.6 NA are used at g- and i-line wavelengths, with typical field sizes of 15 to 20 mm. The 0.6-NA lenses use smaller fields (5 by 5 mm) and 10× magnification to achieve production resolution approaching 0.5 µm.

The imaging results obtained with a 0.45-NA g-line lens in 1.8-µm-thick Shipley 1813 photoresist are shown in Fig. 8.42. The pattern res-

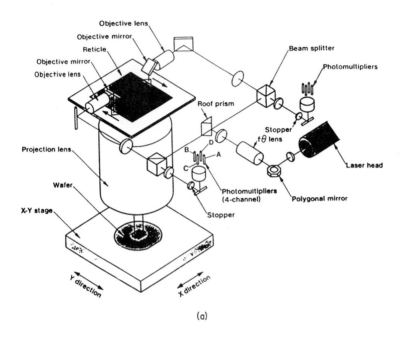

(a)

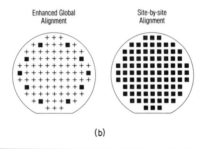

(b)

Figure 8.41 Stepper alignment. (*a*) Optical alignment and imaging paths of a wafer stepper. (*Courtesy of Canon.*) (*b*) Alignment strategies: enhanced global (*n*th die) versus site-by-site alignment. (*Courtesy of Nikon Precision Inc.*)

olution is 0.7-μm lines and spaces patterned equally in *x* and *y* directions. The close-up SEM photo shows the high ratio image obtained with this lens-resist combination. The imaging results with any optical system and depends to a large degree on the optical properties and photochemistry of the resist-developer system. Matching resist sys-

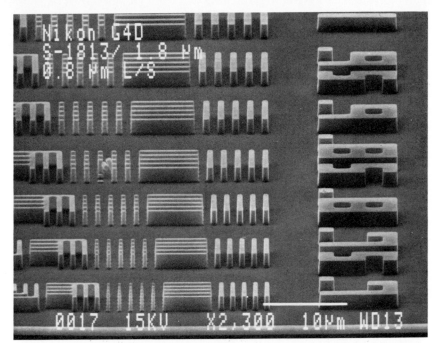

(a)

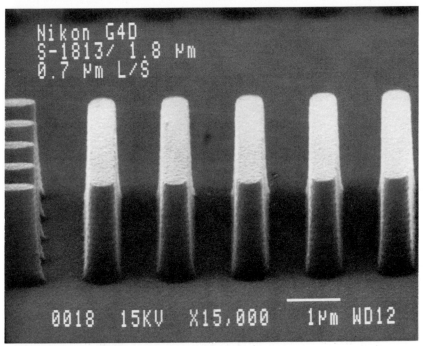

(b)

Figure 8.42 *g*-Line stepper resolution in a single-layer photoresist. (*a*) 1.8 μm of Shipley 1813 positive photoresist imaged with a 0.45-NA, *g*-line lens on a Nikon stepper. (*Courtesy of Nikon Precision Inc.*) (*b*) Close-up of SEM photo in (*a*). (*Courtesy of Nikon Precision Inc.*)

tems to optical imaging tools is a clear and necessary strategy in extending optical lithography to its limits.

Moving to a higher-NA g-line lens further extends the available resolution, as shown in Fig. 8.43. The SEM photo is Aspect System's 812 photoresist at slightly over 1 μm thickness, with 0.5-μm line-space resolution. The lens was a Nikon 0.6-NA g-line lens.

The 365-nm wavelength offers still further resolution improvement. The SEM photograph in Fig. 8.44a shows 0.5-μm line-space resolution with Shipley 6038B resist; Fig. 8.44b shows similar resolution with an Aspect System resist. The lens used in Fig. 8.44a was a Nikon i-line at 0.45 NA; that used in Fig. 8.44b was a Zeiss i-line.

Optical step-and-repeat imaging at 13. Optical projection printing with a 1:1 broadband imaging system has provided a means for very high wafer exposure throughput along with resolution levels needed in VLSI devices. The 1× stepper provides very large field sizes (292 to 450 mm²) with good CD control and submicron resolution. The use of 1× reticles permits rapid reticle turnaround time and multiple (7 or 8) customization layers on one plate.

The optical system that makes this 1:1 lithography practical is a modified Wynne-Dyson. The system, shown in Fig. 8.45a, is a folded

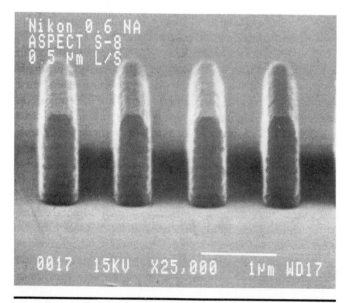

Figure 8.43 g-Line stepper resolution in a single-layer resist: 0.5-μm lines and spaces in Aspect System 812 resist imaged with a 0.6-NA, g-line lens on a Nikon wafer stepper. (*Courtesy of Nikon Precision Inc.*)

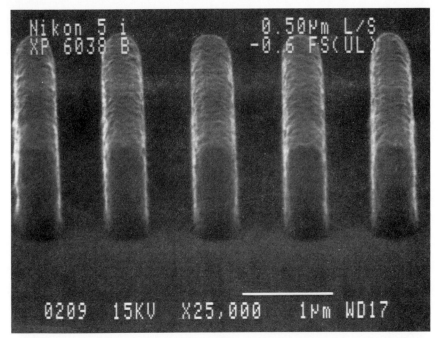

(a)

(b)

Figure 8.44 (*a*) 0.5-μm lines and spaces in Shipley 6038B positive photoresist imaged with a 0.45-NA, *i*-line Nikon lens. (*Courtesy of Nikon Precision, Inc.*) (*b*) Aspect System 9 positive photoresist patterned with a GEA *i*-line (Tropel 2235-1) lens. 0.35 wafer stepper.

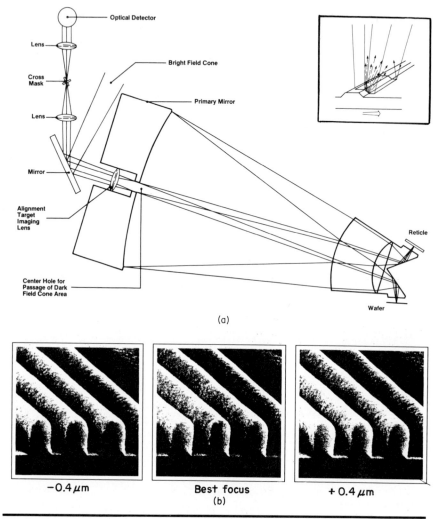

Figure 8.45 1× stepper optical system and imaging results: (a) Wynne-Dyson folded catadioptric system for 1× step-and-repeat imaging; (b) 0.5-μm lines and spaces through focus with single-layer resist. (*Courtesy of Ultratech Stepper.*)

catadioptric lens with symmetry permitting correction of odd-order aberrations such as distortion, lateral color, and coma. The simplicity of the design permits large usable field sizes with excellent correction. This lens system has one-fourth as many optical elements as an equivalent refractive lens for the same parameters. The color correction occurs over a bandwidth of 390 to 450 nm, permitting *g*-line and *h*-line

TABLE 8.5   Total Overlay for 0.75- to 0.8-mm Production and Its Constituents

Machine-to-machine alignment errors, μm	
Alignment system	0.15
Baseline calibration	0
Stage matching	0
Wafer run-out	0
Die rotation	0
Image distortion registration errors, μm	
Relative lens distortion	0.15
Trapezoidal error	0
Magnification error	0
Barometric effect on distortion	0
Reticle overlay error, μm	0.10
Process overlay error, μm	0.10
Root sum square total overlay, μm	0.25 (99%)

mercury arc exposures with minimal standing waves. The depth of focus is 3 μm for 1-μm resolution, 2 μm for 0.7-μm resolution, and 1.5 μm for 0.6-μm resolution.

The unique feature of this lens compared to refractive lenses is that it favors high numerical aperture over large field sizes, whereas in most lenses field size is traded for numerical aperture more severely. Another advantage comes from being a mirror-based lens system, giving low overlay budget figures, as shown in Table 8.5.

Description of the Wynne-Dyson optical system. The lens consists of five optical elements: the primary mirror, two folding prisms, and a two-element achromat. The lens system shown in Fig. 8.45a is folded and symmetric. The two folding prisms split the circular field into nearly rectangular reticle and wafer fields, with 4 cm^2 of usable area. The lens design has zero chromatic aberration at g-line and h-line wavelengths, which supply 33% of the energy used in resist exposure. The lamp efficiency permits 50- to 100-msec exposure time for 1-μm-thick positive resist. Partial coherence of 0.45 is obtained at optical power levels exceeding the levels of other lens systems. The telecentricity of the lens at wafer and reticle planes eliminates magnification errors due to reticle or wafer focal plane positions. A 10-μm focus error causes only 0.02 μm of magnification error. Trapezoidal errors are eliminated by the same lens property. Finally, design distortion of the Wynne-Dyson is essentially zero; there are no pincushion or barrel distortions. Manufacturing variation is the primary source of lens distortion. The resist images shown in Fig. 8.45b provide evidence of the focus tolerance and resolution of the lens system.

Benefits of wafer stepping

1. Stepping each die permits correction for wafer distortion.

2. Vulnerability to dust and related defects is considerably reduced.

3. Resolution capability is extended 30 to 40% beyond 1:1 scanning-exposure aligners by using conventional mercury wavelengths.

4. Stepping permits easier exposure of large-diameter (8- to 10-in) wafers.

5. A chrome- or glass-stepped reticle may be used in place of a conventional mask, thereby increasing the "starting" resolution.

Obtaining submicron line widths in 1-μm-thick steps requires good x-, y-, and z-dimensional imaging, as shown in Fig. 8.50. The coverage of 1-μm steps, particularly polysilicon, aluminum, or similar reflective surfaces, can be difficult in the z direction because sidewall reflections off the polysilicon or aluminum can cause increased resist exposure and narrowing of the lines traversing the steps. Differences in collimation angles of the incident energy, which produce different reflection angles into the photoresist, affect the dimensional control problems associated with reflective surfaces. Dyed resist is used to reduce the problem. Wafer steppers offer both off-axis and through-the-lens alignment. Throughput varies with off-axis alignment. The extension of optical lithography is in the direction of short-wavelength imaging, the topic of the next section.

Deep-UV Photolithography: Resolution-Limiting Parameters

Resolution

The strategy for extending practical resolution in VLSI and ULSI devices obtainable with optical lithography is to reduce the wavelength of the exposing radiation. This is the relationship between wavelength and resolution:

$$R = \frac{K\lambda}{\text{NA}}$$

where R is the minimum practical geometry and K (0.8) is a contrast value tied to the resist material. Assuming an exposing wavelength of 193 nm and an NA of 0.4, resolution in single-layer resists would be calculated as follows:

$$R = \frac{0.8(193 \text{ nm})}{0.4}$$

$$= 0.386 \text{ } \mu\text{m}$$

This resolution can be extended by further reducing wavelength (152-nm fluorine laser), increasing the numerical aperture of the lens, or reducing the value of K. Reducing K to 0.6 by using existing resist chemistry yields

$$R = \frac{0.6(193 \text{ nm})}{0.4}$$

$$= 0.290 \text{ } \mu\text{m}$$

and increasing the lens NA to 0.5 yields

$$R = \frac{0.6(193 \text{ nm})}{0.5}$$

$$= 0.232 \text{ } \mu\text{m}$$

Shorter wavelengths, and therefore better resolution, will be possible only with the development of better transmitting materials for use in lens (all refractive or catadioptric) production.

UV lens materials

Fused silica, calcium fluoride, and barium fluoride are the only candidate materials available, and Table 15.1 shows their optical properties. Fused silica is the most thermally stable, and it is hard enough to polish with relative ease. Calcium fluoride is second to fused silica as a material for achromatizing in a lens system. All deep-uv-transmitting materials also must have resistance to color center formation. That leads to crystal damage in the optic, followed by reduced uv transmission. Special coatings are used to reduce this problem.

Numerical aperture

Increased NA is a lens design function. The Wynne-Dyson lens design (shown in Fig. 8.45a) offers promise as a production lens for wavelengths down to 152 nm, which approach the transmission limit of fused silica.

The lens design used in the experiments described here (Fig. 15.7) is a 52× modified Schwartzchild reflective objective. A numerical aperture of 0.65 is available with this lens. The limited field size and off-axis imaging preclude the use of this design for a production lens but make it ideal for deep-uv resist material and ULSI process research.

Depth of focus

The following Rayleigh criterion is typically used to calculate depth of focus (DoF) for structures at or near the limit of the lens resolution:

$$DoF = \frac{\pm\lambda}{2(NA)^2}$$

Substituting NA out of the equation shows that a reduction in resolution of ½ effectively reduces focal depth to ¼, a square-law relationship. For practical lithography at 0.5 μm, this leaves only about 0.5-μm focal depth for g-line to i-line imaging, but over focal depth, μm, at 193 nm. The total focus budget must be divided, however, among several parameters. The most significant one is wafer nonflatness, which is typically + 0.5 μm for very flat wafers. Tip/tilt errors and wafer topography contribute another 0.5 μm, thereby using the entire budget.

The value of K

The last parameter to further reduce resolution is reducing the K value in the formula. Moving to a "top-layer" imaging resist strategy, where only the upper 1500 to 2000 Å of the resist is used for image transfer, a K value of 0.5 or lower is easily achieved. One such process seeds silicon atoms into the resist layer after excimer laser exposure, followed by RIE pattern transfer. Differential uptake of silicon by the resist makes this possible, and several candidate chemistries based on conventional optical resists (positive novolak type) are under investigation.

Mid- and deep-uv technology

The strategy of using a shorter wavelength to obtain increased pattern resolution is well known. Optical wafer aligners have typically operated on the g, h, and i energy lines of the mercury spectrum, which correspond to 436-, 405-, and 365-nm wavelengths, respectively. In contact and proximity aligners, all of these mercury "spikes" are mixed and the resist absorbs various degrees of the wavelengths. The use of optical step-and-repeat cameras for wafer exposure resulted in a reduction of wavelengths used to the use of 436 and/or 405 nm. These wavelengths are optimized for the mercury energy source (whose primary energy lines are at these points), for the optical elements in the refractive-type stepping system, and finally, for the resist type being used. The longer wavelengths have been ideal, since most optical resists absorb strongly in the 436- to 365-nm range. Also, optical systems for step-and-repeat exposure have avoided shorter wavelengths mainly because of light scattering and absorption within the optical elements. The pressure to improve image resolution has resulted in the development of 365-nm lenses for step-and-repeat exposure.

The logical pathway for reduction of patterning wavelengths in resist exposure is to move gradually below the major mercury lines (436 and 405 nm) typically used. The first area below the standard uv wavelengths to take advantage of better resolution is referred to as mid-uv, the 240- to 380-nm region. The benefit of using shorter exposing wavelengths to obtain increased resolution is based on simple mathematics. In an optical system, resolution is a function of wavelength according to the well-known relationship:

$$V_0 = \frac{2NA}{\lambda}$$

where V_0 = cutoff frequency (normalized to unity), pairs/mm
NA = numerical aperture
λ = wavelength, mm

Numerical aperture is defined as

$$2NA = \frac{1}{f}$$

where f is the f number of the optical system (focal length/effective diameter). Substituting the second equation into the first equation gives

$$V_0 = \frac{1}{\lambda f}$$

Thus it is seen that, in a given optical system, resolution can be increased by decreasing the wavelength.

**Resists for mid- and deep-uv
microlithography**

Assuming that suitable masks (quartz) and imaging equipment for production use are available, selecting a good resist becomes a major factor. In fact, until the resist and imaging source are optically matched for acceptable wafer throughput, a production technology does not exist. The primary resolution range at which mid- and deep-uv imaging are aimed is approximately 0.7 to 0.5 μm. E-beam and x-ray imaging can provide resolution at and below that level, but capital equipment cost and limited throughput are often limiting factors. Mid- and deep-uv imaging offer good throughput with only moderate capital equipment expenditure, a stepping-stone technology toward beam exposure techniques.

Resist selection is based on several criteria, and one of the first areas to study is a comparison of the output spectrum of the energy source with the absorption profile of the resist. Energy sources include filtered mercury vapor, mercury xenon, xenon, deuterium, and excimer lasers. The exposure of the resist should result in uniform energy movement, and thus the limit of absorbance by the resist should be about 0.3 (or 50% transmission). Some of the resists that will be discussed here have higher than 50% initial absorption but undergo bleaching during exposure, a phenomenon that mitigates the high-absorbance requirement and permits good uniform energy distribution in the resist film at the end of the exposure cycle. In fact, if resists do not have fairly strong absorption, they are likely to be too "slow," or lacking in photosensitivity, to allow for good wafer exposure throughput. Figure 8.46 shows three deep-uv resists and their transmission spectra before and after exposure with a 248-nm excimer laser source.

The number of photons per joule at a deep-uv wavelength of 250 nm are half the number at a longer wavelength of 500 nm. Assuming one chemical reaction for each individual photon across the board for re-

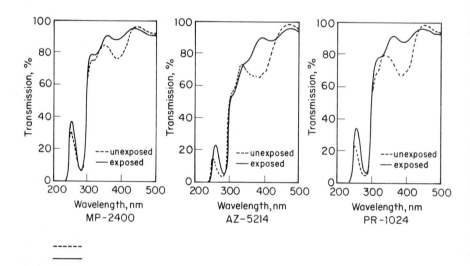

Figure 8.46 Transmission spectra of MP-2400, AZ-5214, and PR-102 before and after 248-nm exposure. (Resist thickness : 0.5 μm.)

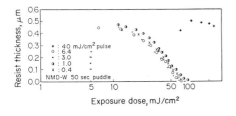

Figure 8.47 Exposure characteristics of AZ-5214 with pulse energy as a parameter.

sists typically used, we can see that, in general, resist sensitivity falls off as we move deeper, or to shorter wavelengths, into the uv region. The problem is one of getting good strong absorption at the strong mid- and deep-uv energy spikes inherent in exposure sources, with little or no absorption at longer, standard uv wavelengths. Figure 8.47 shows a resist sensitivity curve for AZ-5214. Note the reciprocity law behavior based on 248-nm excimer laser exposure.

One deep-uv lamp light source used in deep-uv exposure is a doped mercury lamp produced by Philips (Catalog No. 93146). The output of this source is shown in Fig. 8.48. Note the high-energy peak, from zinc doping, at 214 nm and cadmium and mercury peaks at 229 and 254 nm, respectively. This rich deep-uv output is needed to keep resist exposure times as short as possible. The energy intensity is about 3 mW/cm^2 at a 10-cm distance with a small 75-W (0.83-A) power source. People have doped a mercury-xenon source with zinc and cadmium. The combination of these elements in the excited lamp plasma should render a better strong uv source.

Having determined the exact spectral output of the exposing source, one can now establish the absorption spectra of the resist film.

Exposure mechanics

High absorption of a deep-uv wavelength or range of wavelengths in the short-uv region would seem to favor high sensitivity of the resist, since that is the energy that acts to photochemically convert the resist to a soluble state (in the case of a positive resist) or an insoluble state (in the case of a negative resist). However, this is not always the situation, because high initial absorbance prevents transmission of the reaction-causing energy down through the entire layer of resist. The desired mechanism is one in which the exposing energy is conducted through the film of resist uniformly to the substrate. High absorption prevents that unless, as is the case with Shipley Microposit 2400 photoresist, the resist is simultaneously bleached in the exposing reaction so as to increase its transmission further into the film. This

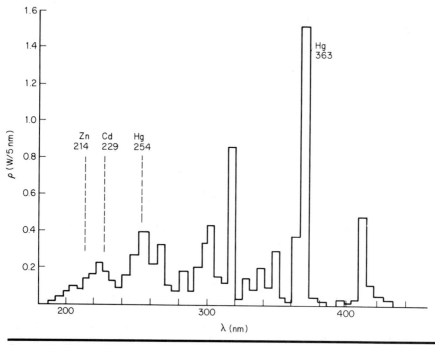

Figure 8.48 Deep-uv light source from Philips used in PMMA and PMIPK exposure tests. (*Courtesy of Philips.*)

self-bleaching phenomenon offsets the energy-restricting absorption characteristic in Microposit 2400.

Optical density, then, plays a key role in exposure mechanics and especially in short-wavelength lithography, in which selective wavelength filtering by the resist is a desired functional characteristic that allows the exclusion of longer wavelengths that detract from image resolution. In the case of most resists, optical densities in the 0.25- to 0.55-μm^{-1} region are desirable. The optical positive resists used for conventional photoimaging, especially all of the novolak resin–based resists, exhibit relatively high optical absorption in the mid- and deep-uv energy regions. Likewise, their transmission is high in the longer wavelength regions, above 300 nm, a factor which makes them quite suitable for the *g*-, *h*-, and *i*-line exposure sources commonly used. The Microposit 2400 also transmits energy in the 300-nm and higher region, so its use in mid- and deep-uv imaging requires special bandpass filters in the optical path to remove those wavelengths.

Actual functional speed, the real yardstick for a process engineer, is how long it takes to expose a wafer with a specified resist coating thickness. This calculation is derived by subtracting the absorption of the unexposed resist from that of the exposed resist. Figure 8.49 shows this "difference curve," along with the spectral absorbance curves. The difference curve is still not a true measure of the sensitivity, since it does take developer solubility effects into account.

Multilayer Resist Processes

Multilayer resist imaging has developed as an alternative to single-layer resist processing mainly because of the problem of ever-shrinking geometries on IC surfaces without a proportional change in the thicknesses at which various resist and dielectric layers are fabricated. The use of high-NA lenses with shallow depth of focus has made submicron imaging in relatively thick resist layers difficult. Multilayer resist imaging helps to solve the problem by providing an initial coating that effectively planarizes the wafer topography. This is followed by a second coating that is typically much thinner and serves as the pattern-forming layer through which the underlying layer is exposed and then developed or etched. In some cases a third, intermediate, layer is used to separate the planarized layer from the top coat mainly because of incompatibility between the chemistries of the top and bottom layers that leads to scumming. Figure 8.50 shows first the single-layer resist example with fine resolution over relatively high topography or step heights and then a comparison of pro-

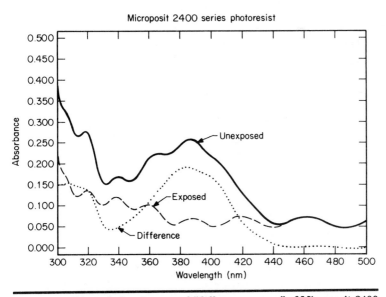

Figure 8.49 Spectral absorbance and "difference curves" of Microposit 2400 photoresist. (*Courtesy of Shipley Company.*)

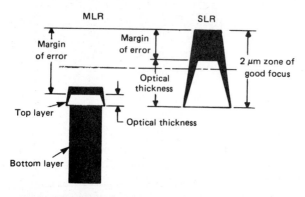

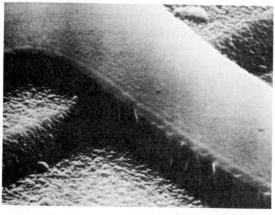

Figure 8.50 Single-layer resist over a high step and focus tolerance for single-layer resist (SLR) versus multilayer resist (MLR).

cess latitude for single-layer resist processes versus multilayer resist processes.

One of the biggest problems with single-layer imaging that is solved by multilayer processes is resist coating thickness variation over steps. This variation means a difference in the amount of exposure energy in thin versus thick areas, and pattern width changes can and do occur as a result of resist thickness variation.

Single-layer imaging also becomes more critical as resist geometries shrink and become more greatly influenced by anomalies either on the wafer or in resist coatings themselves. Multilayer processing avoids this problem by using very thin imaging layers which can be applied with extremely high uniformity. The thick planarizing layer also covers up nonuniformities on the wafer surface that could negatively affect imaging resolution.

Multilayer resist processing has another benefit that has become more important as the amount of reflections incurred in resist exposure increases because more reflective layers are added to IC processes. Many devices, as they increase in complexity, add more metal interconnection levels, a major source of reflection problems. In multilayer processing, the thick planarizing layer absorbs most of these reflections; since it is not used as the master image-forming layer (the top coat is that), the effects of reflections on pattern dimensions are either minimized or completely eliminated.

A key advantage of multilayer processing is using the thin top coat as a means to greatly reduce exposure time of highly capital-intensive exposure aligners. The top coat can be applied as thin as 1000 to 2000 Å without leaving pinholes, and exposure times for that thin a layer are very short. Since the planarizing layer can be batch-imaged in a reactive ion etcher (as will be shown later in this section), wafer throughput is improved.

The overall advantages of multilayer resist processing are summarized as follows:

1. Improved line geometry control

2. Improved image resolution

3. Reduced sensitivity to reflections

4. Reduced exposure times

5. Formation of aspect ratios not possible with single-layer imaging

6. Extension of optical imaging technology

In practice, resists used for single-layer processes may be used for multilayer processing also, perhaps with minor modification. This is an important area of consideration, since research time to develop entirely new resists for a multilayer application could take years.

The disadvantages of multilayer processing are added process steps, which certainly increase costs, and the fact that the processing represents a departure from well-developed process learning curves. Regardless of those difficulties, the pressure to overcome the problems that are solved by multilayer processing continues to increase as IC devices become more complex. An added incentive for using multilayer technology is the extension of existing optical imaging equipment for production of submicron dimensions once thought to be the domain of nonoptical (e-beam, ion beam, x-ray) imaging tools.

In this section we will discuss the major types of multilayer processes and the materials used with each process. The emphasis will be on practical application of each process to an IC production line. The estimated cost impact of each process is also an important parameter, and the cost of alternative processes or technologies must be estimated before a new process is introduced to production.

Two-layer resist process

Polymethyl isophenyl ketone (PMIPK) is an excellent material, commercially available, to serve as a high-sensitivity planarizing layer. In combination with Kodak 809 positive resist, PMIPK can be dyed to optimize its optical performance, absorbing and transmitting at selected wavelengths.

Basic dye properties for the Kodak 809 resist dye are

- Strong 310-nm absorbance
- Remains in the film during softbake
- Will not affect lithographic performance of top layer
- Soluble in the casting solvent for the top-layer resist
- Retains dying properties after softbake (no volatility)

The bottom layer of PMIPK was optimized to give high differential solubility: a 250 Å/sec dissolution rate for the exposed resist and only a 6 Å/sec rate for dissolution of the unexposed resist in the developer. Process parameters using a 2-min methyl isobutyl ketone (MIBK) development time are

1. Coumarin 6 dye, 3% in PMIPK, spin-coated
2. 116°C, 30-min softbake
3. Spin 809 plus 20% BPE dye
4. Prebake at 80°C, 30 min
5. Expose and develop top layer, 1.4-min exposure for 1.5-μm resist thickness
6. 310-nm pattern transfer
7. Redevelop MX 931

PMIPK provides 6 to 7 times the exposure speed of PMMA resist without any sacrifice in differential solubility.

Top-layer transmission is kept to 0.5%, which is important for good pattern transfer. Another key aspect of this process is reduction of reflectivity from the layers and substrate. Figure 8.51 shows an example of an image made with this process.

The dyes used in the bottom layer play a key role in reducing and eliminating interference effects. Wall angles are approximately 85° for the lower layer and 75° for the top layer.

The mask at 0.85 μm gave between 1.0- and 0.7-μm fidelity. The fidelity (mask-to-wafer reproduction) improves as dyes are added, increasing wall angles several degrees.

Process

- Spin on the PMIPK/3% Coumarin 6 dye
- Prebake at 160°C for 30 min to remove all solvents
- Spin on Kodak 809/20% BPE
- Prebake at 80°C for 30 min in a nitrogen atmosphere (high temperatures forms interlayers)
- Expose and develop top layer
- Pattern transfer to the bottom layer with exposure at 310*nm*
- Redevelop in MX 931 to remove all Kodak 809, leaving a residue free PMIPK layer
- Develop PMIPK for 2 min in methyl isobutyl ketone (MIBK)

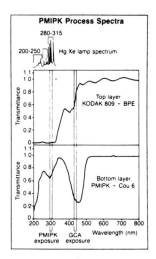

UV Spectra of 2 Layer Resist Process: Kodak 809 + Dye and PMIPK + Dye

Figure 8.51   Two-layer imaging process and resist image.

Polysilanes as multilayer resists

Polysilanes are a class of polymers with a silicon-silicon chemical backbone, as shown in the following diagram:

$$\left(\begin{array}{c} X \\ | \\ -Si- \\ | \\ Y \end{array}\right)_N \qquad \left(\begin{array}{c} X^1 \\ | \\ -Si- \\ | \\ Y^1 \end{array}\right) \left(\begin{array}{c} X^2 \\ | \\ Si \\ | \\ Y^2 \end{array}\right)_M$$

Homopolymer   Co Polymer

Chain lengths from 20 to 1000 silicon backbone units show changes in absorbance, shifting to the blue.

Silicon-containing polymers are desired for high-oxygen RIE resistance, mainly because they leave a layer of SiO_x to protect the underlying layer. In essence, oxygen RIE produces a nearly pure Si_2 layer on the top-layer film.

Polysilanes produce images with very steep wall profiles. Exposure is made at the 313-nm wavelength and contact-printed by using a 10-nm bandpass filter. The developer is an isopropyl alcohol (IPA) mixture. Undercut profiles in the polysilane can be achieved by extending the RIE process, which is used to prepare an image for lift-off processing.

Isopropanol can easily remove the non-cross-linked polysilane layer. The aspect ratio of pattern width to thickness is about 2:1 with 1-μm geometries. Some overexposure to the anisotropic oxygen plasma will result in 0.75-μm resolution, essentially an overetch.

Trilayer process

Trilayer processing was one of the original multilayer processes developed extensively at Bell Laboratories. Trilayer structures are initiated by first applying a planarizing layer to the wafer to cover existing topography. The thickness of this layer may vary according to the height of the steps previously etched. A good rule of thumb is to provide a resist layer thick enough to be 1000 to 2000 Å over the highest step. The second layer in the trilevel process is an inorganic layer, usually a deposited oxide, nitride, or metal layer. The thickness of this layer is approximately 1000 Å; the bottom layer is approximately 2 μm or more. The purpose of the intermediate layer is to serve as a mask for reactive ion etching of the underlying planarizing layer. The main functional requirement, then, is oxygen plasma resistance. The top layer is typically a thin (0.2- to 0.5-μm) energy-sensitive coating (photoelectron, ion, or x-ray resist). The main function of the top layer is to serve as a mask for etching the thin intermediate layer, which in turn is the mask for the bottom layer.

A typical process sequence, shown in Fig. 8.52, is

1. Prime the wafer with HMDS by spin coating or vapor priming.

2. Spin-coat the planarizing layer; use an optical novolak positive resist of PMMA.

3. Softbake the first coat at 100°C for 20 min in a fresh-air-circulating oven or, for a shorter time, on an in-line hot plate surface or in a microwave or infrared baking system.

4. Hardbake the planarizing layer at 200°C for 20 min, or use an energy equivalent.

5. Spin-coat the spin-on glass while using point-of-use filtration to remove any particulates.

6. Softbake the glass-resist structure at 100°C for 5 min (oven) or use an energy equivalent.

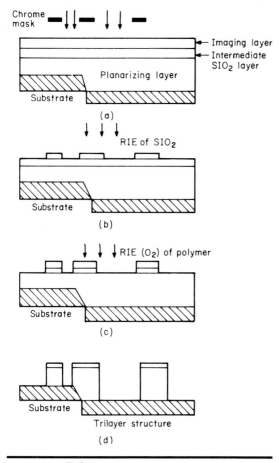

Figure 8.52   Trilayer process.

7. Spin-coat HMDS on the glass layer.

8. Spin-coat the top-layer resist (thin, 5000-Å, patterning layer).

9. Softbake, expose, and develop the top resist layer.

10. Reactive ion–etch (CG_4O_2) down to the planarizing layer.

11. Oxygen reactive ion–etch the bottom layer.

   A key benefit of this variation of two-layer portable conformable mask (PCM) process (Fig. 8.53) is the spin-on application of the ARC, which avoids the process interruption that takes place when a CVD oxide is applied. As with spin-on glass layers, a bake step is used after application, but it too is performed "on chuck" or in-line. Unlike spin-on glass, etching of the layer is avoided by being able to dispense-strip

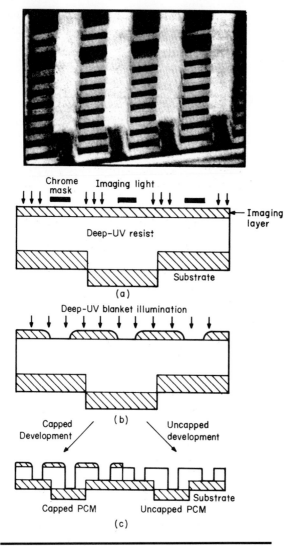

Figure 8.53 Two-layer PCM process: 0.2 μm of AZ-1350J on 1.8 μm of PMMA. (*Courtesy of B. J. Lin, IBM Corporation.*)

the ARC. The ARC also serves to separate the top and bottom resists, eliminate chemical mixing, and eliminate the reflection-interference effect and thereby keep standing waves out of the resist sidewalls.

The ability to pattern sharp images by using deep-uv exposure is a key advantage of the two-layer process, since the edge sharpness of the top layer in two-level, or bilevel, patterning carries through to the silicon substrate interface at which circuit functionality is determined. Even the sidewall roughening of PMMA in the plasma and some roughening effects from reflections are of little consequence, since the top-layer mask, sharp and undisturbed by these phenomena, determines the final pattern shape.

Two-layer PCM with ARC separator

Many modifications of the two- and three-layer (two polymer layers separated by a nonpolymer oxide, metal, or thin polymeric film) PCM process are possible. One that offers flexibility in top-layer resist choice is a process having top and bottom layers separated by a spin-on film of dyed polyimide. This antireflective coating will absorb exposure radiation at mid-uv (313 nm) and deep-uv wavelengths, the regions at which top layer resist exposure occurs. This means that resists such as Shipley Microposit 1470, Hoechst AZ-2400, and Kodak 820 can be used. Since some of these resists can be efficiently exposed at the shorter wavelengths, better top-layer resolution is achieved, especially in deep-uv areas in which ARC absorption is intense. Figure 8.53 shows a typical imaging sequence.

Two-layer process: inorganic resist

The use of two-layer, or bilevel, resist processing has evolved for the same reasons as for trilevel processes. The inhomogeneous exposure of resists that is caused during exposure by reflected light which interferes with itself leaves standing-wave patterns, unpredictable line widths, and line pattern variations when images traverse steps. The line widths particularly affected are at the bottom of steps at which considerable narrowing occurs from standing-wave patterns and inadequate exposure.

There are several bilevel processes, but a notable approach, outlined by K. L. Tai and coworkers at Bell Laboratories, involves a negative-working, photosensitive inorganic resist. The early work on inorganic resists was reported by Yoshikawa in *Applied Physics Letters,* vol. 29, p. 672, and was described as a Ge-Se system with submicron resolution. The work at Bell Laboratories has focused on a silver selenide and germanium selenide ($Ag_2Se/GeSe_2$) system. The lithographic results obtained with this inorganic resist as a top layer and a thick bottom planarizing layer are impressive, including linear patterns that are less than 1000 Å. The Ge-Se resists have shown remarkable resistance to oxygen plasmas. Also, the need to add dye to the bottom resist layer, as in the trilevel process, is eliminated, since the absorbance of Ge-Se is very strong at wafer stepper and other aligner wavelengths. For example, absorbance of 2.5×10^5 cm^{-1} at 400 nm was reported by K. L. Tai. Thus, standing-wave and other reflection effects are eliminated, as they are in the trilevel process, without using dyed resist layers.

The inorganic bilevel resist process begins with the application of the bottom planarizing layers, as do all multilayer processes. Step heights typically run about 8000 to 10,000 Å, and surfaces are often relatively rough. The resists used for bottom layers are Hunt HPR 206, Shipley 1400 Series, Kodak 820, and most of the other optical positive-resist systems, too numerous to cite here. The film thickness used is approximately 2.5 μm or slightly less, depending upon step heights. The positive-resist planarizing layer is baked, as it is in the trilevel process, at approximately 210°C for 2 hr.

The second, or top, layer in the bilevel process is $GeSe_2$, which is evaporated directly onto the top coat resist to a thickness of about 2000 Å. The resist top coat is imaged with a stepper or other type of production exposure aligner.

The exposure dosage for an optical resist layer such as 1350J is 50 to 60 mJ as a matter of comparison. The inorganic resist is then developed, and the wafers are placed in the oxygen reactive ion–etch environment. Figure 8.54 shows both a diagram and a SEM photo of the bilevel resist process results. The degree of resolution and line control and the vertical sidewalls and sharp edges are comparable to the results obtained with the trilevel process. The major single advantage is a simpler process.

Bilevel process: electron beam and deep-uv structures

The various types of bilevel processes are alternatively referred to as dual-level and two-layer, but all share the same fundamental structure, that being a sandwich of two films in which the bottom film is a topography-leveling, or planarizing, layer and the top layer is patterned and serves as a mask for imaging the bottom layer.

In order to extend the resolution of standard optical exposed two-layer structures, shorter-wavelength imaging of the top layer is employed. In the example discussed here, the top layer is patterned by e-beam exposure, and the resulting image, after development, is used as a mask in deep-uv exposure and development of the bottom layer. The wafer is etched as it would be in the more traditional one-layer process.

The two-layer process discussed here was reported by B. J. Lin and coworkers at the Thomas J. Watson Research Center in Yorktown Heights, New York. The top layer of resist in this process is refined to a dichroic layer, meaning that it is able to respond to two different wavelengths of radiation: It passes or transmits one wavelength to the bottom layer and absorbs a different wavelength as part of the imag-

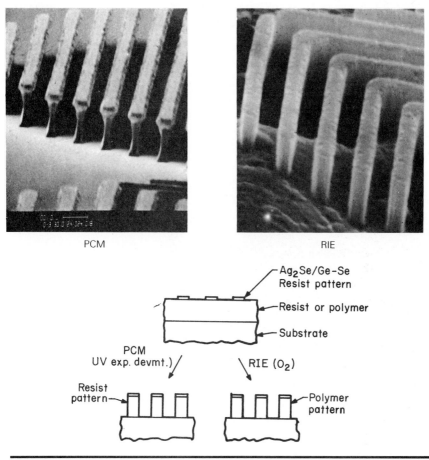

PCM

RIE

Ag$_2$Se/Ge–Se
Resist pattern

Resist or polymer

Substrate

PCM
UV exp. devmt.)

RIE (O$_2$)

Resist
pattern

Polymer
pattern

Figure 8.54 Two-layer RIE using inorganic resist; 0.5-μm lines and spaces on aluminum. (*Courtesy of K. Tai, Bell Laboratories.*)

ing process. Dichroic resist films are especially useful in two- and three-layer resist-imaging processes, since many process engineers will use two or more exposure tools to form the high-resolution patterns needed in VLSI device fabrication.

Figure 8.55 shows a diagrammatic cross section of the hybrid e-beam–deep-uv resist structure. As indicated in the diagram, the top layer may also be exposed with conventional step-and-repeat, scanning, or proximity-type wafer aligner systems. A variation of the process which changes it to a trilevel system involves the application (by evaporation) of a thin (0.3-μm) layer of aluminum. This adds undesirable complexity to the process but does eliminate the requirement that the top layer be optically dichroic. A process summary for the PCM bilevel technique follows:

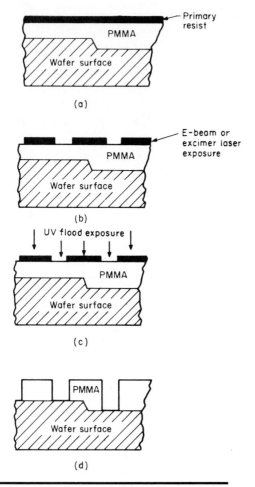

Figure 8.55 Hybrid e-beam–deep-uv structure.

1. Coat PMMA* (Dupont Elvacite-2041 or similar material) to a thickness of 2.0 to 2.5 µm.

2. Softbake PMMA† at 170 ± 10°C for 60 min in a convection oven or similar desolvating environment.

3. Coat Shipley 1350J, Kodak 809, or comparable resist to a thickness of 0.5 to 0.6 µm.[3]

---

*Two separate coatings, one immediately following the other, may be needed to obtain a smooth, uniform coating.

†PMMA with laser dye additions will prevent standing-wave formation in the resist.

4. Softbake 1350J at 80 ± 5°C for 10 min or equivalent.

5. Image 1350J (expose and develop) to form the PCM.

6. Deep-uv flood-expose PMMA though the PCM. NOTE: Descumming in oxygen plasma may be needed to remove interfacial layer; extended top-coat developing or a stronger-than-normal developer may also be useful to completely "chew" through the interfacial layer.

7. Develop PMMA by using chlorobenzene-xylene to retain the top layer for image fidelity comparisons from layer to layer or for lift-off; use methyl ethyl ketone–methyl isobutyl ketone to remove the top layer while the PMMA is being developed. This developer is more practical and simpler to use than the former developer.

Bilevel process with two optical resists

In a bilevel system, there are two resists and almost all of the two-layer processes discussed involve at least one relatively uncharted resist type. For example, PMMA combined with 1350J was cited earlier, and PMMA does not have the learning curve or database that exists with 1350J. The process we will discuss here utilizes two well-known and production proven materials, the 2400 and 1350J resists. In this particular bilevel approach, these two optical positive resists are patterned in the portable conformable mask style, the bottom layer (2400) being used in a deep-uv flood manner. The top layer (1350J) is imaged with conventional optical step and repeat, scanning projection, or nonoptical exposure with an e-beam. The primary benefit from this process is being able to use resists that have good lot-to-lot consistency (quality control as provided by the manufacturer and established functional parameters including dry etch resistance). This permits rapid implementation of a new imaging process with a fairly high degree of confidence.

The actual process step for this optical resist bilevel scheme is:

1. Wafer surface clean and dehydration bake.

2. Coat approximately 1.8 μm of 2400 photoresist. Spin-coat at 3000 rpm for 15 sec; use a mixture of 1 part 2400 and 1 part 2430.

3. Softbake at 90°C for 45 min in a fresh-air oven.

4. Generate buffer layer by CF_4 plasma treatment. Use 0.8 torr and 50 W of rf power for 45 sec in a plasma etch system.

5. Coat Microposit 1300-19 to a thickness of 0.55 μm by spinning at 3500 rpm for 25 sec. If e-beam exposure is desired, reduce coating thickness to 0.2 to 0.4 μm.

6. Softbake resist at 90 ± 2°C for 45 min in a fresh-air oven.

7. Expose the top resist layer at 435 nm (wafer stepper) at a minimum energy dose of 30 mJ/cm^2. E-beam exposure is performed at 20 μC/cm^2 at 10 keV for a minimum energy dose of 30 mJ/cm^2.

8. Develop the top layer of 1300-19 by using Microposit 351 developer at a 1:5 makeup (1 part 351, 5 parts deionized water). Develop for 40 sec at 21°C. Reduced exposure levels are possible by making the 351 developer more concentrated (makeup at 1:3.5). Follow with a thorough water rinse and dry.

9. Blanket deep-uv expose at 320 nm through the portable, conformable 1400 mask.

10. Remove the buffer layer in an oxygen plasma for 2 min at 0.55 torr and 50 W of rf power.

11. Develop 2400 photoresist in Microposit 2401 developer made up of 1 part 2401 to 4 parts water. Develop for 30 to 90 sec, depending upon resist thickness.

12. Follow with substrate patterning step, i.e., ion implantation, ion etching, or other processing.

Contrast enhancement layer process

The contrast enhancement layer (CEL) imaging technique moves even closer to being a straightforward production-viable process than the simplest bilevel schemes described thus far. In a CEL process, a 1000- to 3000-Å-thick layer of a photobleachable organic film is spin-coated directly onto the bottom planarizing layer, which is a standard optical positive resist. The two layers are then exposed to standard optical aligner wavelengths, step-and-repeat or scanning, and the top layer bleaches out during exposure. This bleaching increases the contrast of the bottom resist layer considerably by effectively becoming a high-contrast mask in intimate contact with the bottom resist. In essence, the results are very similar to patterns obtained when hard-contact printing is used. Most lithographers regard the resolution and vertical sidewall performance of hard-contact printing as the ultimate when compared to all of the other off-contact exposure image results. The CEL offers this same image quality, obviously without the disadvantages (mask wear, particle entrapment, and subsequent wafer damage) of hard-contact printing. The actual threshold energy dose of the resist is increased by the presence of the CEL layer, which effectively raises the contrast of the resist.

As the energy from exposure enters the top CEL layer, the bleaching process begins. The bleaching continues until it reaches the bottom layer of resist and then exposure of that layer is initiated. The threshold energy dose is quite predictable and therefore provides a useful means of establishing good pattern dimension control and improved exposure latitude.

Simplicity is the major benefit of the CEL process. There is immediate compatibility of the process with standard wafer coating and exposure equipment. In fact, all of the standard optical resists are probably compatible. The CEL layer also provides another benefit by reducing scattered light and poorly collimated light. After exposure, the CEL coating must be removed in a special solvent, followed by developing of the resist below. Figure 8.56 shows both a cross section of the CEL structure and SEM photos of the results obtained.

Resist planarization properties

A major objective of all multilayer imaging processes is planarization of the wafer surface followed by imaging on the planar layer. Even in processes in which planarization is not complete, such as the single-layer double-softbake techniques, getting the resist to image as if the surface were planar, as if the step heights and resulting reflections and thickness differentials were absent, is still the goal. A key area to study for a better understanding of how to control planarization or, at best, improve wafer topography resist coverage is planarization modeling. Figure 8.57 shows a typical etched step cross section overcoated with a single layer of resist. The various sections in which polymer thickness changes are greatest or are changed by resist coating are noted. This particular description is from the experiments of L. K. White of RCA Laboratories, Princeton, New Jersey. The formula used in this work is based on the various key sections around a wafer topographical feature in which overcoated layers produce various polymer thicknesses. The formula White used is

$$H_1 - H_2 = T_{\text{bot}} - T_{\text{top}}$$

where H_1 = height of the topographical feature
H_2 = amount of polymer film thickness
T_{bot} = thickness of the polymer from the substrate to the top without influence of the step (the thickness taken from a resist supplier spin speed versus thickness chart, called spun-on thickness)
T_{top} = polymer thickness on the top of the topographical feature

Note that in this process T_{max} equals the area of greatest polymer thickness, always taken at the base of a step, and T_{min} equals the area of least polymer thickness, always taken at the edge of a step.

The wide variety of polymers, resists, and solutions now used in multilayer and single-layer resist imaging suggests the need to plot

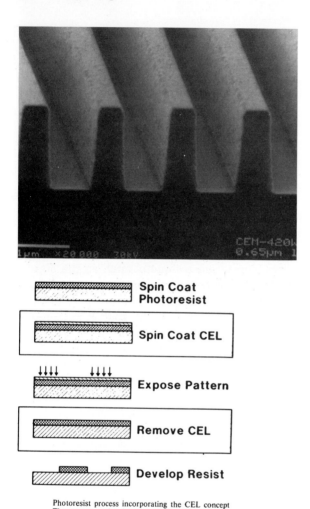

Photoresist process incorporating the CEL concept
The nonconventional steps are outlined in boxes.

Figure 8.56 CEL process and resist image: 0.65-μm lines and spaces in CEM-420WS resist on 1.4-μm-thick layer of Aspect System 8 photoresist imaged with an Optimetrix 8010 stepper using an *h*-line lens at 0.32 NA. (*Courtesy of Paul West, General Electric.*)

these various resist thickness changes as a function of polymer type and differences in the thickness variation across the area of a topographical feature. The area of greatest concern is the difference between T_{max} and T_{min}, since it will determine the greatest excursions in exposure dose and resulting differences in both pattern dimension width and resist sidewall angle. The data shown in Fig. 8.58 plots thickness difference from T_{top} and T_{bot} versus polymer thickness for

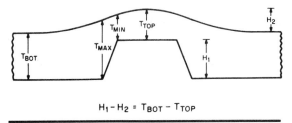

$$H_1 - H_2 = T_{BOT} - T_{TOP}$$

Figure 8.57 IC topography cross section and resist thickness variation.

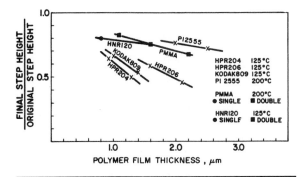

Figure 8.58 Resist planarizing model.

several polymer types. The way to identify a good planarizing resist is to determine which resists deliver the lowest thickness aspect ratios over a given topographical feature. On the chart in Fig. 8.58, the lower-left corner represents the resists with minimum total height above a step at the lowest polymer thickness. Following this type of coating thickness analysis on steps, planarization constants can be established for various step heights and for various resists. Any new fabrication process should consider such an analysis before committing a given resist system to production.

Overall, the best planarizing materials are the diazoquinone novolak resin–based resists such as HPR 204, Kodak 820, and Microposit 1400. Fortunately, these are the very same materials that fit into the multilayer imaging schemes. These resists have all of the other critical properties needed to generate submicron pattern resolution over wafer topography by using standard optical processing and wafer exposure equipment. These resists have very special optical properties required for multilayer processing, including transmission and absorption in very specific wavelength regions for PCM masking and sensitivity to the exposure tools. These resists are the workhorse

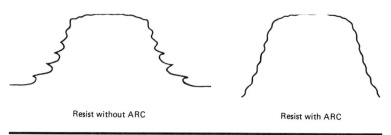

Resist without ARC

Resist with ARC

Figure 8.59   Resist profiles with and without antireflective coating.[29]

materials that deliver economic submicron lithography to IC manu-facturers.

Antireflective coatings

VLSI technology for submicron processes requires special attention to management of critical dimensional (CD) control to a degree well be-yond that of only slightly larger 2-μm processes. The dimension is scaled in half, but the degree of difficulty in terms of critical dimen-sion control increases in the range of 5 to 10 times. One technique used to help control CD parameters is the antireflective coating. This is a thin film, in the range of 100 to 250 nm thick, that is typically applied to metallization or other highly reflective films onto which the 1-μm imaging scale dimensions are to be placed. These coatings can be applied to within 10-nm coating uniformity, and the best method for coating these spin-applied solutions is hotplate softbake followed by resist coating. This method of bake ensures uniform ARC thickness wafer-to-wafer and within one wafer. Typically, the photoresist is ap-plied without a primer when the ARC solution is employed.

After both the ARC and resist are applied and baked, the composite layer is exposed, with the ARC absorbing 95% of the wavelengths used on a GCA (436-nm) stepper. A spray develop followed by a plasma descum ensures clean removal of both resist and ARC, leaving vertical sidewalls. Figure 8.59 shows the dramatic imaging difference obtained with the use of an ARC. The solution referenced here was an ARC obtained from Brewer Science, Rolla, Missouri, and used in a process described by M. Listvan, M. Swanson, A. Wall, and S. Campbell of Sperry Corp.

REFERENCES

1. W. S. DeForest, *Photoresist: Materials and Processes,* McGraw-Hill, New York, 1975.
2. J. Pacansky and J. R. Lyerla, "Photochemical Decomposition Mechanisms for AZ Type Photoresists," *IBM J. Res. Dev.,* vol. 23, no. 1, 1979, p. 42.

3. L. F. Thompson and R. E. Kerwin, "Polymer Resist Systems for Photo and Electron Lithography," *Ann. Rev. Mater. Sci.,* vol. 6, 1976, p. 267.
4. Technical Literature on Positive Resists (D1300, D1400, D111), Shipley Co., Newton, Mass., 1980.
5. F. J. Loprest and E. A. Fitzgerald, "The Photochemical Speed of Positive Photoresists," *Photographic Sci. Eng.,* vol. 15, no. 3, 1971, p. 260.
6. J. M. Shaw and M. Hatzakis, "Performance Characteristics of Diazo-Type Photoresists under E-Beam and Optical Exposure," *Kodak Interface Conf. Proc.,* 1977, p. 119.
7. E. B. Hryhorenko, "A Positive Approach to Resist Process Characterization for Linewidth Control," Eastman Kodak, Rochester, N. Y., 1980.
8. D. F. Ilten and K. V. Patel, "Standing Wave Effects on Photoresist Exposure," *Image Tech.,* February–March 1979, p. 9.
9. Technical Literature on Micralign Projection Printers, Perkin-Elmer Corp., Norwalk, Conn., 1980.
10. D. L. Markle, "A New Projection Printer," *Solid State Tech.,* June 1974, pp. 50–53.
11. R. Krause, Shipley Company, private communication.
12. J. Roussel, "Step and Repeat Wafer Imaging," Society of Plato-optical Instrumentation Engineers, *SPIE,* vol. 135, 1978, p. 30.
13. M. Kaplan and D. Meyerhofer, "Response of Diazoquinone Resists to Optical and Electron-Beam Exposure," *RCA Rev.,* vol. 40, 1979, p. 166.
14. J. G. Skinner, "Some Relative Merits of Contact, Near-Contact, and Projection Printing," *Kodak Interface Conf. Proc.,* 1973, p. 53.
15. M. C. King, "Principles of Optical Lithography," in N. G. Einspruch (ed.), *VLSI Electronics Microstructure Science,* vol. 1, Academic, New York, 1981.
16. M. Hohga and I. Tanabe, "Fabrication of High Precision Fine Pattern Photomasks and Evaluation of Photoresist Processing," Hitachi, Musashi Works, Japan. (*Kodak Interface 1977*).
17. D. Elliott, *Microlithography,* McGraw-Hill, New York, 1986, fig. 4.23, p. 137.
18. Technical Literature on the Micralign Projection Printer, Perkin-Elmer Corporation, Norwalk, Conn.
19. Technical Literature on the SRA 100 Projection Printer, Censor, Vaduz, Liechtenstein.
20. Ref. 19.
21. I. Higashikawa, et al., "Recent Progress in Excimer Laser Lithography," Toshiba Corporation, VLSI Research Center, Kawasaki, Japan.
22. B. J. Lin, "Multilayer Resist Systems and Processing," *Solid State Tech.,* May 1983, p. 105.
23. M. Watts, "A High Sensitivity Two Layer Resist Process for Use in High Resolution Optical Lithography," *SPIE,* vol. 469, March 1984.
24. Ref. 22.
25. K. Tai, AT&T, Bell Laboratories, Murray Hill, N.J.
26. P. West and B. Griffing, "Contrast Enhancement—A Route to Submicron Optical Lithography," *SPIE,* vol. 394, March 1983, p. 33.
27. L. White, "Planarization Properties of Resist and Polyimide Coatings," *J. Electronic. Soc.,* July 1983, p. 1543.
28. Ref. 27.
29. M. Listvan, M. Swanson, A. Wall, and S. Campbell, "Multiple Layer Techniques in Optical Lithography: Applications to Fine Line MOS Production," Sperry Corporation, Eagan, Minnesota.

Development

Introduction

Resist development has steadily progressed from a manual immersion process to a highly mechanized lithography step with the aid of laser end-point detectors, automatic titrators, and computer modeling programs that describe the developing process. The use of analytical instrumentation to profile and measure resist-developing parameters has increased steadily, thereby removing much of the uncertainty and unpredictability often associated with this process step.

Providing good management of the developing process has not been a function of computer modeling programs and instruments alone. Many new pieces of software-intensive wafer-handling and -developing equipment have been commercialized. Developer-processing equipment includes many hardware devices to implement the programmed software. Examples include temperature-controlled chucks to adjust and control the temperature of the solution and pressure-adjustable jets and spray nozzles to control solution spray intensity and uniformity. Special developer containers, such as plasma bags, are used to insulate positive-resist developers from oxidation in the air. Metering devices are used on in-line wafer-processing systems to regulate the amount of solution aspirated or dispensed. Hardware options also include timing functions that are software-regulated and that allow a flooding of the developer solution on the wafer surface followed by wafer spinning.

Developing equipment and solution chemistry combine to manage a process in which in-line normality controllers or specific-gravity monitors are used to control developer solution strength by injecting small

---

Reference 9 was a source of background information for this chapter.

amounts of developer concentrate. A constant developer strength can therefore be maintained in a production process, a key factor in controlling resist line geometries. Refinements in hardware continue to make the wafer-to-wafer developing process more repeatable. This increasing level of control allows lithographers to begin to manipulate the pattern sizes by changing any of a number of variables in the equipment, software, or actual developer solution. In that way, developing is evolving to a level of importance to process engineers much the same as exposure did in the past. Developer process manipulation is nearly as useful a way to change resist pattern sizing and sidewall slopes as is exposure, although the tuning leverage is less. Real-time software control of the resist-developing process will permit better predictability and control of this critical lithography step. Complete characterization and control of resist developing will permit higher levels of automation in IC manufacturing, a needed step to reduce handling-related defects and increase device yield.

Developer Mechanisms

Negative photoresists

Developing negative photoresist involves chemical removal of the unexposed portions of the photoresist coating. Typical negative resists in the liquid phase consist of a photoinitiator, a resin, and a solvent system. Assuming that the sensitizer, or photoinitiator, remained in solution (no crystallization or precipitation) through the coating and softbake stages, we begin with a photoimaged wafer of uniform consistency. The exposure has caused the photoinitiator to transmit photons in the exposing wavelengths to the polymer, resulting in an insoluble film. The exposure-initiated insolubilization of negative resists generally occurs as a cross-linking or chain growth within the resins of that particular resist. The unexposed negative-resist areas are quite soluble in solvent developers and are readily removed. They remain only slightly polymerized after exposure, and like positive resists, the differential solubility that exists between exposed and unexposed areas is the basis for image formation. Unique to negative resists, however, is the reaction that takes place between the resist and its developer, a reaction quite different from that of positive resists.

When wafers are sent into developing cycles, the exposed and the unexposed negative-resist areas are affected by the developer. The exposed areas, having undergone polymerization, are highly resistant to the solvent developer, yet a considerable amount of swelling takes place. This occurs as the solvent molecules attack and penetrate the molecules of the resist that have been cross-linked by exposure. As a result, some of the exposed negative resist is removed and the remaining area is distorted because of absorption of the solvent developer. In the unexposed negative resist, the solvent developer removes the resist by a process of penetration, swelling, and dissolution. The solvent molecules tie themselves to the polymer molecules and, by overcoming the forces of attraction holding the polymer together, separate the polymer in small sections from the film. This erosion process takes place even after the unexposed resist is separated from the bulk of the coating, so that a complete chemical breakdown is effected. The surrounding and separating of these relatively small polymer areas is optimized in the formulation of proprietary developer solutions. The size of the larger polymer molecules in the exposed areas is differentially so great that by the time developer molecules have surrounded the polymer molecules and are ready to break them loose from the bulk of the film, most of the unexposed polymer areas are dissolved and the wafer is nearing the postdevelopment rinse.

Most negative-resist developers are mixtures that include a solvent with high-solubility parameters for the unexposed resist and "buffer" solvents with much lower solubility parameters to control the action of the primary solvent.

Photoresists with aqueous-based developers are preferred to avoid image swelling that becomes an unacceptable condition in submicron lithography. The ability to reverse-tone positive resists is an alternative to using negative resist (see "Image Reversal," Chap. 8).

Positive photoresists

The mechanism for developing positive photoresists is often described as a "surface-limited dissolution process." In this sense, it is similar to the mechanism just described for negative resists. The major difference between the two lies in the greater selectivity achieved by positive-resist developers. Recalling the exposure mechanism (Chap. 8), the wafers enter the developer having exposed areas converted to a base-soluble carboxylic acid, while the unexposed areas remain the same, i.e., a relatively insoluble diazo inhibitor in the phenolic resin complex. Deforest[1] discusses two slightly different chemical reactions postulated by Sus and Levine for the development of positive resists. In the earlier Sus reaction, postulated in 1944, a carboxylic acid structure is formed during exposure, a product of a reaction between a ketene and a water molecule. The proposed reaction is shown in Fig. 9.1. The carboxylic acid is reacted with the basic or alkaline developer and thereby neutralized. In the Levine reaction, shown in Fig. 9.2, note that the ketene formed is relatively unstable and reacts with the inhibitor or sensitizer to form a lactone. A "ketene" is a compound

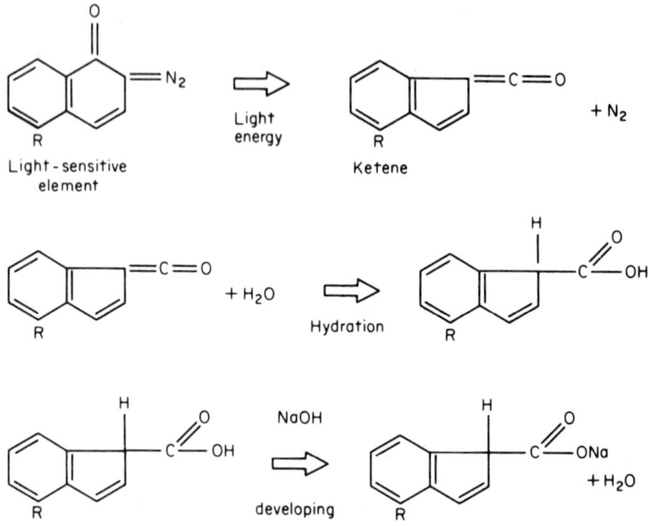

Figure 9.1 Sus reaction for positive photoresist.[1]

having a carboxyl group and carbon atoms with double-bond attachments. Ketenes are so short-lived that they are difficult to isolate during the exposure of the resist. A "lactone" is a ring-structured compound, and the developer breaks this structure and forms a hydroxyl group and carboxylic acid. In both proposed reactions, ketenes are formed and a carboxylic acid is ultimately neutralized by the mildly alkaline developer. The surface of the positive resist is slowly penetrated by this isotropic reaction. A micrograph of the exposed positive photoresist after partial development is shown in Fig. 9.3.

Another reaction that occurs in positive-resist development also is described by Deforest.[1] This is a coupling reaction wherein the diazide, phenols, and amines join to form an azo linkage. This reaction occurs immediately as the imaged wafer enters the developer. The visible evidence of this coupling reaction is the "red cloud" that is observed at the surface of the wafer when developing in the immersion mode. The reaction is shown in Fig. 9.4.

The exposed positive resist then is reacting its carboxylic acid groups into amine or metal salts in the presence of the developer, and the salts are rapidly dissolved away into the developer solution. Since no such groups are formed in the unexposed areas, they remain largely unaffected by the developer. The dissolution of the salts from the exposed resist areas causes the developer normality to drop, since the active ingredient, often sodium or potassium hydroxide, is consumed in the reaction. The monitoring of developer normality is used

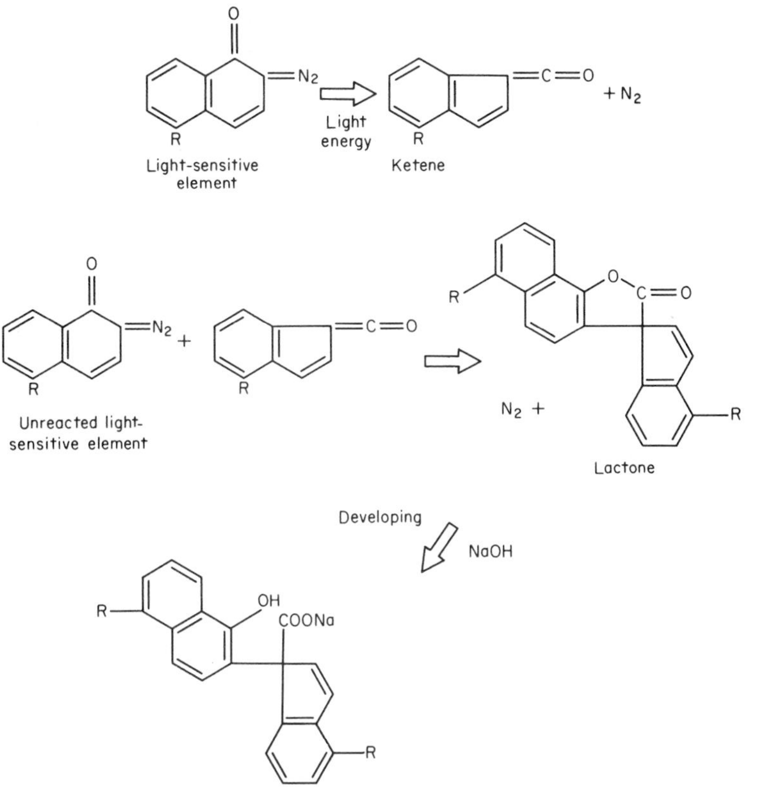

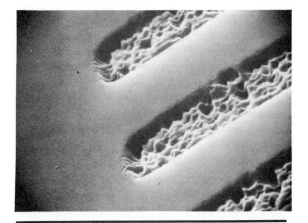

Figure 9.2 Levine reaction for positive photoresist.[1]

Figure 9.3 Micrograph of the surface of a partially developed positive photoresist.

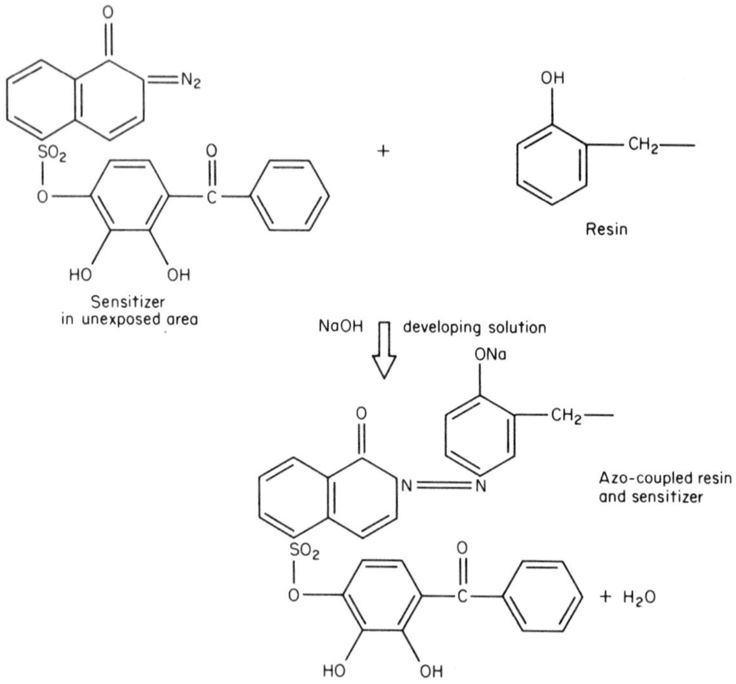

Figure 9.4 Reaction of sensitizer molecule with the resin in the unexposed areas.[1]

as a control test for developer bath depletion; if it goes below the normality point specified by the supplier, the bath is discarded. On in-line puddle or spray systems, the normality of the system solution should be checked daily.

The main functional difference between positive- and negative-resist development reactions is that the resulting image with positive resists is not physically or chemically distorted, as it is in negative resists by solvent-developer absorption. Control of photoresist geometries is so critical that fractions of a micron of dimensional change can, by accumulating through the imaging process, result in an out-of-spec overlay budget. Solvent-developer absorption is one of the reasons why positive resists are predominant in VLSI/ULSI technology.

Development approaches. Developing exposed (positive) or unexposed (negative) resist from wafers and masks can be accomplished by a variety of methods. The type of development technique used depends upon the complexity of the pattern geometry, size of the substrate, the production rate required, and the composition and configuration of the substrate. Resist coatings can vary in thickness from 0.1 to 3.0 μm, all used to fabricate the micron and submicron geometries used in advanced IC fabrication processes.

VLSI wafers are typically processed with high-volume production equipment, either in-line systems developing one part at a time or batch systems in which several cassettes of wafers are developed at once. Silicon ribbon, solar cell substrates, and other microelectronic parts to be imaged require special development approaches. In this section we will review the standard approaches, including immersion, spray, puddle, and some less frequently used techniques for special applications.

Immersion. Immersion developing is the simplest and oldest technique for resist processing. It involves placing the exposed substrates in a bath of made-up solution and agitating them gently in order to remove resist, thereby allowing fresh developer to enter the pattern area constantly. In research and prototype circuit areas, immersion developing is desirable because it can be performed in a shallow glass tray, beaker, or other equally simple equipment. Temperature and other parametric control is also simple because of the small volume of developer used and the size of the container. Immersion developing is easy to monitor on this scale.

The mechanism of development in immersion is cellular. The gradual erosion through the thickness of the resist layer is illustrated in Fig. 9.3. This particular example is a 1-μm optical positive resist. Negative resists behave similarly except that, instead of exposed areas dissolving away, unexposed areas are removed. Agitation during immersion is very important to keep the chemistry of the solution uniform and therefore maintain a uniform developing rate.

Immersion developers should be covered when not in use to maintain bath volume and strength. Solvent evaporates from negative-resist developers, and carbon dioxide in the air reacts with positive-resist developers to form useless carbonates.

The key parameter to control in immersion developing is bath concentration, since the active ingredients determine the rate of development and the concentration decreases in proportion to the amount of resist dissolved in the bath. Temperature can be maintained at 21°C ±1°C (or ±0.5°C to ±0.1°C for submicron geometry processes) with a constant-temperature bath. Developing time for VLSI applications, using approximately 1-μm-thick coatings, is about 60 sec in immersion, and developer rate curves are useful in predicting the amount of time required to dissolve through a given thickness of resist under a specific set of process conditions. Providing good process control of development time, bath concentration, temperature, and agitation will enable a process to generate good process yield.

Application of immersion developing beyond laboratory and prototype operations is rare but not without example. A few sizable wafer fabrication lines use large immersion tanks for batch immersion of one or more cassettes at a time. Since up to a few hundred wafers can be developed and rinsed simultaneously, production rates are excellent. Cassettes are loaded into a programmable system that provides control over key variables.

Immersion developer characterization. Characterization of immersion developing is performed in one of several tests, the most popular being the characteristic curve. Characteristic curves can be plotted in several different ways, but generally they express the percent of resist thickness remaining versus log exposure. Figure 9.5 shows a characteristic curve for Shipley Microposit MF-312 used in the immersion mode. The resist thickness change as a function of developing time for various exposure doses tells us much about the contrast, solubility, speed, and other critical properties of the resist-developer system. The resist used for these data was Microposit 1400-31 at a thickness of 1.2 μm. The developer temperature was maintained at 22.5°C. A variety of contrast tests with this resist-developer combination showed that the contrast tests of the resist over this developer temperature range remained fairly constant. Since much better exposure throughput can be obtained with the higher developer temperature, without sacrificing contrast, the higher temperature was chosen.

The one concern in immersion development is a loss of contrast or differential solubility caused by poor efficiency of removal of exposed resist. Differential solubility is defined here as the difference between the solubility of the exposed resist and that of the unexposed resist. Note in Fig. 9.5 that solubility of the resist is high at very low exposure doses, which is evidence of a lower contrast situation or poor differential solubility. Figure 9.6 shows the same plot or characteristic curves except that a different developer, the Microposit MF-314, was

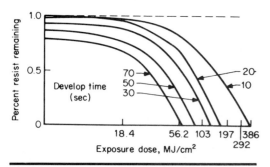

Figure 9.5 Characteristic curves for MF-312 developer in the immersion mode.[10]

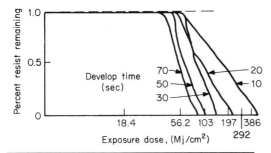

Figure 9.6 Characteristic curves for MF-314 developer in the immersion mode.[10]

used. The resist solubility is considerably less in this case, especially at the lower exposure doses. The dissolution of Microposit 1400-31 by the MF-314 metal-ion-free developer begins much later in the exposure process. For example, dissolution begins at exposure doses close to 50 mJ/cm, and the resist developed with MF-312 starts dissolving immediately. The 10-sec development time even showed some dissolution at low exposure doses with MF-312.

The good resist retention at low exposure doses with MF-314 is indicative of good contrast. Also, the very sharp profiles of resist dissolution for all developer times are evidence of the high contrast, which is especially important in high-resolution lithography. At submicron imaging levels, the stray light and spreading of projected images is "edge sharpened" by a high-contrast resist system. The rate of development should always be measured over a range of bath temperatures and at various developer concentrations in an attempt to achieve maximum differential solubility or contrast.

Puddle development. Puddle developing (or static flood developing) is a technique whereby a small amount of developer is dispensed onto the wafer or other surface. The amount of developer used is just enough to form a meniscus on the surface so that surface-edge tension keeps the developer from running over the side of the wafer. This technique essentially puts immersion developing on line, or in line, so that cassette-to-cassette automated wafer-handling systems can be used and thereby eliminate the need for the operator to play a direct role in processing the parts. The puddle of developer formed on the exposed resist surface immediately begins acting on the resist. The dwell time is predetermined by the length of static immersion time required to dissolve away the nonimage resist areas. When the developing time is complete, the wafer is sprayed, by the dispense head, with deionized water to quench the developing action.

Special developer formulations are used for puddle developing, in

addition to the standard positive- and negative-resist developers originally formulated for immersion. The main requirement of puddle developers is good operation at ambient temperatures, since they are
dispensed from canisters stored below the wafer-handling system. The
resist used in puddle development processes must be well baked to
avoid excessive unexposed resist thickness loss, and the developer
should wet the surface of the resist readily and uniformly so that the
dissolution of resist proceeds at a constant rate across the wafer surface. Differential solubility testing is especially important in puddle
developing in which the substrate cannot be agitated or the solution
moved during the dissolution reaction.

One chemical system for puddle developing is the Shipley
Microposit MF-314 metal-ion-free developer and Microposit 1400
photoresist system. The characteristic curve data for this resist-
developer combination are shown in Fig. 9.7. The improvement in contrast between these curves and the ones for MF-312 and MF-314,
shown in Figs. 9.5 and 9.6 for immersion developing, is considerable.
In a direct comparison between MF-314 for standard immersion and
for puddle development, note the better profile for the puddle-
developed samples. In positive resist developing, the by-products of
the reaction, including dissolved resist, actually accelerate the rate of
developing. This means that the time to reach the substrate is less, yet
the attack on the unexposed resist is actually less because fresh developer does not reach the unexposed areas. Puddle developing, by
keeping the wafer motionless, prevents solution movement which normally carries away dissolved resist products.

The spray-developing method, however, cannot be matched for the
amount of contrast it imparts to the resist image. Figure 9.8 is a comparison of the three most commonly used developing methods versus
the contrast obtained in the resist image. The large difference in contrast between immersion and spray developing is especially important

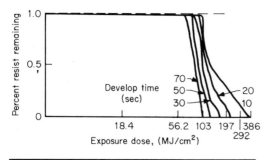

Figure 9.7 Characteristic curves for puddle-
developing MF-314 developer.[10]

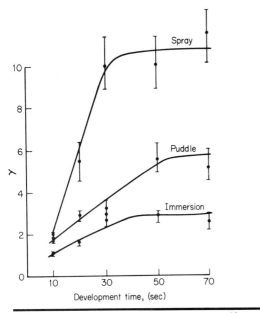

Figure 9.8   Contrast versus developing mode.[10]

to submicron lithography in which several degrees of resist sidewall angle are critical in resolving an image dimension and holding the more important critical dimensional tolerances. In the future, more attention will be paid to improving the efficiency of developing mechanisms and the equipment that delivers this chemistry.

Optimizing the puddle-developing process becomes increasingly important with smaller pattern geometries. Considerations for rapid and complete wetting of the wafer, especially one of the larger sizes, must be taken. The surface-wetting properties can be modified by changing the wetting contact angle of the developer, increasing its tension for surfaces, or lowering its internal surface tension. Surfactants and wetting agents can be added to a developer, or their existing concentration can be increased. Since dissolution rates on properly exposed resist are high, the entire surface should be contacted at once so that some areas do not get overdeveloped.

The physical aspects of puddle development are just as important as the chemical aspects. Puddle developing, by showing each wafer a "fresh" puddle, is highly repeatable and uses less developer than spray techniques.

Spray-puddle developing. There are several varieties of developing, and spray-puddle developing differs from flood-spray developing in two important ways. First, the solution is applied to only one wafer at a time and is not recirculated. Second, the solution is allowed to puddle with the wafer in a stationary position in between two spinning steps. The actual sequence is approximately as follows:

1. Initiate spin cycle, 200 to 4000 rpm (1 to 2 sec).
2. Spray dispense developer solution (1 to 2 sec).
3. Reduce spin speed to 100 to 200 rpm while spraying.
4. Stop spinning; developer meniscus forms.
5. Puddle develop for 20 to 30 sec, depending on resist type, developer strength, and other variables.
6. Decelerate to low-speed spin (100 to 200 rpm) while spraying developer.
7. Initiate water rinse; overlap with the developer spray for 2 to 3 sec to prevent developer from drying on the wafer.
8. Spin dry.

The sequence is used for several types of metal-containing and metal-free developers for positive optical resists such as Microposit MF-314 Developer, CD-31. This specific developer is used at 20 to 25°C and is controlled to +1°C measured at the surface of the wafer. Temperature control is increasingly important as spray pressure increases because of evaporative cooling effects. That is why temperature of the solution should be measured at the wafer surface.

Most of the in-line track wafer-handling systems have sufficient software flexibility to accommodate the variety of process changes described in the above procedure; they offer a high level of automation for spray-puddle developing. The developers used for this method must have the same capability as straight puddle developers in delivering high-contrast images by providing good differential solubility, which is largely a developer chemistry function. The general rule for developer attack on unexposed resist for a well-controlled process is that it should not exceed 200 to 250 Å.

The spray-puddle technique tends to conserve on developer usage, since the spray cycles before and after puddling are short and serve mainly to first wet the wafer for uniform initial dissolution and then to remove developed resist after the static puddle cycle. The solution movement after static developing may also serve to remove any monolayers, veils, or thin skins of resist that sometimes form in the developing process.

Resist system solubility parameters. Positive optical and negative optical resists for IC fabrication are able to form images by some type of process differentiation. Resists are classified in several ways; most are polymer-based systems using solvent carriers. The chemical basis for image formation is generally either photopolymerization by cross-linking (negative resists) or photosolubilization by ring breaking (positive resists). There are several other types of chemical reaction categories, but they constitute the chemical reactions used in most VLSI resist systems.

Since the bulk of wafer imaging for advanced IC processes is performed with optical positive resists, we will use such systems as the example of how image formation occurs. However, the same basic principles work for almost all other types of resists—positive, negative, and of varying chemistry.

The components of a positive optical resist system are easily broken into three parts: the resin(s), the photoinhibitor (sensitizer), and the solvent system (carrier). Each of these component categories is responsible for delivering specific properties to the functional behavior of performance of the entire system. The solubility of each component, as it exists in the film through the imaging process, is shown in Table 1.2. First, we see the dissolution rate of the cast-resin film. This is a rather high figure, especially compared to the next step, in which the photoactive compound, or dissolution inhibitor, has been added. Extremely low solubility is very necessary for the preservation of good resolution in overdevelopment situations, for high differential solubility, and for maintaining the integrity of the resist surface to provide maximum etch protection. The solubility parameter of an unexposed film of resist in the developer is often more important than the rate of dissolution of exposed resist. For example, if the development time needs to be long, the developer attack on the unexposed area will be minimal for a low-solubility, unexposed resist layer.

During exposure, the resist is reacted in order to greatly increase its solubility so that exposed resist, dissolving at several hundred times the rate of the unexposed resist, quickly forms the image. An extra margin of 10 to 25% overdevelopment is commonly used to ensure that all of the exposed resist is removed. The higher the differential insolubility, the more latitude the process engineer has in controlling the imaging process.

Some processes cut back on the amount of exposure given a resist and typically add 20 to 50% more time in the developer. Some processes will elevate the developer temperature to achieve better exposure throughput. In any one of the cases, good developer resistance on the unexposed resist is needed. In Chap. 3, the statement is made that good developer resistance is a function of complete softbaking. When solvents are left in the resist film, the solubility rate of the unexposed film increases. Thus, the unexposed resist solubility can be increased by additional baking until photosensitivity begins to be degraded, which, of course, would then begin to degrade the solubility ratio or contrast.

1. Cast a film of the desired resist by spin-coating onto a wafer(s) or similar substrate.

2. Softbake to remove residual solvents. Complete desolvation is necessary to minimize developer attack caused by the presence of solvent.

3. Measure the thickness of the resist at several representative points around the wafer or mask.

4. Immerse the wafer in the developer for the standard developing time for a normally exposed sample.

5. Expose other parts prepared (coated, baked, and measured) in the same way.

6. Develop the exposed samples by using a laser end-point detection system or similar in situ method (an example follows the list) for determining exactly the point of complete development.

7. Calculate the amount of attack on the unexposed resist by remeasuring the sample. This figure divided by the amount of resist thickness of the coating developed is the solubility ratio. Figure 9.9 illustrates a method for calculating the solubility ratio of a resist-developer system.

Spray developing. Spray developer processes are as varied as the types of developer chemistries and the equipment used to process through those chemical types. One example of a production spray process used for 1.5-μm geometries is shown below.

This process includes the ranges in bakes, rinse overlap, and other variations likely to occur among different processes.

1. Resist coating thickness of 1.2 to 1.5 μm.
2. Softbake in-line infrared for 6 to 8 min.
3. Expose (resists are Kodak 820, Hunt 204, Shipley 1400).
4. Develop
   a. Type and concentration, Microposit 351, diluted 3:1 with deionized water or Microposit MF-312 diluted 3:1, 2:1, or 1:1 with deionized water.
   b. Time, 10- to 20-sec spray or puddle with 1- to 4-sec overlapping spray rinse. Rinse time 15 to 30 sec, followed by a spin dry.

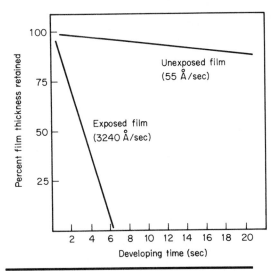

Figure 9.9 Resist developer contrast relationship (differential solubility ratio).

c. Pressure, 20 to 25 psi of nitrogen pressure or 5 to 10 psi of cannister pressure.

d. Temperature, 21°C ± 1°C.

These parameters should be regarded as guidelines and should be adjusted as required to meet specific process needs. Spray developing has distinct benefits and characteristics that separate it from puddle and immersion developing. Spray developing is rapid, both in wetting the resist and in total time to dissolve away exposed positive resist or unexposed negative resist. It tends to be more repeatable than the other types of developing, including combinations of spray with static programs. A summary of image repeatability results of spray, spray-puddle (static), and spray-static-spray programs run on a GCA positive-resist-module shows that the average standard deviation from a 2.5-μm feature for the spray-static-spray program is 0.05 μm, 0.10 μm for spray static, and 0.043 μm for the straight spray. There are separate wetting cycles or overlap steps that can be used on in-line systems. The use of several different developer concentrations offers the possibility of changing exposure times. Figure 9.10 shows the GCA DSW exposure value, in seconds, plotted against development time, in seconds. The spray-static program for the 5:1 makeup of Microposit 351 gives a slightly faster exposure, which is explained by acceleration of developing at the resist interface by developer by-products as opposed to fresh developer. At the 3.5:1 makeup, all spray

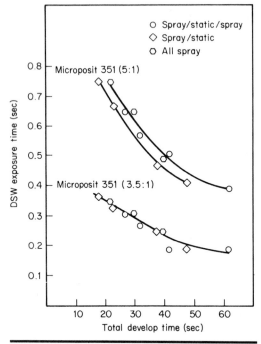

Figure 9.10 Exposure energy versus developing time.[12]

cycles resulted in equivalent exposure time. One important aspect of spray developing is the option to use low spray pressures to avoid air entrapment and foaming of the solution in the spin bowl. In many spray-developing systems a 360°C exhaust is used to remove vapor and condensation. A special liquid feed system eliminates the most common problems associated with spray developing, including salt caking or developer salt crystal buildup in the nozzle, along with postcycle dripping. The low-pressure spray greatly reduces evaporative cooling, thereby reducing the temperature sensitivity of the process. A flood rinse cycle removes developer by-products.

Development rate monitoring. The principle of single- and multiple-channel developer rate monitoring (DRM) is illustrated in Fig. 9.11. The change in resist thickness as development proceeds is based on the change in light interference in the resist layer. The change in the laser output signal varies sinusoidally with resist thickness change. The modulated photodiode signal is recorded on a strip chart, and the signal magnitude is dependent on the phase difference between the

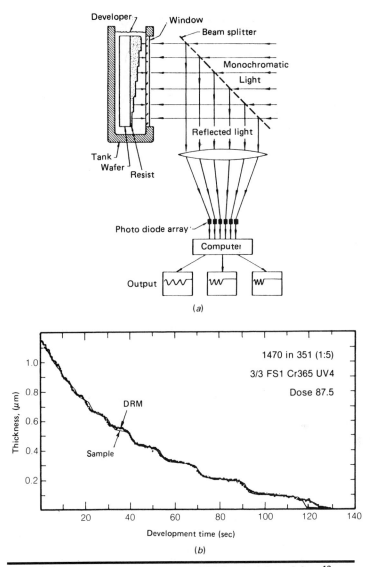

Figure 9.11 Developer rate monitoring (DRM) principle and data.[13]

two output beams, which is in turn dependent on the optical thickness of the resist. The amplitudes A_1 and A_2 are measured by using the standard I_{max}/I_{min} calculations. The change in signal intensity ΔI allows for the calculation of optical path variation Δnd. The final computation for resist thickness at any point is given by the known developing time figure, the refractive index of the resist, and the laser wavelength.

One of the advantages of this approach to characterizing developer processes is data accuracy because of the relatively high density of data points derived from the measurements. The signal is plotted, in millivolts, on the ordinate and the corresponding development time, in seconds, on the abscissa. Note the flattening out point after approximately 25 sec, indicating complete development. When the thickness calculations are completed as described earlier, resist thickness (removal) can be plotted versus development time. This plot is also called a characteristic curve, since it correlates to a given resist exposure dose. The results of this type of data on developing rate allow process operators to know exactly when developing is complete and thereby to more efficiently protect resist geometries from overdevelopment; in addition, the effect of exposure on developing rate is determined by this method. A typical plot of this type of development rate data made for one fixed exposure dose using a positive optical resist is shown in Fig. 6.18. One aspect of development rate monitoring that relates to the high data point density is sensing and recording small exposure excursions within the thickness of the resist layer. For example, standing-wave patterns are very precisely mapped on samples that are exposed to narrow energy bandwidths; broadband mercury or similar sources produce a wavelength-averaging effect that washes out most of those exposure and corresponding thickness excursions.

Multichannel development rate monitoring. A device developed at Perkin-Elmer Corporation takes the data relationship shown above in the single-channel DRM and expands it into 15 separate channels. Each channel corresponds to a different exposure level, which provides rapid characterization of a resist-developer system. The device naturally saves many hours of resist system evaluation time and allows for a significant amount of data collection on a single silicon wafer; it further reduces test variability by having a common substrate for all exposure data points. Most IC facilities require complete evaluations of lithographic properties of several similar but competitive resist systems. The number of available resists for advanced IC device fabrication keeps increasing, which makes the evaluation task more costly and time-consuming. The need for a 15-channel development rate monitoring system is thus obvious, and its availability answers a key industry need.

One other aspect provided by the multichannel DRM system is standardization. For many years the semiconductor industry relied on heuristic approaches to lithography. In the absence of good analytical equipment tailored for measuring lithographic chemical relations, that led to the impression that microimaging technology is a "black art," and the industry has not yet shaken that image. Resists and their developers have been evaluated and optimized by more diverse test methods than there are companies that make ICs. The DRM is one of several tools that permit IC industry standardization in a resist sensitivity definition, contrast determination, computerized modeling of resist systems, and software for wafer-handling systems that allows for real-time adjustments to wafer-imaging processes to compensate for process variations. The quantitative techniques are removing the art from lithography and replacing it with scientific, objective methods. More important, the industry technical experts are beginning to be able to speak a common language to express their findings, thereby allowing more meaningful use of new research and process engineering findings.

With resist thickness decreasing constantly, the output signal of the system registers the resist thickness at the specific point of measurement. Multichannel development rate monitoring uses multiple parallel data streams to process the signals. This results in a profile from each channel plotted on a single graph. Developing times are shown, in seconds, on the abscissa and resist thickness on the ordinate.

Developer Parameters

Time

Since positive photoresists are widely used in VLSI lithography, they will be referred to primarily as the developing process is described. Developing is a "surface-limited etching reaction," so it is necessary to determine the developing rate of a photoresist and calculate the time required for complete removal of, in this case, the exposed positive resist. "Development time" is the time required to completely remove or dissolve away the resist from the nonimaged areas. Increased exposure results in reduced inhibitor concentration and increased dissolution rate of the positive resist, as shown in Fig. 9.12. The measurement of development rate is determined by measuring the dissolved film in situ until none remains. The in situ optical monitoring device is shown, along with data from the tests in Fig. 9.13. This was a precursor of the DRM shown in Fig. 9.11. As the exposure level increases, so does the rate of dissolution of the resist in the developer. Thus one key function of development time is exposure time.

Softbake temperature also affects developing time, because increased softbake reduces solvent content and thereby "hardens" the resist to developer penetration, all for a constant exposure level. Thus there are predictable changes in development rate as a function of

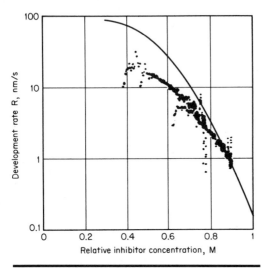

Figure 9.12 AZ-1350J development rate curve (1:1 AZ developer at 20°C, 100°C softbake).[2]

softbake temperature. Since softbake is typically optimized first on the basis of solvent removal (Chap. 7), this becomes constant when the proper developing time is calculated.

Developer concentration is also a major determinant of development time. Increasing developer concentration greatly decreases development time, which time is also a function of developer agitation, resist thickness, and developer temperature. The major variables affecting development time are:

1. Exposure time

2. Developer concentration

3. Developer temperature

4. Developer agitation method

5. Photoresist thickness

6. Photoresist softbake temperature

The above parameters are optimized to provide maximum resolution, CD control, and process latitudes. Equipment constraints generally dictate that development time extend only as long as is necessary to process an economical number of wafers per hour. Resist thickness is fixed as the minimum needed to protect the underlying surface from breakdown during etching while allowing for some loss in developing. Agitation also is fixed by the type of equipment used. Exposure time

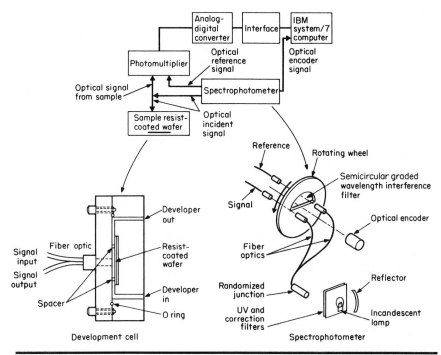

Figure 9.13 Diagram of in situ resist film thickness measurement apparatus.[2]

and developer concentration are frequently played against each other to balance the needs of the process relative to throughput and resolution.

The primary requirement of establishing photoresist development time is complete, residue-free removal of the nonimage resist. This means keeping wafers in the developer long enough to prevent underdevelopment, which will leave films in the areas to be etched and subsequently prevent complete etching. Keeping the wafer in the developer too long also will cause problems. A negative resist that is overdeveloped will swell excessively, and a relatively large percent of the exposed (polymerized) resist will be removed. That will reduce the plasma or wet-etching resistance and cause pinholes. A positive resist that is overdeveloped also will undergo unwanted loss in some areas (unexposed portions), which will reduce the processing resistance. In both cases, extended development time will change the geometries of the final image, often away from the desired size.

Since development time is a major variable in achieving good reproduction of photomask geometries, it must be carefully determined and monitored in process. Even after the developing cycle is complete, developing action continues as long as there is solution on the resist coating. Quenching of the developer action is important for that reason, and a dilute aqueous acidic spray is recommended for positive-resist developers. A small addition of a dilute acid to the rinse water will suffice. Negative-resist development can be followed by a rinse in a solvent to ensure clean development. The effect of the solvent action in the resist geometries must, however, be included in process optimization.

In general, development times should be long enough to allow precise time control, i.e., 20 ± 1 sec, because shorter times will reduce the process latitude and increase the margin of error in developed geometries.

Another aspect of calculating the proper time is the wetting ability of the developer. Some photoresist developers do not readily wet the wafer or mask surface, and variability in wetting will mean variability in the amount of development for a given time cycle. In some cases, wetting agents must be added to commercially available products to keep development times at a minimum.

In summary, development time is a critical parameter that must be closely controlled in any IC lithography process.

Temperature

Developer temperature is another critical imaging parameter in photoresist processes. The chemical action of the active ingredient in the developer is a direct function of developer temperature, and varying temperature will produce a change in resist image size. In that respect, developer temperature is really more critical than developer time.

Developers for positive and negative resists have varying degrees of activity according to their chemistry. Some metal-free developers for positive resists, for example, operate more efficiently at below ambient temperatures, whereas most developers are optimized for about 21°C. Thus each developer used should be tested at several temperatures to determine its point of best differential solubility, i.e., the temperature at which there is the greatest difference between the rate of removal and the rate of attack on the resist image.

Increasing developer temperature increases the rate of attack on the resist image. Figure 9.14 shows the relationship between line width change and developer temperature for various exposure settings. A test of this type should be run for any resist-developer combination so that predictable geometry changes are known and developer temperature control can be properly specified. In Fig. 9.14 the line width change is ¼ μm for a developer-temperature shift of ±2°F.

Most resist suppliers specify a developer temperature of 21 ± 1°C,

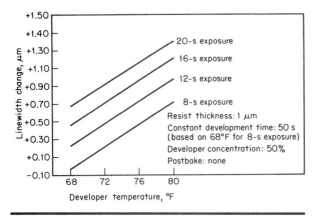

Figure 9.14 Developer temperature and exposure versus line width change.[3]

but most IC manufacturers prefer to use a slightly higher temperature, usually about 23°C. There are at least two good reasons for the difference. First, the higher temperature provides an increase of a few percent in effective photospeed, and increasing exposure throughput is always a major objective as long as it does not seriously affect image and etched resolution, which the 2°C will generally not do. Second, it is easier to heat and control a developer at 23°C, or above clean-room ambient, than it is to cool and heat the same bath and keep it at 21°C, or at clean-room ambient. The main precaution is that certain developers, especially organic metal-free types, may perform more efficiently at subambient temperatures because their differential dissolution ratios will be poor at elevated temperatures.

The majority of wafers are developed in automatic or semiautomatic equipment by utilizing nozzles for spray or dispense application. When aspirated spray developing is used, a drop in developer temperature of up to 8°C can occur. In the pressure-dispense spray, the evaporative cooling effect is less—about 2 to 3°C. In both cases, a heated nozzle may be available as an option to compensate for this effect. Some spray equipment manufacturers use a combination of aspirated spray and pressure-dispense developer application. The pressure-dispense system, in which the developer solution is pumped out in a stream to form a puddle on the resist before or during wafer spin, has the advantage of a small amount of evaporative cooling. Aspirated spray developing, in which the developer is broken up into a fine mistlike spray, provides increased uniformity of development at the cost of increased evaporative cooling and reduced developing action. Thus a combination of heated nozzles, pressure dispense, and aspirated spray (modified nozzles) balances the overall need for uniform developing application at a suitable temperature. This compromise leans in the direction of the aspirated spray because less solution is likely to run down the motor shaft and cause corrosion. In addition, the yield per gallon of solution is greater.

In general, a high degree of developer temperature control is required, especially in ICs with geometries at sub-half-micron levels. In those applications, ±0.1°C is suggested. For 2-μm geometries and larger, ±0.5°C is adequate.

Concentration

Varying the concentration of positive-resist developers is a method of increasing exposure throughput, and this illustrates the role of concentration in photoresist developing. Figure 9.15 shows the variation in both resist window size and exposure requirements for three concentrations of AZ-351 developer used with AZ-1370 and AZ-1350J photoresists. Note the wide range of exposure times made possible by simply changing concentration. The most concentrated bath (3×, or 2 parts water to 1 part AZ-351 developer) provides the shortest exposure times, but note the sensitivity to window width dimensional change as a function of exposure time variation. The next concentration, a 4.5× bath, provides consideralby more developer latitude but also increases exposure time. The 6× bath gives the greatest latitude and the longest exposure time. These data are based on tests run with a Micralign projection printer, and they show that the greatest control of resist line geometries is achieved with the most dilute developers. Increased developer concentration can be used to increase exposure throughput as long as the loss of latitude is tolerable.

The effect of increased developer concentrations on the surface of the AZ-1300-type resist is illustrated in the three micrographs in Fig.

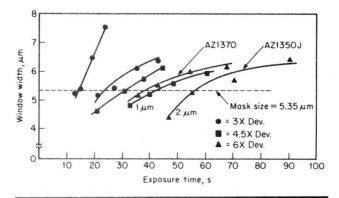

Figure 9.15 Exposure versus line width for three developer concentrations.[4]

9.16. The third micrograph (6× developer) displays a smooth finish on the resist; the 4.5× micrograph appears somewhat "grainy" because the developer has removed several hundred angstroms of unexposed resist and also has highlighted the standing waves on the resist image sidewalls; the 3× micrograph clearly shows both a highly textured surface and an observable reduction in photoresist thickness. Although the higher concentrations of positive-resist developer may remove over 1200 Å of unexposed resist, there generally remains enough resist to protect the wafer against pinholing in etching. Any pores or micropores that might have been created by the concentrated developer could possibly be flowed shut in the postbake step (Chap. 10).

Increased developer concentration is used with positive resists to reduce exposure time, and it has only one major negative side effect: loss of image dimension latitude. As long as IC geometries permit this variability, concentrated, or "high speed," developer makeups will be usable. A need for future developers is rapid dissolution rate of exposed resist with reduced attack on unexposed resist, or greater contrast, to get greater wafer throughput without sacrificing geometry control.

The constant change in developer concentration that occurs in im-

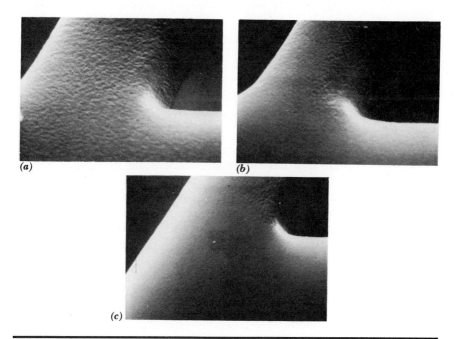

(a)   *(b)*

(c)

Figure 9.16 Developer concentration versus resist surface attack. *(a)* 3× developer. *(b)* 4.5× developer. *(c)* 6× developer.[4]

mersion developing makes spray or puddle-spin developing, in which a supply of fresh developer is maintained, more controllable. The in-line spray or puddle developing system permits variations in concentration, temperature, and pressure. Fresh solution is pumped directly onto the wafer surface.

Changes in positive-resist developer concentration are also caused by exposure to air. The alkaline developers often contain sodium or potassium compounds, which, when contacted by carbon dioxide in the air, combine to form a relatively inactive carbonate. Thus the active ingredient may be depleted if the developer is exposed to air. Immersion developer baths should be covered when not in use, and spray developer sumps can be blanketed with a layer of small-diameter ($\frac{1}{4}$ to $\frac{1}{2}$ in) polyethylene or polypropylene balls. These are also useful to blanket heated or solvent solution to minimize evaporation.

Agitation

Keeping a continual supply of fresh developer solution on the exposed photoresist surface should be the objective of the agitation method of developing. There are several techniques for meeting the objective, including movement of the resist-coated wafer, bubbling gas through the developer, spinning the wafer while dispensing a stream of developer on it, and spraying developer on the wafers, generally also while spinning. In immersion developing, the wafer movement creates eddies in the developer solution that carry away the solution as it dissolves the photoresist. Immersion developing with gentle part agitation is extremely reliable if a mechanical agitator is used. With manual systems, operators use different types of individual "strokes" when developing a cassette of exposed wafers, and these differences may affect critical line tolerances.

A variation of part agitation is bubbling nitrogen gas through the developer to create solution movement at the wafer surface. Nitrogen is used instead of air because carbon dioxide from the air will deplete the active ingredient in positive-resist developers. A bubble developing system generally includes a gas flow meter, time and temperature controllers, and a plate that rests at the bottom of the stainless steel tank and contains numerous holes through which gas flows. Dispensing developer solution onto the wafer while the wafer is spinning, called "pressure-dispense developing," is another way to provide developer solution movement. In some cases, the developer may be dispensed onto a stationary wafer so that it completely covers the surface. This is called puddle developing. A programmed delay of several seconds allows the developing action to begin, and wafer spin then commences. Another variation is to simply dispense the developer while the wafer is spinning, generally at 500 rpm. Spray developing, in which the solution is aspirated, provides very uniform developing action, but it requires temperature correction to compensate for evaporative cooling effects. Spray agitation breaks the developer up into many small droplets and provides a scrubbing action that efficiently removes the dissolving resist material. Spray and puddle spin agitation are production techniques, and both methods provide good wafer throughput.

Metal-free developers

Since most developers for positive resists contain free metal ions that may contaminate wafer surfaces if not properly rinsed away, organic sources of alkalinity are provided in what are commonly called "metal-free or low-metal-content developers." Since all major negative resists use solvent developers, metal ion contamination from developers is not an issue with negative resists. Early attempts at low-metal developers were mixtures of conventional alkaline inorganic solutions with isopropyl alcohol. These solutions reduced the differential solubility parameters for the positive resists, and better alternatives were researched. Several low-metal organic developers for positive resists are now available, and most have less than a few ppm of total metals. The lower the metal content, particularly the sodium level, the less concern there is for mobile ion entrapment within the matrix of the device that could alter its electrical parameters and cause rejects. Most low-metal developers work well in a spray mode, yet all tend to be aggressive on the unexposed positive resist; they remove from 10 to 20% of the coating during a typical cycle when the developer is used in the greatest dilution recommended. Metal-free and low-metal developers are important as device geometries get smaller. Moreover, the cost of rinsing to resistivity with expensive deionized water increases as better rinsing is required on the advanced devices. A typical qualitative spectrographic analysis of a low-metal-content developer is given in Table 9.1.

An alternative to a low-metal developer is a no-sodium developer, in which potassium hydroxide is substituted as the source of alkalinity.

TABLE 9.1 Typical Spectrographic
Analysis of Low-Metal-Content
Developer

Metal	Content ppm
Aluminum, silicon	0.3–3.0
Calcium, copper, sodium	0.1–1.0
Tin, titanium	0.03–0.3
Chrome, magnesium, lead	0.01–0.1
Gold, silver, barium, nickel	0.003–0.03
Potassium	Less than 1

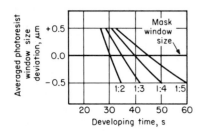

Figure 9.17 Photoresist window size deviation versus developing time.[5]

A developer of this sort will have about the same type of functional performance as a sodium-based developer. Metal ion contamination from photoresist developers will become more of a concern as IC density increases, and this should result in increased activity in formulating new and improved low-metal developers.

Developer Optimization

Concentration versus latitude

Although a large number of variables affect the development process, only a few are critical when an optimization scheme is established. For example, we have shown how softbake changes the developing time; yet softbake parameters are preestablished by exposure, CD control, latitude, and adhesion constraints and should not enter into developer optimization. The same is true of resist thickness, which is preestablished on the basis of resolution pinhole and process resistance needs. Even developer temperature is preset according to its point of best differential solubility with the resist. Agitation, another potential variable, is established by the configuration of developer equipment. The main parameters left to be optimized, then, are developer concentration and development time–exposure time relationships. The developer concentration can be selected on the basis of the required line geometry control. Figure 9.17 shows variation in resist line geometry as a function of developer concentration; as expected, the latitude increases as developer concentration decreases (at the cost of longer developing time).* The more critical the line tolerances, the more dilute the developer should be. Conversely, larger geometries with larger tolerances can be processed through concentrated or

---

*Process conditions: 1-μm-thick AZ-1370 photoresist; softbake, 90°C for 20 min; exposure, 5 sec at 10 MW/cm² with a mercury vapor lamp. Development conditions: immersion at 20°C in AZ-351 made up as shown in the figure.

high-speed developers. Thus, in optimizing the developer step, one should make several different dilutions and plot the tradeoff between line control and developing time for a fixed exposure time. There usually are some equipment constraints that establish a maximum practical time for economic wafer throughput, and they will dictate the maximum dilution tolerable. The maximum concentration is determined not only by resist geometry control but also by differential solubility. At some point, developer concentration is such that attack rate and damage to the resist image are more significant than the line control. For that reason, more dilution is required.

Development time versus latitude

Having established concentration on the basis of acceptable wafer throughput, line control capability, and resist image integrity, the relationship of development time to exposure time should be plotted. This is the most critical tradeoff of all, since exposure throughput is nearly always given priority. However, there is, as mentioned, also a requirement for wafer developer throughput. Process operators should plot line deviation versus development time for several exposure times. Balancing exposure time and developing time against line width change essentially completes the developer optimization process.

Selecting the exposure and developing times presupposes the ability to control the time itself. For example, an exposure time of less than 8 sec may create a mechanical control problem wherein the accuracy of the equipment in providing identical dosages may be questionable. The same is true of developing times shorter than 10 sec; times that short are subject to image size change as a function of delays between the developer and rinse cycles, unless they are overlapped. In many cases, the developing parameters are set entirely as a result of production requirements by first determining throughput and then "backing into" the other parameters. As long as attention is ultimately given to the variables that are yield-related, the ordering or sequence of steps in process optimization is somewhat arbitrary. Optimization should build enough latitude into a process to permit reasonable excursions in time, temperature, and concentration without adversely affecting CD control.

Developer Equipment

Batch type

Equipment for developing resists has become increasingly sophisticated in order to meet the requirements of better process control. Developing equipment for batch processing was the first type used, originally with immersion agitation. A number of companies still use immersion batch developing, but commercially available equipment is very scarce and many companies develop their own simple designs. Immersion batch-developing systems have been largely displaced by in-line spray or puddle-spin equipment. In-line filters are used to remove any aggregates or particulates from the developer bath before they contaminate the wafers or block the nozzle. An example of one of these resist particles is shown in Fig. 9.18. These "platelets" of sheared-off sections of resist films are caused by poor softbaking, overdeveloping, or partial exposure at a point somewhere beneath the surface of the resist film, where interfering light waves create an exposure maximum (or standing-wave node). In addition, since defects are a function of the amount of wafer handling, any system that can develop, rinse, and dry in the same physical chamber is preferred to one in which wafer cassettes are transferred between these operations.

In-line type

In-line track developing removes the concern over variation in developer strength by using fresh solution for each wafer. Consumption of developer will increase, but the added cost is small relative to the assurance of uniform solution strength. In-line developer tracks typically provide temperature control to 0.2°C by using heated spray heads. Nitrogen is used to atomize the developer in the nozzle as well as create a blanket over the solution in the sump to protect it from carbon dioxide absorption. Developer flow rate, time, and temperature

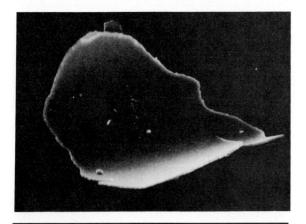

Figure 9.18 Micrograph of a positive-resist platelet.

are controlled by a microcomputer, as are the cycle parameters and wafer transport. Typical output of a track system may vary considerably depending on resist thickness, developer type, and other parameters, but it is generally about 100 wafers per hour per track. A developer track system is shown in Fig. 9.19. In-line systems can provide pressure dispense, atomized spray, puddle spin, or a combination of those forms of developer solution movement. Occasionally, mechanical failures occur in process or wafers are inadvertently moved against each other or into portions of the track.

Development Trends

Smaller IC geometries have created a greater dependence on metal-free or low-metal developers for positive resists. This has resulted in increased usage of low-metal positive-resist developers in spray systems, which eliminate salt formation or "salt-caking" on nozzles. Metal-free developers generally attack unexposed positive resists more readily than the standard inorganic developers.

Smaller geometries have created the need for increased control of the developing process and equipment, especially for positive resists.

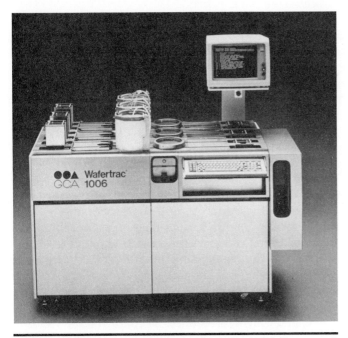

Figure 9.19 Track developing system. (*Courtesy of GCA Corp., a unit of General Signal.*)

Greater temperature and bath concentration control are among the areas getting the most attention.

The use of oxygen plasma as a developing medium offers at least a reduction in wafer handling if wafers can be kept in a plasma chamber through the developing, etching, and resist removal stages.

Plasma-developing resists are still being researched and developed, and it may be some time before we can truly assess their capability for production use.

Computer Modeling of the Development Process

Computer modeling and process simulation have been applied to major microlithography process steps ranging from softbake through expose, develop, and etch. Each step has a sizable number of variables each of which contributes to the final IC dimension. Dimensional control is the major component of manufacturability and yield in IC fabrication. The need for analytical models that can first predict and then provide a means of controlling the lithography process steps is a critical one. In the following paragraphs, a line width control model for the developing process step is described.

Figure 9.20 shows the aerial image profile of a 1-mm space (top half) and the development contours resulting from gradual dissolution of the exposed resist in the aqueous-based developer. The resist used is AZ-1370, and the developer is AZ312 MIF developer diluted 1:1 with deionized water. The intensity profiles and resultant developer contours are different for variously sized geometries. Each profile has a characteristic shape determined by the proximity (adjacency) effects in exposure and development, reflections determined by the substrate, light, resist solubility rate, standing-wave intensity, developer agitation conditions, and many other process variables.

The development model incorporates, hopefully, all of the variables that contribute to line width in this step, including parameters outside the developing step. For example, softbake temperature must be part of the calculation, since the log solubility rate is roughly proportional to the reciprocal of the bake temperature. All of the rates of change are incorporated into a model, and then on-line experiments are carried out to evaluate the model against actual process conditions. Once verified, the model can be used as part of the process of setting specifications and control procedures to enable better yields to be realized on existing products and controllable process conditions for

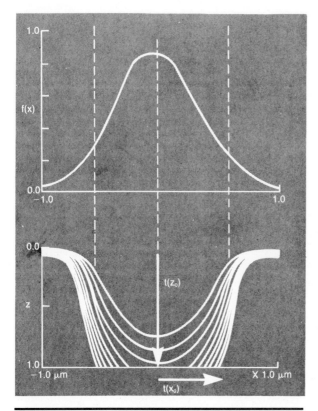

Figure 9.20 Sample simulations of a 1-μm aerial image (top half) development contours (bottom half). (*Courtesy of M. P. C. Watts, AZ Photoresist Products.*)

new devices with increased component and line width density and resolution. Computer modeling and process simulation technology are slowly finding their way into production processes, where their benefits can be realized.

REFERENCES

1. W. S. DeForest, *Photoresist: Materials and Processes,* McGraw-Hill, New York, 1975, p. 53.
2. K. L. Konnerth and F. H. Dill, "In-situ Measurement of Dielectric Thickness during Etching or Developing Processes," *IEEE Trans.,* vol. 22, 1975, p. 440.
3. E. B. Hryhorenko, "A Positive Approach to Resist Process Characterization for Line-width Control," *Kodak Interface Conf. Proc.,* 1980.
4. D. J. Elliott, "Increasing the Functional Speed of Positive Photoresist," *Solid State Tech.,* September 1977.
5. Technical Manual, Shipley Company, Newton, Mass., 1980.

6. Technical Literature on the FSI Single Cassette Positive Photoresist Developing System, FSI Corporation, Chaska, Minn., 1978.
7. Technical Data Sheets on System 4000 Developing System, Kasper Instruments, Sunnyvale, Calif., 1980.
8. F. H. Dill, W. P. Hornberger, P. S. Hague, and J. M. Shaw, "Characterization of Positive Photoresist," *IEEE Trans.*, vol. 22, no. 7, 1975, p. 445.
9. Castillo, Shipley Company, private communication.
10. V. Marriott, "High Resolution Positive Resist Developers: A Technique for Functional Evaluation and Process Optimization," *SPIE Proc.*, Spring 1984, paper 394–18.
11. R. Leonard, G. Sim, and R. Weiss, "Automated In-Line Puddle Development of Positive Photoresists," *Solid State Tech.*, June 1981, p. 99.
12. Wafertrac Application Note, GCA Corporation, 1982.
13. A. McCullough and S. Gringle, "Resist Characterization Using Multichannel Development Rate Monitor," Perkin-Elmer Corporation, Norwalk, Conn., 1984.

10

Postbaking

Postbaking is used as a postdevelopment step to increase the resistance of the resist to the etching process. Baking resists after developing or exposure is also used to increase the ion implant or metallization resistance of resists. Postbaking often causes resists to flow, and plastic flow is used in some processes to seal micropinholes in the resist coating, thereby reducing the incidence of pinholes in oxide after etching. In short, "postbaking," or "hardbaking," is a process step wherein the photoresist, x-ray, or electron-resist images are thermally treated to improve their resistance to a subsequent process step. Deep-uv curing is another postdevelopment treatment to stabilize resists in thermal or corrosive etch environments. In this chapter we will examine the physical and chemical changes that resists undergo in postbaking, the establishment of postbake parameters, and the effects of postbaking on resist removal and etchant undercutting.

Postbaking Mechanisms

Physical changes

Many positive and negative resists contain thermoplastic resins that flow at various temperatures. In some cases, multiple resin system resists will have a flow range wherein maximum movement of the resist images will occur. The plastic flow range or glass transition temperature should be determined for any resist used in IC manufacturing, and several tests will provide these data. Differential thermal analysis (DTA) is the most common way of determining the plastic flow properties of a resist. The data from DTA will show the temperatures at which resist images distort and how much movement occurs per degree Celsius. Such data can be generated for an individual resin or for a resin system.

Another way of measuring thermal plastic flow in resist images, as it would occur during postbaking or any step in which the resist is subjected to high temperature, is through SEM analysis. In this case, the resist images are coated in an evaporator with a thin (approximately 1000 Å) film of gold or a palladium-gold alloy. Resist images may be looked at with an SEM without the metal coating, but the pictures taken will not have the same detail or contrast. TEM analysis also provides a great deal of detail from samples, but sample preparation is more complex. Preparing the sample for this type of analysis involves scribing and breaking the wafer at a 90° angle to a series of long lines (resist images) and spaces. This will render a sample specimen with several different cross sections of resist lines on a small portion of the wafer.

Nitrogen freezing can be used to make negative resist more brittle (before the wafer is broken) and provide a clean break of resist and wafer. Positive resists are generally brittle enough without freezing and will break cleanly with the wafer. Samples postbaked at various temperatures and prepared in this way are shown in Fig. 10.1. The figure shows 1-μm-thick images in positive (AZ-1370) photoresist, and each has been postbaked at an increasingly higher temperature to see the effects of temperature on image distortion. Note the increasingly rounder corners produced by higher temperatures, forming a mushroom shape over the etched silicon dioxide (1-μm-thick) layer.

The main concern over plastic flow in postbaking is its effect on the final critical dimensions after etching. Since most of the dimensional change occurs at the top or surface of the resist image, it is of little concern. Some studies have indicated that this rounding does change the wetting characteristics of the resist in an etchant.

One aspect of plastic movement that may be of concern is shrinkage of the resist at the resist-oxide interface. SEM studies have shown that, with positive resists, a small amount of shrinkage which might affect critical dimension tolerances takes place. However, since higher postbake temperatures, which produce increased shrinkage, also provide reduced undercutting, the two may have a cancellation effect relative to dimensional changes.

Chemical changes

Postbaking of resist images inevitably removes some residual solvent left after softbaking. Although the amount is generally less than 3 to 4%, the more complete drying adds to the process resistance of the resist images and thereby enhances performance. Complete solvent removal in softbake is especially important if wafers are then placed in

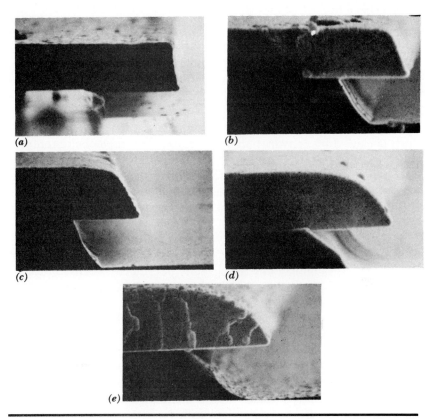

Figure 10.1   Postbake temperature versus thermal flow (1350J): (*a*) 100°C; (*b*) 110°C; (*c*) 120°C; (*d*) 130°C; (*e*) 140°C.

an ion implanter, sputter etcher, or any vacuum environment in which the presence of solvent in the film could result in solvent burst effects in the image. Heat generated in these processing environments drives remaining solvents from the film. If solvent retention is considerable (greater than 6%), distortion of the image may result. Solvent "popping" or "burst" are terms often used to describe this phenomenon.

Another physical change that occurs is the buildup of stress in the film. Tensile stress in resist films changes according to the level of postbake. Deckert reported a tensile stress of approximately 10^8 dyn/cm^2 in Waycoat IC and Shipley AZ-1350J films prior to postbake. After a 20-min hotplate bake at 120°C, the samples were measured again and found to have at least twice as much tensile stress. After a 30-min delay at room temperature, the samples were measured again and found to have about the same tensile stress as before the postbake. Tensile stress has been implicated as a cause of resist lifting during silicon dioxide etching.

The chemical changes occurring among the resol-type polymers, as reported by Jinno et al.,[1] are shown in Fig. 10.2. These reactions confirm empirical studies that show a shrinkage of the photoresist images at postbake temperatures above 100°C. The conventional positive-working resists exhibit thermal plastic deformation at about 140°C, which is caused by dehydration of the resins. When postbake temperatures reached 200°C, line patterns underwent gradual shrinkage, which was caused by a change from resol to novolak polymers and a decrease in resin volume. The dimensional change in the OFPR-2 resist images is plotted as a function of temperature in Fig. 10.3.

Resist Adhesion Factors

Photoresist adhesion is a key lithography parameter because of its direct effect on the dimensions of etched images. There are many factors that influence resist adhesion and its relationship to etchant undercutting, and there are factors other than resist adhesion that directly affect etchant undercutting. Since resist adhesion is generally measured as a function of lateral undercutting during etching, undesirable undercutting is almost always blamed on adhesion failure in the resist. Dry etching with an anisotropic etch system permits low hardbakes and greatly reduces lateral etching. Good priming and dehydration baking result in vertical etch profiles and easy resist re-

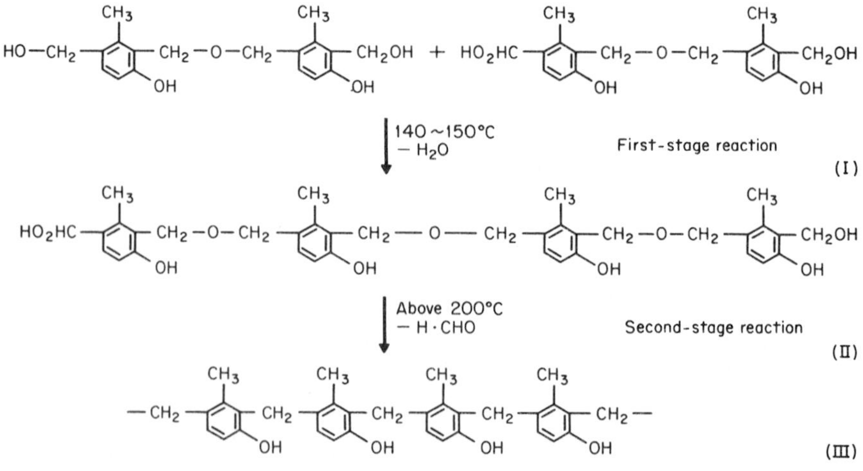

Figure 10.2   Reaction model of base resin polymers in postbaking.[1]

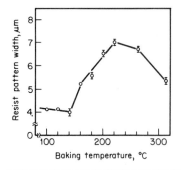

Figure 10.3 Postbake temperature versus line width for OFPR-2.[1]

moval. In some cases, nonresist factors may be the causes of undercutting. For example, etchant composition can be varied, and a variety of undercut profiles can be generated to correspond with the compositions. The question is: Did the adhesion value of the resist change with etchant composition change, or was the change in slope or undercut due to varying attack rates of the etchant on the resist-substrate interface? Should adhesion be defined purely as a value of physical attraction or expanded to include relative chemical resistance to various etchant compositions? Table 10.1 addresses the subject from the standpoint of factors influencing the etchant undercut and methods to control the amount of undercut.

Postbake Optimization

Adhesion versus temperature

Postbake studies have shown that increasing postbake temperature does not always increase adhesion. Figure 10.4 shows that, for AZ-1350J positive resist, the maximum postbake temperature should be 170 to 180°C, beyond which point adhesion values are reduced presumably by rupturing of the bonds between the resist and the underlying silicon dioxide. A postbake of 120°C is considered optimum for positive-resist adhesion to silicon dioxide, and 150°C is optimum for aluminum with the same positive-type resists. Negative resists adhere better to oxides by nature (more chemical-bonding sites) and require lower bake temperatures for the same adhesion values. Some positive-resist formulations have markedly better oxide adhesion than other positive resists and can provide excellent etching results even without a primer. Primers are still used because they remove microparticulate contamination immediately before the photoresist is

TABLE 10.1 Factors Influencing Etchant Undercut and Methods of Control

Factors influencing degree of resist undercutting during etching	Associated problem	Suggested methods to reduce or control the amount of etchant undercutting
Substrate type:		
Highly doped surface	Rapid etch rate, high undercut rate	Dehydration baking; priming to bond resist stronger at substrate-resist interface; high-temperature postbake to increase adhesion energy; special resist with comparatively better adhesion to doped surfaces
Extremely smooth surface	Resist lifting because of inadequate bonding sites	Priming to increase bonding; dehydration baking; chemical microetch to improve mechanical-chemical bonding
Granular surface	Penetration of etchant along grain boundaries or surface topography variations	Conversion coatings or primers to retard etch penetration at resist-substrate interface; microetching to "level" surface topography
Resist type:		
High adhesion to oxides	Metal adhesion marginal	Modification of metal-etch composition; bake metal slightly to promote slight metal-oxide growth (i.e., on aluminum)
High adhesion to metals	Oxide adhesion marginal	Dehydration bake; priming
Substrate cleanness or surface condition	Particulate contaminants under resist film causing high stress in resist film. Contamination deposited with metal or oxides during evaporation, sputtering or CVD. Moisture contamination	Brush, scrub, or spin clean with primer or solvent immediately before coating resist; clean evaporators; remove source of contamination; prebake at greater than 300°C for 10 min in air or nitrogen
Resist contamination	High incidence of resist dewetting	Replace resist; filter resist through (or let stand overnight) in diatomaceous earth or filter aid
Etch type or etch condition	Etchant contains chemicals that attack the resist. Excessive etch concentration, power, or temperature	Reduce the percentage or eliminate completely the acid in question (nitric acid attacks many positive photoresists); predetermine the optimum operating conditions of the etchant
Postbake conditions	Postbake too high: ruptures bonds between substrate and resist and increases the amount of undercut. Postbake insufficient to resist etchant and minimize amount of undercut. Brittleness in resist because of film stress	Reduce postbake to optimum level on the basis of undercut; postbake in air instead of nitrogen to derive oxidative resist cross-linking that increases etch resistance; also, postbake may be too low in temperature; run a matrix of bake times and temperatures and measure undercutting after etching

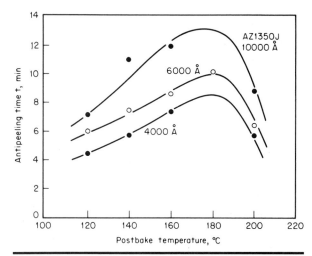

Figure 10.4 Antipeel time versus postbake temperature.[2]

applied. The optimal postbake or hardbake temperature is a function of the resist chemistry (available bonding sites).

Postbake versus chemical resistance. Postbake temperature also determines the amount of chemical resistance. The 180°C maximum postbake for AZ-1350J shown in Fig. 10.4 was determined by placing test samples in a mixture of hydrofluoric, nitric, and acetic acids after postbake and measuring the time until the resist pulled off. This is a good chemical resistance test; it shows the increased chemical resistance of thicker coatings because the etchant takes longer to permeate the thicker resist coating. Note that the cause of resist lifting in this case has little to do with the absolute adhesion of the resist to the substrate; rather it has to do with a chemical breakdown of the resist film and, finally, the attack of the etchant on the resist-oxide interface.

Adhesion versus postexposure baking

Some studies have shown that baking the resist after exposure and before development improves adhesion both in developing and in etching. This technique involves a second standard softbake of 5 to 10 min at 90°C and is especially useful on highly doped phosphorous oxide and polysilicon surfaces. In these cases, if wafers through the developer can be successfully processed without any photoresist lifting, plasma etching can have typically good results. Thus, when adhesion is extremely difficult with positive or negative resists, as it is on quartz, the added bake is useful to get images successfully through a developer. Plasma etching is recommended in these cases because wet etching will generally cause resist lifting. The postexposure bake not only bonds the resist more solidly to the oxide or semiconductor surface but also increases its chemical resistance to the developer.

Some low-metal organic developers are particularly aggressive toward positive resists and will remove up to 2000 to 3000 Å of unexposed resist and induce pinholes in the remaining coating thickness. The postexposure bake will reduce chemical attack by these developers and, if raised to the bottom end of the resist's melt temperature, may close micropores in the resist. This level of baking may affect the developing rate; if so, baking to close pinholes via resin flow should be delayed until the standard postbake.

Another application for postexposure baking is discussed by Walker.[3] The postexposure bake is described as eliminating the standing-wave latent image by causing a redistribution of the inhibitor or sensitizer in the exposed areas where the sine-wave pattern occurred. There is a shift of the inhibitor from the regions of low intensity to those of high intensity.

Postbake time versus thermal flow

Microwave baking of silicon wafers suggests a method of reducing the problem of thermal flow in postbaking. Since the thermal flow range of most resists, particularly positive resists, is relatively narrow, raising the temperature in postbake very rapidly would permit only a very short time for resist resins to be in the critical flow range. Microwave energy offers the possibility of driving wafers (for postbaking) through their glass transition temperatures before any significant movement or distortion takes place. Similarly, any baking energy source, such as an intense and highly focused exposure source, that will permit a very rapid temperature rise is desirable. The greatest advantage of these rapid-heating sources, and the reason for the popularity of microwave heating, is the speed at which they get the job done and permit increased wafer throughput. The only concern over microwave energy is the possible frequency-related damage it could do to devices in the wafer structure. Answers to these questions should be provided by the equipment manufacturers. Infrared energy is another means of achieving very rapid wafer temperature increase to minimize the bake duration at resist flow temperature ranges.

Postbake versus resist removal

Higher postbake temperatures mean more difficult removal. Negative resists are almost universally more difficult to remove than positive resists after any type of postbake, primarily because they result in a high concentration of cross-linked polymeric species. Positive resists can be baked up to 175 to 180°C and still be removed in hot aqueous alkaline solutions; negative resists require more aggressive resist strippers for the same level of postbake. The main point is to only postbake as high as is truly necessary to obtain the desired etch resistance.

Image Hardening

Several techniques besides postbaking have been used to render resist images more durable in subsequent wet- or dry-etching processes, including ion milling and reactive ion etching.

One approach is to subject the image to a 50-keV ion beam (arsenic) for a dose of 5×10^{15} cm^{-2}. (That is as referenced in U.S. Patent No. 4,201,800 (IBM), issued May 6, 1980.) A hardening solution of sodium,2-diazo,1-oxynaphthalene,4-sulfonate also may be used; the resist image is immersed in the solution for 2 to 5 min at room temperature.

Resist surface cross-linking also will increase the resistance of an image layer, and a short plasma gas exposure will cause the resist to link. All the preceding techniques greatly increase the solvent resistance of the image and permit a second resist layer to be applied without redissolving the first.

Prior to sputtering, with positive resists as masks, excess nitrogen (which causes outgassing problems in the sputtering chamber) is removed by exposing the resist for up to 8 times the minimum exposure time. This reacts the PAC and not only eliminates outgassing but also makes resist removal much easier.

Most of the preceding treatments will result in a tough resist image that requires the use of harsh removal techniques after etching. The safest and most aggressive dry removal technique is a 30-min exposure to oxygen plasma at 300 W in a reactive ion etch chamber.

Deep-uv resist stabilization

The density of pattern elements and the critical dimensional tolerances to which they must be held in advanced lithography make standard thermal movement of these images a major concern. Working at resolution levels above 2-μm resist, thermal flow in high-temperature processing was too small a percentage of the total dimension to enter the line width control equations. At the 1-μm level and below thermal movement, either shrinkage or plastic flow can account for up to 25% of the pattern element size and can exceed the total critical dimensional tolerance to stabilize the image after patterning and before postbaking or hardbaking and before subsequent elevated temperature processing.

One of the techniques used to stabilize resists, especially optical positive (novolak-based) resists, is deep-uv radiation. The technique is used for stabilizing not only patterns already formed but also the flood-exposed planarizing layers of PMMA through a preformed mask, commonly referred to as the portable conformable mask process. Following the flood exposure, the wafers are developed in a PMMA developer.

Photostabilization provides the resist with greatly increased chemical resistance to plasma aluminum etches and other corrosive dry-etch environments including ion milling and sputter etching. Many of the chemical constituents in reactive ion etchers will attack positive resists.

The use of photostabilization permits high-temperature postbaking prior to etching. The ability to give the resist high-temperature (up to 200°C) postbakes (or hardbakes) without thermal distortion allows positive optical resists to be used in moderate- to high-energy direct ion implantation. Resist used as an implant mask without this treatment will crack badly and distort. Direct implant masking with a resist "stencil" eliminates an oxide-etching step, thereby reducing process complexity and increasing yield potential.

High-temperature hardbakes, because of photostabilization, are used to improve resist adhesion to such surfaces as quartz and highly doped glass. In addition high-temperature baking may be used prior to process steps that require *complete* resist desolvation. In standard softbaking, a small percentage of the resist solvent system stays in the film.

Microwave-powered deep-uv flood exposure systems are a common type of equipment for photostabilization resists. The basic components of a microwave-powered deep-uv system are a magnetron, a power supply, and an optics system. The operating principle is one of magnetron-generated microwaves heating a bulb filled with the appropriate elements which then emit the desired radiation. In the example shown in Fig. 10.5, the bulb is a 19-mm sphere of quartz. Once filled with argon, mercury, and plasma-stabilization additives, it can be heated. The magnetron, at 1400 W of output power, beams its microwaves directly into the bulb cavity and causes the gas mixture to heat. Heated argon gas causes the mercury to vaporize, thereby producing the intense uv radiation. Efficiencies of this method are high: Approximately 80% of the power is actually coupled to the bulb. The energy distribution is about 25% in the 200- to 400-nm region, 20% visible light (400 to 700 nm), and the balance radiated heat energy.

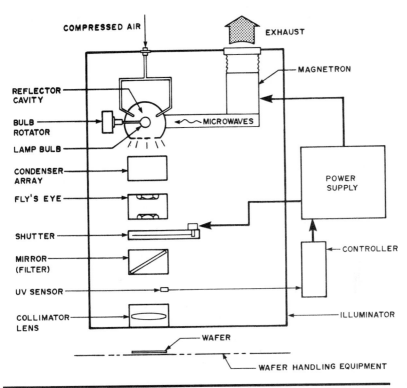

Figure 10.5 Diagram of a deep-uv flood exposure unit.[5] (*Courtesy of Fusion Semiconductor Systems.*)

The temperature of the bulb surface is about 850°C, air-cooled as the system runs.

One of the primary requirements of deep-uv flood exposure equipment is the constancy of output in both energy intensity and wavelength uniformity. The life of a microwave-energized bulb is typically approximately 500 hr, and both power and spectral output are reasonably stable throughout the bulb's life. One example of a test to measure this important parameter is to plot the spectral output both before and after sizable production running time. Figure 10.6 shows a typical deep-uv flood lamp output spectrum. In the range of 200 to 250 nm, the total radiated power is given at approximately 100+ W, or a conversion efficiency of approximately 10% if the bulb is rated at 1200 W. Cadmium or some other element is added to shift the spectral distribution of these lamps, depending on the resist system in use.

The balance of the system includes optics to smooth out intensity variations and provide collimation to eliminate lateral exposure under a mask. Note the condenser array that is used to first capture the

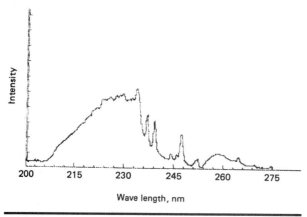

Figure 10.6 Spectral output of a deep-uv flood lamp.[5]

maximum amount of energy from the cavity surrounding the reflector. A fly's-eye integrator is used to make the beam uniform. A mirror in the optical path selectively removes, by attenuation, nonfunctional wavelengths outside the deep-uv region. For example, the attenuation with a PMMA resist would be set at 240-nm and longer wavelengths. Resists can be changed to suit different process needs, which brings about the need to change the mirror. The last element in the optical path is the collimator which projects the image energy onto the resist-coated (and sometimes patterned) wafer. The energy sensor can be set at the maximum absorption point of the resist in use to monitor the intensity of the system at the most critical wavelength and change or integrate exposure *time* so that a constant dose is received by the resist. Since the power is stable in this type of system, there is little concern for the shift in wavelength.

The optical performance of these systems is quite good: beam collimation half angles of approximately 40° and 5% energy intensity uniformity in the critical wavelength range. Overall, the energy measured at the wafer plane during deep-uv flooding is telecentric (equal angular ray distribution across a surface).

Ultraviolet stabilization with the flood exposure systems requires up to 3-min exposure depending upon the type of resist, thickness of the coating, degree of stabilization required, percent of photoactive compound in the resist, and process temperature. Some resist films can be sufficiently stabilized for further thermal processing in as little as 30 sec. A general test for whether or not a resist is stabilized is to achieve image stability after a 200°C, 30-min oven bake. Images exposed to deep-uv flood units long enough to pass this test will likely pass most other thermal process environments without harmful distortion.

The 200- to 300-nm uv energy achieves photostabilization by causing chain cross-linkages in the resist layer. The higher-molecular-weight groups formed by this cross-linking are much more resistant to the thermal energy of ion implantation and other high-energy and high-heat processes. The change in the flow of a resist layer by exposure to this high-energy deep-uv radiation is dramatic. Figure 10.7 shows examples of resist before and after deep-uv flood exposure, exposed to the same level of thermal energy. Note that the edges of the nondeep-uv flooded image are rough, probably from nonuniform flowing of the resin. The many advantages of deep-uv flooding, at 220 and 260 nm with mercury-zenon lamps and at other wavelengths with different sources, promise to establish deep-uv flooding as a key process tool in production VLSI fabrication. Improvements in intensity will be needed to increase wafer throughput and reduce exposure times.

Thermolysis

Baking conditions vary considerably in single and multilayer processes, and special mention of high-temperature baking conditions, called thermolysis, is appropriate in this section. The ranges in the following list are independent of chemical activity; they represent the three basic areas in which thermolysis reactions occur in positive novolak-type resists:

70–90°C	Desolvation reaction
70–150°C	PAC reactions
130°C and up	Resin reactions

About 8% nitrogen and 2% photoactive compound volatilizes at approximately 120°C, and the primary thermal reaction occurs in positive resists. The small amount of diazo-oxide PAC decomposition does not change the resist's properties significantly. These reactions form different products from those formed by standard photolytic reactions.

In vacuum or inert atmospheres in which water is not present, different reactions occur. In those cases, carboxylic acid products, which require water for formation, will not form. Without water present the ketene reacts with the next most likely compound, a novolak resin. An ester is the result of this reaction, and the solubility is decreased significantly.

Phenolic groups are responsible for the resist's solubility, and tying up or depleting them can provide undesirable process variations. Another product formed from the diazo-oxide, besides ketene and subse-

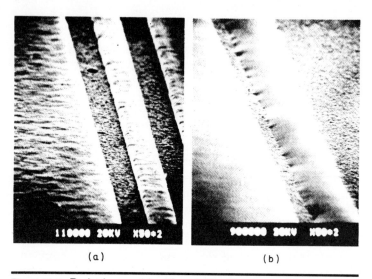

(a)   (b)

Figure 10.7 Resist images (a) with and (b) without use of deep-uv flooding after developing.[5] (*Courtesy of Fusion Semiconductor Systems.*)

quent groups such as carboxylic acid, is diazo dye. This is a low-temperature reaction that occurs in almost all diazo-sensitized resists. The dye inhibits light absorption in the resist and changes exposure parameters. Fortunately, only less than 1/10% dye is formed, which reduces critical impact of this chemical reaction. Plasma and removal resistance increase as the thermolysis reactions, which are time-and-temperature related, become more intense. High temperatures cause increased thermal flow, after which cross-linking occurs. Homolytic cleavage leads to a stabilized radical. Ideally, flow would occur *before* cross-linking, but that is not the case. Reactive sites are removed by the high-temperature baking, making decomposition more difficult because there are more bonds to break.

Surface effects, aside from the bulk effects cited above, also play a key role in resist performance. The cross-link density of the surface can change dramatically depending upon the oven conditions. These surface effects result in film nonuniformities. Polymerizing plasmas are a major source of surface reactions that cross-link the resist. Ion implant also causes severe hardening and cracking of resists. Ions decelerate in the resist and deposit energy as heat and cross-link the resist.

REFERENCES

1. K. Jinno, Y. Matsumoto, and T. Shinozaki, "Baking Characteristics of Positive Photoresists," *Photographic Sc. Eng.*, vol. 25, no. 5, 1977, p. 290.

2. H. Yanazawa, T. Matsuzawa, and N. Hashimoto, "Chemical Characterization of Photoresist to Silicon Dioxide Adhesion in Integrated Circuit Technology," Kodak Interface Seminar, 1977.
3. E. J. Walker, "Reduction of Photoresist Standing-Wave Effects by Post-Exposure Bake," *IEEE Trans.*, vol. 22, no. 7, 1975, p. 464.
4. Don Johnson, "Lithography Process Optimization," Microlithography Consulting, 1984.
5. J. Matthews, M. Ury, A. Birch, and M. Lashman, "Microlithography Techniques Using a Microwave Powered Deep UV Sources," *SPIE,* vol. 394, March 1983.

The etching process in IC fabrication transfers the pattern made in resist lithography to the underlying layer, which is typically a semiconductor film of silicon dioxide, aluminum, selicide, silicon nitride, or polysilicon. Etching is performed with wet or dry (gas plasma, ion, or reactive ion etch environment) methods. In this chapter we will discuss in detail the wet- and dry-etch technologies used for pattern transfer. Major emphasis is placed on accurate transfer of the resist pattern geometry so as to conserve IC area. Precise control of dimensions, along with operator safety ecology and cost efficiency is a key aspect of etch technology. Etching wafers with high uniformity and control has led to the increased use of dry-etch technology. Films less than 100 Å thick must be etched with high uniformity and repeatability. In the following sections all of the issues needed in providing appropriate etch processes to fabricate ULSI devices are discussed.

Wet Etching

Silicon etching

Semiconductor silicon, including single-crystal and polycrystalline material, is typically wet-etched in a mixture of nitric and hydrofluoric acids. The chemical dissolution of silicon occurs as nitric acid forms a layer of oxide (by oxidation) on the silicon and hydrofluoric acid dissolves the oxide away. In this way, the etchant gradually dissolves the silicon.

Nitric and hydrofluoric acid silicon etchants can be used to perform a variety of special silicon-etching applications depending on the ratio of the etch components, which can also include water or acetic acid as a diluent. Figure 11.1 shows etch rate plots for two ratios of the basic HNO_3-HF etchant mixture. Note the lower rate of etching with the

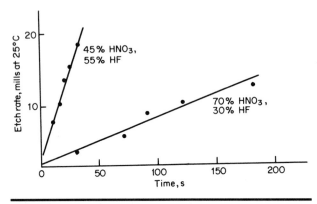

Figure 11.1 Etch rate versus time for two etchant compositions.[1]

etch rich in nitric acid, indicating that the rate-limiting reaction is the dissolution of the oxide by the relatively small percent of HF in the bath. The more equal ratio of 45% HNO_3 to 55% HF allows more efficient use or balance of the function of each component—nitric acid to form the oxide and hydrofluoric acid to dissolve it away. In addition to this basic silicon-etching mechanism, numerous submechanisms are at work in the etching process; they depend not only on the etchant-component ratios, but also on crystal orientation, dopant type and level, etch agitation, and so on. Examples of the influences of dopant type and agitation on etch rate are plotted in Fig. 11.2 for doped oxides of silicon. The addition of catalysts ($NaNO_2$), water, or acetic acid add other variables to etch performance.

The variety of permutations of silicon etchants and their behavior is considerable. Table 11.1 identifies some silicon-etching applications and the suggested etchant composition.

Controlled silicon etching.   The silicon etchants based on nitric and hydrofluoric acids are largely isotropic and, in most cases, do not allow a high degree of control over their operating ranges. Since there are many applications in which it is desirable to either make the etching stop at a specific point or etch a specific shape or structure in the silicon, controlled etching is of considerable interest.

Anisotropic silicon etching is possible with etchants based on the following:

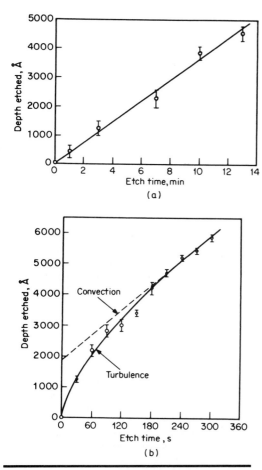

Figure 11.2 Etch time versus depth for doped oxides: (a) boron-doped oxide; (b) phosphorus-doped oxide.[2]

Primary component	Hydrazine
	Ethylene diamine
	Potassium hydroxide
Complexing agent	Isopropyl alcohol
	Catechol
Diluent	Water

The mechanism of silicon etching is through the formation of hydrous silica and subsequent dissolution by a complex that is formed. The etch rates of this type of etchant, as described by Declercq, Gerzberg, and Meinol, are greater along the ⟨100⟩plane than along the ⟨111⟩plane. In etching {100} oriented surfaces, the etch has a useful

TABLE 11.1   Silicon Etching Applications and Suggested Etchant Composition

Application	Etchant composition
Junction etching for device stability and improvement of electrical properties.	High nitric acid percentage will permit smooth surface after etching with minimal attack on metal; suggested etchant: 1 part HF to 7 parts HNO_3 to 1 part $HC_2H_3O_2$ (by weight).
p-Layer etch (to reach n layer with adjacent aluminum).	Galvanic effects from the aluminum are minimized by high nitric acid etch; selective etching of the p layer desired without attacking n layer: suggested etchant: 1 part HF to 8 parts HNO_3 to 1 part $HC_2H_3O_2$ (by weight).
Highlighting structural crystal detail.	Avoid smudging of crystal detail by using slow etch (minimize catalyst coupling); suggested etchant: 1 part HF to 1 part HNO_3 to $HC_2H_3O_2$ or nonautocatalytic etch ($NAOH \cdot H_2O$); or ½ part HF to 2 parts HNO_3 to 7½ parts acetic acid.
Chip thinning.	Close thickness-tolerance control essential; leave a smooth surface with square edges and corners; remove large amount of material; suggested etchant (two stage): (1) 4 parts HNO_3 to 4 parts HF to 2 parts $HC_2H_3O_2$; (2) 7 parts HNO_3 to 2 parts HF to 1 part $HC_2H_3O_2$.
Polycrystalline silicon.	Highly controlled rate, slow etching, residue-free or insoluble residue-free, and good etch-bath stability to keep rate fairly constant: 1. Fast etch: 7.5 mL ethylenediamine, 2.4 g pyrazine, 2.4 mL water; use at 115°C. 2. Slow etch: 6 g pyrazine per liter of ethylenediamine, 7.5 mL ethylenediamine, 1.2 g pyrazine, 1 mL water; use at 115°C. 3. Shipley Remover 1112A.

self-stopping property at the {111} intersection planes where the {100} bottom layer ends.

Hydrazine-water–based silicon etchants are recommended for several different applications, as shown in Table 11.2.

These etchants should be used at 100°C, and special care should be taken to ensure that only clean, dry wafers enter the etch. As the amount of water in these etches increases up to about 30%, so does the

TABLE 11.2   Several Applications for Hydrazine-Water–Based Etchants

Application or objective	Etch composition
V-Groove rings (high wall quality) V-MOS walls (high wall quality)	65% hydrazine (N_2H_4), 35% water (H_2O)
Two-layer etching (good bottom-layer quality) Sensors, electrodes (smooth surfaces)	70 to 80% hydrazine (N_2H_4), 20 to 30% water (H_2O)

amount of undercutting at a 20°C etch temperature. Water concentrations above 30% cause undercutting to increase again, making 30% H_2O to 70% N_2H_4 an optimum point. This relationship varies, however, as a function of temperature and concentration. There are a variety of optimum points for each balance of water and hydrazine at specific temperatures. From the work by Declercq et al., the curves for these relationships can be obtained.

The overall guidelines for etching silicon in hydrazine-water mixtures to obtain anisotropic quality structures are:

1. Minimize etch evaporation and obtain a constant balance of etch components by using in a system that monitors and replenishes automatically (controller). This technology is readily available for simple in-house control systems to satisfy the application.

2. Avoid drag-in of organic or inorganic surface contaminants and residues that may introduce additional etch complexing or disrupt the normal complexing reaction critical to anisotropic etch quality.

3. Preselect the specific H_2O to N_2H_4 balance according to the application, since each mixture will produce varying degrees of etch structure, surface finish, etch rate, and undercut profile.

4. Use a temperature-controlled bath to maintain close etch rate and profile control for all the possible etch combinations.

Silicon dioxide etching

The most commonly etched material in IC fabrication is silicon dioxide. Overcoming the problem of IC surface area consumed by undercutting (lateral etching) has been the challenge of both wet- and dry-etch technology. In the wet and dry etching of silicon dioxide, controlling the lateral etch is also important in all fabrication steps in which another material is to be deposited over the etched structure. When silicon dioxide, silicon nitride, polysilicon, or aluminum is applied over etched oxide structures, the layer must completely cover the corners of the silicon structures, and coverage of the corners is dependent on the slope of the etched silicon dioxide. The role of the silicon dioxide sidewall slope angle in obtaining adequate coverage of a subsequent deposited layer is critical and should be a highly controlled parameter in the IC fabrication process. Thus there are two etching objectives that tend to work at cross-purposes: (1) minimizing the amount of undercut to conserve IC real estate and (2) providing sufficient etched sidewall sloping for subsequent deposited layer protection. In the wet- and dry-etching sections of this chapter we will discuss the techniques necessary to achieve this balance.

The technology that has gradually changed silicon-etching technology from an art to a science, from relatively uncontrolled isotropy to controlled anisotropy, also has fed into silicon dioxide production etching techniques.

The basic etching control parameters for wet-etching doped and undoped silicon dioxide are

1. Concentration

2. Time

3. Temperature

4. Agitation

The concentration of hydrofluoric acid to ammonium fluoride may vary from 4:1 to 10:1, depending on the level of dopant and respective etch rate, since increasing levels of boron or phosphorous dopant increase the rate and, along with it, the undercut ratio. The most common mixture range for etching undoped silicon dioxide is 5:1 to 7:1, and most wafer fabrication processes stay close within this range to achieve an etch rate of 900 to 1500 Å/min. Thus the etch concentration feeds directly into the second control parameter, time, by setting the rate of silicon dioxide dissolution. Etch time is always so related to the layer of silicon dioxide thickness to be etched that reasonable amounts of overetching will not take the device out of specification. For example, etching a layer thickness of 1 μm with an etch concentration yielding a rate of 1000 Å/min gives process operators relative freedom for overetching; a 30- to 60-sec etch time past etch layer completion will remove an additional 500 to 700 Å laterally, an amount of control consistent with 4- to 5-μm etch geometries. Processes with 2- to 3-μm geometries naturally call for greater etch time control, which is partially achieved by reducing etch concentration. Figure 11.2 shows a relationship for silicon dioxide time versus etch depth. Note that the etch rate is generally not dependent on any specific point within the silicon dioxide layer but is relatively continuous throughout.

The temperature of the silicon dioxide etchant is probably the most critical parameter to control, since it is the most sensitive factor in relation to undercutting. In measuring the temperature dependence of undoped versus doped silicon dioxide, a change is seen in the shape of the two curves. Undoped silicon dioxide undergoes a geometric change in reaction rate for an algebraic change in temperature; doped oxides have an etch rate–temperature relationship that is nonlogarithmic and basically a function of the dopant concentration with the oxide. Figure 11.3 shows these relationships for both doped and undoped silicon dioxide.

The fourth major silicon dioxide etching parameter is solution agi-

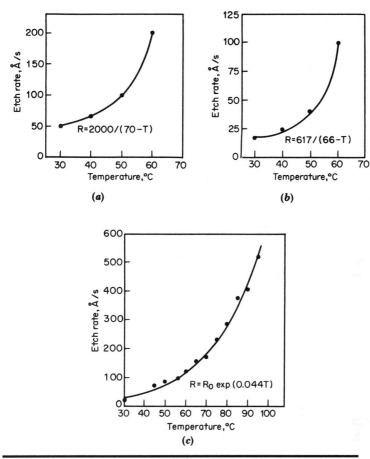

Figure 11.3   Etch rate versus temperature for doped and undoped silicon dioxide: (a) phosphorus-doped oxide; (b) boron-doped oxide; (c) undoped oxide.[2]

tation, and this also can produce significant changes in the way silicon dioxide etches. There are several different forms of wet silicon dioxide etchant agitation, including thermal convection, spray, mechanical (wafer), bubble, and ultrasonic agitation. The etch rates for these forms of agitation differ; they are plotted in Fig. 11.4. The lowest rate, as expected, is with thermal convection. Convection movement occurs naturally as a result of the heat gradients within the etch bath as well as the chemical potential differences that occur while the etch dissolves away the oxide. Although this is the simplest to implement of all wet immersion techniques, it also generates the greatest lateral etching or undercut. Part movement or mechanical agitation of the wafer increases the etch rate by about 25% and reduces the undercut

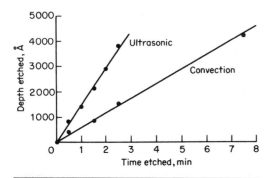

Figure 11.4 Etch rate versus agitation for undoped oxide.[2]

obtained in simple immersion with no artificial agitation. Mechanical agitation tends to deepen the etch solution and move it away from the wafer surface as the silicon dioxide is dissolved, thereby bringing fresh etchant into contact with the oxide.

Ultrasonic agitation, while producing almost twice the etch rate of thermal convection, has the undesirable side effect of also making the photoresist coating lift from the oxide. Therefore, it is seldom used. Bubble agitation, in which an inert gas is pumped through the solution to provide solution movement, produces about a 20% increase in etch rate compared with thermal convection, but it does not produce the uniform surface etch achieved with the spray and mechanical techniques. Consequently, like ultrasonic etching, it has not become a popular approach. Spray etching offers the solution movement advantages of all the other approaches, with primary improvement in reducing lateral etching or undercut. The etch rate with spray etching is considerably greater than with other forms of agitation as well.

In all the various forms of etch agitation, the nature of the silicon dioxide removal is the same. The etchant dissolves the oxide by creating numerous single cells or pits in the layer, and as dissolution progresses, the cell units increase in size until they overlap and combine to form a smaller number of individual cells. This type of material removal, called "domain etching," is similar to positive-resist developer dissolution in positive-resist layers. A replication of a wafer section illustrating the nature of domain etching is shown in Fig. 11.5. This phenomenon occurs not only with all the forms of agitation cited but also with all types of silicon oxides.

Wet etching of silicon dioxide therefore has four primary control parameters that must be characterized and monitored according to the needs of a specific IC process. The most sensitive of the parameters is temperature, followed by agitation, concentration, and time in declin-

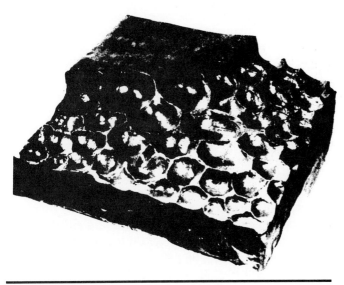

Figure 11.5   Domain nature of SiO$_2$ etching.[2]

ing order of control difficulty and effect on the etching result. The doped oxides etch according to the level of dopant; they etch most rapidly near the surface (phosphorus doping) or near the silicon oxide interface (boron doping). Undoped oxides etch uniformly through the layer thickness. All forms of etch agitation increase the etching rate, as does any increase in temperature. Agitation also results in reduced undercutting and smoother oxide surfaces and edges.

Slope control.   Changing the ratio of ammonium fluoride (NH$_4$F) to hydrofluoric acid (HF), along with temperature change, will permit slope control over the etched oxide. A 7:1 NH$_4$F/HF ratio will produce steep sidewalls at 25°C, whereas 30:1 ratios at 55°C will produce nearly flat, highly undercut profiles. These represent the extreme cases, and etch temperatures and dilution ratios in between will permit a variety of slope angles.

Silicon nitride etching

Silicon nitride (Si$_3$N$_4$) is wet-etched in solutions of either phosphoric acid (refluxing) or aqueous hydrofluoric acid. A major problem for etching silicon nitride in phosphoric acid is that the phosphoric system must be closed to permit the refluxing action to occur. U.S. Patent No. 3,859,222 describes the addition of fluoroboric acid, in various ratios, to the phosphoric acid, which eliminates not only the closed-system disadvantage but the other major problems of this etching approach as well. Notably, photoresists must be very highly postbaked to withstand the acid effect of lifting the resist. The 170 to 190°C phosphoric acid presents a safety problem; it will *not* etch silicon dioxide; and applications requiring etching of both layers mean a two-step etch.

The etchant described in U.S. Patent No. 3,859,222 utilizes a ratio of from 1 to 6 parts fluoroboric acid to 100 parts phosphoric acid at 105 ± 5°C. This etchant is not as highly corrosive to photoresists, is safer to work with because the operating temperature is lower, and it will etch through adjacent layers of silicon nitride and silicon dioxide. An etch ratio of 1:1 for each layer is obtained by using 100 parts phosphoric acid (by weight) to 1 part fluoroboric acid at 105°C. Increasing the etch rate of this etchant for silicon dioxide is done only by increasing the ratio of fluoroboric acid. Increasing the rate of etching of the silicon nitride is accomplished by increasing the temperature of the etchant up to 110°C. The rates of etching are thus variable, and a rate of 100 Å/min is a typical dissolution speed. The nitride thickness used in the patent description was 300 Å, and the oxide was about 1400 Å thick. Other sources of the fluoroborate are the fluoroborate salt of ammonia and sodium fluoroborate.

The mask for this etchant is either a negative or a positive photoresist postbaked at about 140 to 160°C (Chap. 10). During etching, the color changes in both the silicon nitride and the silicon dioxide layers will serve as a guide to the depth of the etch. See Chap. 5 for oxide color/thickness charts. When the oxide and nitride layers are etched to the silicon, complete etching is checked with an ohmmeter, followed by resist removal. In MNOS device processing, the etched structure would be metallized with aluminum alloyed to form a metal nitride–oxide–silicon structure. If a layer of nitride were etched to form a pad mask (dielectric protection as well as a moisture and dust barrier, an ideal application for silicon nitride because of its density), the silicon nitride etchant could not attack the aluminum metallization layer over which it is deposited. The high density of nitride makes it suitable in passivation layers or where high dielectric properties are required. In masking nitride, many processes use silicon dioxide as the etch mask when the etchant does not etch silicon dioxide. The oxide pattern is formed with a positive or negative resist; the oxide is etched; and the resist is removed. After nitride etching, the oxide etch mask can be removed by an etchant, such as buffered HF, that does not react with the nitride layer. An oxide etch mask (shown in Chap. 1) is used to etch materials whose etchants are not readily compatible with the resists that provide the required resolution. Polysilicon, single-crystal silicon, and some metals are masked first with an oxide, and the oxide is then removed or left on, depending on the process.

Aluminum etching

The aluminum metallization layers are generally etched in solutions of phosphoric acid, nitric acid, acetic acid, and water. The etching rate and quality will depend primarily on the etch composition and use parameters (temperature, agitation, and time), the type of resist used, and the impurities or alloyed metals in the predominantly aluminum layer. Aluminum etchants are composed of a large percentage of phosphoric acid and small amounts of the other materials mentioned. A typical etch composition is 80% phosphoric acid, 5% nitric acid, 5% acetic acid, and 10% water. The etchant is used in immersion or spray modes with uniform agitation at about 45°C. The etch rate of pure aluminum at this temperature is about 3000 Å/min and will, of course, vary as small amounts of copper or silicon are added. The copper is used to improve the ductility of the aluminum and prevent microcracks in the layer; silicon is added to promote alloying of the aluminum at the contact holes and thereby ensure good ohmic contact.

The mechanism for the chemical etching of aluminum is similar to that of silicon in that an oxide is created by an oxidant and is then reduced or dissolved by other components of the etchant. The nitric acid in aluminum etchant forms an aluminum oxide. The phosphoric acid and water dissolve the material, and the rate of dissolution increases as the percent of water is dropped and the bath is agitated. Thus, conversion of the aluminum to a type of oxide takes place simultaneously with a dissolution process, and the activities often compete with each other as their respective reactions progress.

Positive resists are generally used to mask aluminum metallization layers for two reasons: (1) their compatibility with the aluminum etchant is good, as is their adhesion to the aluminum (the two factors are interactive), and (2) the reflectivity of the aluminum often causes metal bridging when negative resists are used. "Metal bridging" is the shorting of two aluminum "traces" or circuit lines by an unwanted connecting piece or bridge formed by a negative resist. This occurs when light reflections during exposure of the resist polymerize resist between two intentionally exposed areas and the resist remains to protect the underlying metal from etching, thereby forming the bridge. Positive resists are not as sensitive as negative resists to the partial exposures caused by light reflection. Therefore, they are used on metal layers, especially when the line geometry spacing is close, making the potential bridging distance very close.

Spray etching.   Spray etching overcomes some of the isotropy problems in immersion etching by directing the etch at the wafer and dissolving through the oxide or metal layer at a much greater rate than the lateral etch rate. In recirculating sump systems, the sheer size of the sump permits almost complete uniformity of etch rate despite changes in concentration as production flows through the machine.

Dry Etching

Requirements for a dry-etch process

The lithography process sets the stage for etching by defining where structures are to be placed in the various semiconductor layers. The major requirement for the etch process is accurate and repeatable *pattern transfer* into the film layer beneath the resist mask. The pattern transfer occurs with some degree of etch bias or undercutting. Figure 11.6 shows the differences in etching openings (windows) and islands (doors) in a semiconductor layer. Maintaining dimensional control is a critical aspect of etching technology and is one of the reasons why dry etching is displacing wet etching.

Accurate and reproducible pattern transfer is only one of a long list of VLSI/ULSI requirements in dry-etching technology. The following list outlines these criteria for a typical production process.

1. *Fidelity/repeatability.* Reproduction of the mask dimensions must be accurate and repeatable. Process CD specifications identify mask-wafer bias needs; wafer-to-wafer reproducibility should be less than 3.5%.

2. *Uniformity.* Etch uniformity across the wafer must be high (typically less than 5%).

3. *Selectivity.* Etch selectivity across the wafer must be high; it varies from 7:1 up to 30:1. The higher the selectivity or contrast the better.

4. *Slope angle control.* Sidewall angle control (0 to 45°) should exist to permit controlled-slope etching. Various slope angles are needed for different steps of the process.

5. *Operator and environmental safety.* The etch process must be compatible with operator safety, plant safety, and environmental safety regulations. This includes all chemical, radiation, and mechanical hazards to both system users and the use environment.

6. *Radiation damage.* Radiation used in the etch process must not damage the devices.

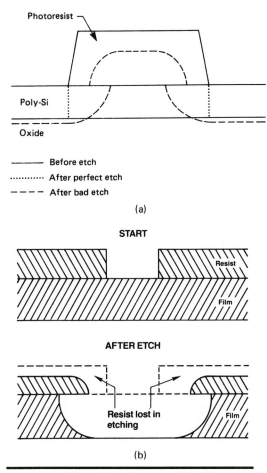

Figure 11.6 Wet-etch undercut profiles: dimensional control problem in (a) door etching[10] and (b) window etching.[11]

7. *Thermal distortion.* Thermal effects encountered must be low enough to be compatible with resist thermal flow properties as well as within wafer distortion limits to meet overall critical dimension specifications.

8. *Throughput and capital expense.* Wafer throughput must be high enough to satisfy the economic requirements and production demands of the overall process. An ROI analysis of the capital equipment costs will indicate needed throughput.

9. *Cleanliness.* Etch process equipment must be noncontaminating

to meet the specifications for "particles added per pass," the accepted method for determining cleanliness levels of systems. A minimum number of particles of a given size will be accepted.

10. *Mask removal.* The mask used to block the etch specie(s) must be completely and easily removed after etching. Ease of removal is determined by the defect impact and process time and cost needed to remove the etch mask.

Several mechanisms are used to etch away semiconductor films. They range from purely chemical means such as wet immersion etching in an acid bath (SiO_2 etching in buffered HF, for example) to purely physical means such as sputtering. A combination of physical and chemical means results in mechanisms that combine the features (and problems) associated with both approaches. Reactive ion etching is an example of this combination. Etching mechanisms can be categorized by the level of excitation energy, pressure of the system, and amount of physical or chemical activity taking place in the etch reaction. Table 11.3 indicates the features of various types of dry etching.

Plasma etching

Plasma etching uses molecular gases that contain one or more halogen atoms. In a plasma, species of these become reactive with semiconductor layers and form volatile compounds. The inert molecular gas in a glow discharge generates ions, radicals, and reactive atomic species that combine with oxides, silicon, and metals to produce a volatile by-product. Under an operating pressure of 10^{-1} to 10^{+1} torr, these species are absorbed on the semiconductor surface, the reaction occurs, and the by-products that are absorbed subsequently diffuse in

TABLE 11.3   Types of Dry Etching

	Plasma etching	Reactive ion etching	Sputter etching
Relative excitation energy	Low	Medium	High
Relative chance of radiation damage	Low	Medium	High
Relative selectivity	High	Medium	Low
Undercut, directionality	Isotropic	Directional from quasi-isotropic (slope) to anisotropic (vertical profile)	Directional from sloped to vertical sidewalls
Pressure	Greater than 100 mtorr	Approximately 100 mtorr	Less than 100 mtorr

the etch chamber environment. The relatively high pressure of plasma systems is offset by a relatively low excitation energy of the plasma (partially ionized ions, electrons, and neutrals). The glow discharge is simply a plasma that is created by the pressure, energy, and temperature conditions in the chamber.

Plasma etching includes such dry-etch subtechnologies as reactive ion etching, plasma etching, and reactive sputtering plasma-assisted etching. These descriptions address the chemical mechanism; still other names for dry etching, such as parallel-plate etching and barrel etching, address the equipment used. To this complexity of chemistry and equipment is added the third dimension of parameter variations, each of which will greatly influence the final etched result. These key parameters include configuration of the electrodes, etch rates used, slope angle profile of the etched film, pressure in the etch environment, and excitation energy. Parallel-plate reactors are used for most of the advanced IC device processes. The various types of films to be etched and the respective gas plasma used are listed in Table 11.4. Note that CF_4, $CF_4 + O_2$, and CCl_4 are able to etch most of the oxides, nitrides, silicides, and combinations of materials used. This presents an easier path to optimizing a resist process to be compatible with the plasma chemistry. The plasmas used are necessarily compatible with the resist in the sense that they react only slightly with the resist in comparison with their reactivity with the film being etched. There have always been problems with etch-resist compatibility in terms of undercutting and resist attack, but processes are then adjusted to overcome most of them. The reason for the trend to dry plasma etching is, therefore, not better resist compatibility, but better etch uniformity. The films being etched are much thinner every year, just as the pattern dimension widths are narrower. The allowable nonuniformity

TABLE 11.4   Common Gas Plasmas and Corresponding Materials to Etch

Plasma type	Materials to be etched
Silicon (Si)	CF_4, $CF_4 + O_2$, CCl_2F_2
Polysilicon (Si)	CF_4, $CF_4 + O_2$, $CF_4 + N_2$
Silicon nitride (Si_3N_4)	CF_4, CF_4O_2
Silicon dioxide (SiO_2)	CF_4, CF_4O_2, CCl_2F_2 (C_3F_8 (diode system), $CF_2F_6 + H_2$ (diode system), HF (selective)
Molybdenum (Mo) and molybdenum silicide ($MoSi_2$)	CF_4, CF_4O_2
Tungsten (W) and tungsten silicide (WSi_2)	CF_4, CF_4O_2
Gold (Au)	$C_2Cl_2F_4$
Titanium (Ti) and titanium silicide (TiS_2)	CF_4
Tantalum (Ta) and tantalum silicide ($TaSi_2$)	CF_4
Aluminum (Al) and alloys	CCl_4, $CCl_4 + Ar$, BCl_3

or percent overetching allowed thus decreases, and the need for selectivity between the material being etched and the underlying substrate increases.

Dry-etch uniformity

The early barrel plasma reactors were slightly better in uniformity overall, but across a single wafer the percent deviation in patterns attributable to the etch process was nearly the same as for wet etching. Only with the advent of the planar reactor did etch uniformity across a single wafer improve significantly from 15 to 20% to less than 10% while batch-to-batch uniformity improved from 15% to 5% average variation. The next technology advance to improve uniformity of wafer etching came with the use of reactive ion species in the etch process which, combined with the equipment configuration advances gained in moving to parallel-plate reactors from barrels, provided a new uniformity capability of 2 to 5% across a wafer, across a lot of wafers, and from one lot to another.

Classification of plasma gas reactions

Inside a plasma reaction environment the gases react to generate the etching species, which in turn determine the rate and nature of the etching process. Plasma processes are very diverse and differ according to two basic criteria: the nature of the discharge (inert or reactive) and the amount of bombardment. Inert gas discharges at low-energy ion bombardment are used for cleaning applications in which small amounts of wafer surface soils need to be knocked off the wafer. These applications are run with the wafer on the ground plane electrode.

The same inert gas discharge used with high-energy ion bombardment with the wafer on the target electrode is sputter etching. There are four basic categories of reactive gases: volatile gas surface products and involatile gas surface products, each at low and high ion bombardment levels. Volatile gas surface products at low ion bombardment levels constitute plasma etching and oxygen plasma ashing of organics, all with the sample at ground. At high ion bombardment levels, the process is called reactive sputter etching or reactive ion etching, all with the wafer on the target electrode.

The involatile gas surface product reactions at low ion energy levels (sample on ground plane) are called plasma anodization (O_2 plasma), and their counterparts at high ion energy levels are reactive sputtering reactions, again with the wafer on the target electrode. These general categories are organized in Table 11.5.

TABLE 11.5   Plasma Reaction Categories

	Low-energy ions	High-energy ions
Volatile reactive gas discharges	Plasma etching and ashing (O_2)	Reactive sputtering, reactive ion etching
Involatile reactive gas discharges	Anodization (O_2 plasma)	Reactive sputtering
Inert gas discharges	Cleaning applications (light soils)	Sputtering (heavier soils, very thin films)

Plasma equipment configurations

Plasma etch systems vary widely according to the type of reaction used. Earlier barrel etch configurations stacked wafers vertically and bled gas through the tube-shaped chamber with rf potential on top and bottom surfaces. Sleeves were added to protect the wafers from energetic ions and electrons and to equalize the activity of plasma reactions and increase etch uniformity. The planar plasma reactor uses two electrodes on either of two plates, depending on the energy level. Basic planar plasma has the rf potential opposite the sample; reactive sputtering and reactive ion etch configurations place the wafer on the target (rf) electrode. There are also etch configurations in which rf is applied to a gas mixture to extract special species while wafers are positioned downstream.

Wafers kept on the grounded electrodes are subjected to minimal radiation damage; placing them on the powered electrode involves much higher ion bombardment levels and an increasing potential for radiation damage to the device. In all of the various types of equipment approaches, the concept is basically the same: Subject the wafer to a field of chemically active species that will combine with the film to be removed and volatilize it away. Second, the role of the plasma in any etching chamber is to create energetic electrons and ions that will also react (physically) with the wafer surface. Many researchers have demonstrated the role of energetic ions as catalysts for more gas-surface chemical reactions, behavior that has served a primary role in reactive ion beam etching (RIBE). In standard plasma chemical reactions, without the benefit of the physical action of highly energized particles, the process is isotropic and lacks directionality. Highly energized ions create anisotropy and add new chemical reactions. The flux of energetic particles will also accelerate the rate of a standard plasma chemical reaction and thereby increase the wafer throughput. Energetic ion-enhanced plasma chemistry is technically what separates planar plasma etching processes from reactive ion etching.

Glow discharge

The glow discharge aspect of a plasma etch process is the glowing plasma of ions, electrons, and neutrals. Light electrons absorb more energy from the field than ions, and the resulting strong electron currents give rise to floating potential negative charges on electrically isolated objects of 5 to 50 V in excess of the plasma potential. In sputtering and reactive ion etching environments, it is the ion bombardment that plays a major role; in plasma etching, only relatively low intensity ions are present. Ion bombardment energy is largely a function of the sheath voltage developed between the surfaces exposed in the reactor and the plasma itself. The dc sheath voltage is formed on the electrodes of etch systems; it stems from the large rf electron currents that flow in that area. Thus the electrode area is the main variable to change when moving from low ion intensity to the large intense ion bombardment required in sputtering or ion milling. Ion milling and sputtering equipment arrive at high ion intensity by reducing the area of the target electrode considerably with respect to the grounded counterelectrode. In reactive ion etching, the amount of ion bombardment is still much greater than in conventional plasma etching. Planar electrodes in most systems are usually sizable, so that relatively high energy environments are generated even when the wafers are placed on the grounded electrode as they often are in RIE.

In planar electrode or barrel reactors, the amount of energy, in the form of energetic ions, will vary widely, but the function of the glow discharge remains constant. The discharge is a source of chemically reactive elements or molecules generated by the ionizing impact of electrons. Essentially, the discharge performs three functions:

1. It creates ions responsible for etching.

2. It controls energy and ion bombardment flux.

3. It generates active chemical species by dissociation under electron impact.

An example of how the glow discharge works in this regard is shown in the following equation, in which relatively inert CF_4 gas is energized by electrons and dissociated into the active fluorine species that will dissolve away semiconductor layers. The reaction is as follows:

$$CF_4 + e \cdot CF_3 + F + e$$

The process of dissociation occurs in the glow discharge, in which are produced the chemically active species that then move to combine with oxides, silicon, nitride, metals, and other layers being etched.

Other conditions (pressure, temperature, physical configuration, etc.) are then regulated to determine the outcome of the etch.

Plasma gases

Plasma reactions in semiconductor device fabrication have become widely used in many different parts of the process. Plasma cleaning, for example, has proved to be an excellent method for in-line, ultraclean processing as opposed to using wet chemical dips. Plasma surface cleaning on reworked wafers, using oxygen to strip monolayers of organic resists, is one application. Plasmas are, of course, widely used as a way to enhance the deposition of dielectrics, as in plasma-enhanced chemical vapor deposition (PECVD). Plasma chemistry is most popular, however, as the medium for dry-etching patterned layers of oxides, silicon nitride, polysilicon, aluminum and its alloys, and metal silicides.

The gases used in plasma etch reactions are part of a chemical family called halogenated hydrocarbons. These complex molecules are basically hydrocarbons that have had some or all of their hydrogen atoms replaced by halogens: fluorine (F), chlorine (Cl), bromine (Br), or iodine (I).

The number of halocarbons (shortened name for halogenated hydrocarbons) is considerable, and the number used in semiconductor processing increases as the number of dielectrics and alloys used in devices increases. Also, many mixtures of halocarbons, commonly called "azeotropes," are used in dry etching. One definition of an azeotrope is "a mixture of two or more halocarbons that exhibit equivalent liquid phase composition and equilibrium vapor." Having identical equilibrium plasma etching provides azeotropes with a certain behavioral predictability when the stochiometry, rates, and other aspects of plasma reactions are calculated. Plasma gas blends are commercialized by many of the dry-etch product manufacturers.

The formation of halocarbons from parent hydrocarbons, as mentioned above, is by simple replacement of hydrogen atoms with halogens. Figure 11.7 shows a couple of typical examples.

Increases in environmental legislation, concern for employee safety, and IC densities have all favored dry-etching technology. Actually, dry etching comprises several distinctly different technologies: (1) plasma etching, (2) reactive ion etching, (3) sputter etching (also called ion milling or ion beam etching), and (4) reactive ion beam etching. All these techniques provide higher resolution potential by overcoming the problem of isotropy discussed under "Wet Etching." Some lateral etching may take place with these dry techniques, but the amount is small. Another major benefit of dry etching is the elimina-

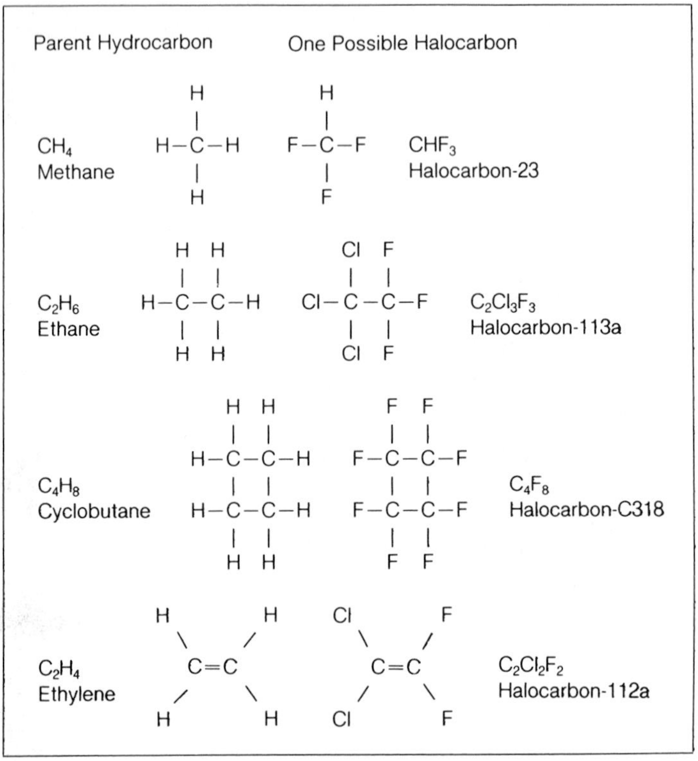

Figure 11.7   Halocarbon formation reactions.[12]

tion of spent acid disposal problems. Dry etchants do not attack resists as readily as wet etchants do, which allows the use of lower postbake temperatures and sometimes simplifies resist removal after etching. The uniformity of etching is good but not radically better than that of wet etching, and dry-etching equipment manufacturers have worked to improve this aspect of the process. In terms of overetch control, dry etching is superior, because the energy source can be shut off and etching ceases immediately; end-point detection provides a high degree of etch control. There are some remaining problems with dry-etching techniques, but the problems solved are far more significant; therefore, dry etching will continue to have an increasing share of IC wafer etching.

Barrel plasma etching

A barrel plasma etching system is composed of the following major components:

1. Radio frequency (rf) power source

2. Plasma etch chamber

3. Gas flow control system

4. Vacuum system (exhausts by-products of reaction and maintains vacuum)

A typical barrel plasma-etching operation involves the loading of the etch chamber with the wafers followed by vacuum pulldown. The gases are then bled into the chamber and the rf power is turned on, resulting in a glow discharge that produces the reactive etch specie or plasma. A "gas plasma" is simply a form of the specific gas mixture that contains atomic and ionized species when the rf energy couples with it. The plasma is really a breakdown process caused by the rf energy breaking the bonds of the supplied gas molecules into atoms of the same molecules. In the example of etching silicon dioxide, the CF_4 molecule supplies a fluorine atom, which, along with atomic oxygen, forms an OF etching specie. The etching reaction is as follows:

$$\begin{matrix} SiO_2 \\ SiO_2 \end{matrix} + F/O \xrightarrow{H, \text{ and } C} \begin{matrix} SiF_2\uparrow \\ SiOF_2\uparrow \\ Si_2OF_6\uparrow \end{matrix} + O_2\uparrow + F_2\uparrow + CO_2\uparrow + H_2O\uparrow$$

The vacuum level for this process is about 200 μtorr. A constant etch rate is maintained by metering in a predetermined amount of the etching species. All the volatile products of the reaction are removed by the vacuum sump. In the barrel (or tunnel) etching approach, the reactant etching species is random in motion and tends to produce isotropic etching. A diagram of the barrel etcher, with its external electrodes and the perforated metal cylinder, is shown in Fig. 11.8. The cylinder separates the plasma chamber into two regions; outside the cylinder is the discharge region where the ionization takes place. The cylinder wall protects the inside from ions that detract from etch uniformity and do not increase etch rate. Only free radicals of the ionized mixture enter the cylinder. The perforated cylinder permits each uniformity of $\pm 1\%$ across wafer surfaces. Barrel plasma etchers are relatively inexpensive and provide excellent wafer throughput compared with parallel-plate etchers.

Doped and undoped oxides of silicon etch at various rates in plasma as well as in wet etching. In Fig. 11.9, the etch rates of these oxides are shown as functions of rf power. Phosphorous oxide etches fastest,

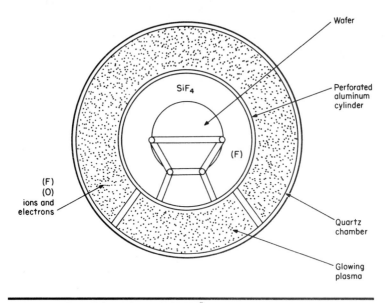

Figure 11.8 Barrel-type plasma reactor.[5]

followed by undoped oxide and boron-doped oxide. The rf power level also can be used to adjust the slope of the undercut as well as the rate of etching. In addition to doping level, etch rate is a function of the density of the material being etched; the denser materials have the lower etch rates.

Another component in the calculation of etch time in plasma processing is the exposed area to be etched. This calculation is plotted for silicon dioxide in Fig. 11.10. An exposed area of about 30% will result in an etch rate of 750 Å/min. These data are based on an 8-in barrel system using an rf power generator with a 500-W output.

Etching silicon nitride is especially attractive in a plasma medium, recalling the problems of hot phosphoric–wet silicon nitride etching. In addition to silicon nitride, oxynitride ($Si_xO_yN_z$) and nitride hydrides ($Si_xH_yN_z$) are found in nitride films. An active fluorine-based plasma is effective in removing all these species according to the following reaction:

$$
\begin{matrix}
Si_3N_4 \\
\text{or} \\
Si_xO_yN_z + F/O \rightarrow \\
\text{or} \\
Si_xH_yN_z
\end{matrix}
\qquad
\begin{matrix}
SiF_4\uparrow \\
SiOF_2\uparrow \\
Si_2OF_6\uparrow + \text{volatiles} \\
(N_2, O_2, F_2, CO_2, H_2O)
\end{matrix}
$$

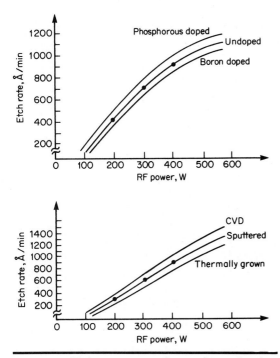

Figure 11.9 Etch rates versus rf power for semiconductor oxides.[6]

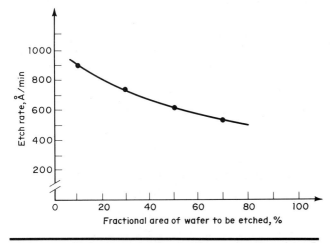

Figure 11.10 Etch rate versus fractional area of oxide opening.[6]

The binary mixture of oxygen and tetrafluoromethane (CF_4) in a simultaneously inductively coupled rf discharge is a patented technique. The preceding multipath approach involves parallel reactions occurring at once to produce the volatiles shown. The temperature of the wafers in plasma etching of silicon nitride with the perforated cylinder insert is about 100°C, and etch uniformity is ± 1%. The greatest gain in plasma etching compared with wet etching is reduction of process steps. The elimination of three individual process steps is possible, as shown in Fig. 11.11.

In summary, the barrel plasma etcher is characterized by a circular quartz or rectangular metal reaction chamber in which the etching specie is broken down on the reactor walls. An aluminum cylinder with perforated sides is commonly placed inside the reaction chamber to shunt the rf field and deactivate the atomic oxygen, ions, and electrons from the glowing plasma. The protection afforded by the cylinder or tunnel protects the wafers from edge heating caused by the ionizing radiation, but the etching species (atomic fluorine) passes easily through the holes in the tunnel to etch the oxide. Variables in the barrel plasma system are gas composition, rf power, wafer spacing, gas flow rate, pressure, uniformity of etch layer thickness, and wafer temperature. Turning the end wafers in the cassette inward will protect them from the larger amount of reactive etch specie in the plasma. Improved etch rate and uniformity are possible by putting fewer wafers in the cassette (spacing every other one) or loading wafers back to back in the holder. Placing wafers in an oven at about 100°C to preheat and then loading them into the barrel will improve etch uniformity. Despite all these techniques, barrel etching still poses uniformity problems that are solved only by overetching. But overetching can be used only to a certain resolution level, beyond which single wafer etching must be used. In a single-wafer etch system, process conditions can be optimized for each individual wafer. Table 11.6 gives relative etch rates for plasma barrel etching of several materials.

Planar plasma etching

Planar plasma etching is separated from barrel plasma etching in both configuration of equipment and chemistry of the etching reaction. Figure 11.12 shows a typical configuration of a planar plasma etcher. In the planar system, internal electrodes are used and the wafers lie directly in the plasma discharge, where they are continually exposed to the reaction products. A complex fluorocarbon gas (C_3F_8) provides a reversal of etch ratios compared with barrel plasma etchers. It is postulated that silicon is being etched by atomic fluorine and silicon dioxide and silicon nitride are etched by atomic fluorine *and*

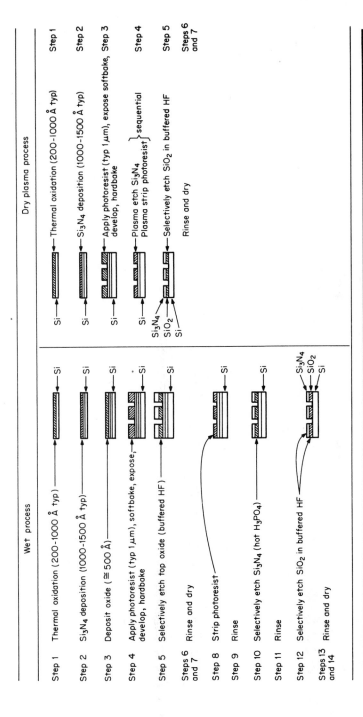

Figure 11.11 Wet-etching versus dry-etching process steps. (*Courtesy of LFE Corp.*)

TABLE 11.6   Barrel-Etching Rates

Material	Relative etch rate
Thermal silicon dioxide	1.0
Deposited silicon dioxide	2.0 to 3.0
Phosphorous-doped SiO_2	2.0 to 3.0 plus 10%
Boron-doped SiO_2	2.0 to 3.0 plus 10%
Si_3N_4	4.0
Oxygen-rich SiO_2	2.5
Silicon-rich Si_3N_4	5.0
Single-crystal silicon	10 to 12.0
Polysilicon	8 to 10.0
Amorphous silicon	8 to 10.0
Tantalum	5.0
Tungsten	10.0
Molybdenum	5.0
Titanium	0.8

intermediate species resulting from the breakdown of C_3F_8. These intermediates passivate and protect the silicon from the atomic fluorine and add selectivity to the etch. Since this same etching species will not produce the same reaction in a barrel plasma etcher, it is believed that the wafers, because they are in the center of the barrel etcher, do not participate in the intermediate reaction that wafers in a planar etcher do.

Minimum etch rates of 1000 Å/min can be achieved with the planar system for silicon dioxide, silicon nitride, polysilicon, and aluminum, and these rates are greater than those achievable with barrel systems. Etch uniformity in planar systems is improved through substrate movement and focusing techniques; it increases as pressure decreases. By having wafers transported through a parallel-plate system, increased uniformity is achieved in comparison with stationary barrel systems. A key application for planar and all single-wafer systems (RIE, RIBE) is removal of silicon nitride over silicon, and the thin

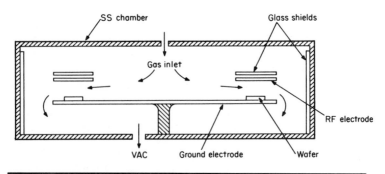

Figure 11.12   Planar-type plasma reactor.[5]

layer of silicon dioxide under the silicon nitride also must be removed. Planar systems permit removal up to the silicon, where etching stops, whereas barrel etching involves the use of a wet-etch dip to strip the silicon dioxide.

The slopes of etched sidewalls are generally near vertical with planar systems, especially if the resist process is adjusted to provide proper slopes before etching. Ideally, near-vertical resist images will provide enough protection from lateral etching to yield near-vertical (80°) slopes on etched oxide-metal sidewalls.

Positive resists offer special benefits in planar etching, wherein resist attack is somewhat greater than in barrel etching. The C_4F_8 and CF_4 plus 8% O_2 gases will penetrate a 6000-Å film of negative resist in less than 2 min, and thicker positive-resist coatings with better resolution can be used to mask the etching action. Thin resist coatings, typically needed with negative resists to provide high resolution, break down at high current levels and low pressures. Only in applications in which thin layers (2000 to 3000 Å) of silicon nitride or other materials are etched can a negative or thin coating of positive resist be recommended. When contact holes are to be etched in thick (> 1.2 μm) phosphorous vapor-deposited oxide (VAPOX), a thick, high-resolution resist is necessary. In summary, planar plasma etching is used for all major IC etch applications.

Since nonuniform wafer-heating effects are reduced in single-wafer systems compared with barrel etching, etch uniformity is better. Etch rates of all commonly used semiconductor materials are high (especially silicon dioxide). The etchant (C_4F_8) will nearly stop when it reaches the silicon layer, making this technique ideal for contact hole etching in thick silicon dioxide or phosphorous VAPOX. In the SEM of a dry-etched oxide with high aspect ratios, shown in Fig. 11.13, a resist (positive type 1350H) is imaged over the etched steps.

Aluminum etching. Dry etching of aluminum is performed typically with gases containing chlorine, which include $SiCl_4$, BCl_3, and CCl_4 or combinations of them with Cl_2. A freshly coated aluminum film, if not passivated with aluminum oxide, will react with chlorine to form a volatile $AlCl_3$. The initial stage of etching will result in the removal of a thin (50-Å) layer of native oxide if the etch provides a short sputtering or chemical reduction mechanism, since the oxide will not be reactive with Cl or Cl_2. The other aspect of initiating an aluminum etch is water removal, because moisture can retard the etching action. Wafers should be completely dry and preferably be sent through a vacuum load lock before entering the etch chamber to ensure complete dryness. Etching of aluminum begins only after complete removal of oxide and moisture.

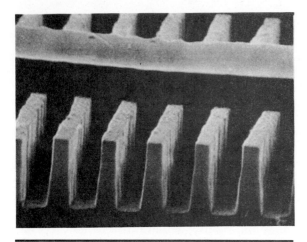

Figure 11.13   Dry-etched silicon dioxide, 0.5 μm wide and 1.8 μm thick, with 1350H positive resist imaged over steps. The resist in the background is 2.5 μm thick. (*Courtesy of H. Lehmann, RCA, Zurich*)

Polymer formation is another problem to avoid, usually by using recombinant gas mixtures of BCl_3-Cl_2 or CCl_4-Cl_2. The BCl_3-Cl_2 mixture has proved best for elimination of polymer deposits in the etch system. These gases provide highly anisotropic etch profiles with reactive ion or plasma etch systems. Anisotropic etching is critical in metallization pattern etching because of the need for high density and close conductor path spacing in ULSI devices. Even with multilayer metal devices, aluminum conductor path spacing requires nearly vertical wall profiles after etching. These same gases work well with aluminum-silicon alloys; they reduce the silicon to a volatile chloride, even at 3% silicon levels. The silicon keeps the aluminum from spiking through thin junctions on the device.

Copper also is used in aluminum alloys, usually in the 2 to 5% range, to reduce electromigration of the aluminum. Copper will not etch in the above-named etches unless sufficient ion energy is present to sputter it off. Copper residues left after dry etching can be chemically wet-etched in a separate operation. After etching, post-etch corrosion (caused by hydrochloric acid formation) often results because there is moisture in the system. Wafers should, therefore, be resist-stripped immediately after etching or subjected to a fluorocarbon plasma that converts the active chloride residue to inactive fluorides. Aluminum with copper is especially reactive in this regard.

Selectivity in aluminum etching varies with both the etch gas and the underlying material. Selectivity over silicon dioxide is 20:1 to 25:1

with a 500 Å/min etch rate in $BCl_3 + Cl_2$. Selectivity over silicon is only about 4:1; with resist, it is about 6:1 or 7:1. That is true of both pure and alloyed films. Figures 11.14 and 11.15 show examples of dry-etched aluminum. Note the nearly vertical sidewall angles and sharp edges (Fig. 11.15) as well as the multilayers of etching.

Parallel-plate plasma etching helps solve other aluminum-etching problems such as undercutting and "necking" of aluminum lines over steps and gas bubbles that cause metal bridges. In barrel aluminum etching, hygroscopic compounds (Al_2Cl_6) cause water vapor to condense on chamber walls, and the water vapor subsequently reacts with the aluminum to form stable products that reduce the etch rate. Also, water vapor is corrosive to the system by virtue of other by-products formed.

In planar plasma aluminum etching, a dry nitrogen purge in the chamber and a liquid nitrogen trap reduce the water vapor and, by the nature of the process, etch uniformity is improved. Some systems passivate the metal surfaces to prevent corrosion and reaction with the etch gases.

In dry aluminum etching, wafers can be given a 60-sec oxygen clean, which removes contaminants from the system while drying the wafer. Carbon tetrachloride and helium form the basis of the etching plasma. The final product is aluminum chloride, a dry white powder that is later cleaned from the exhaust line of the reactor. The last step

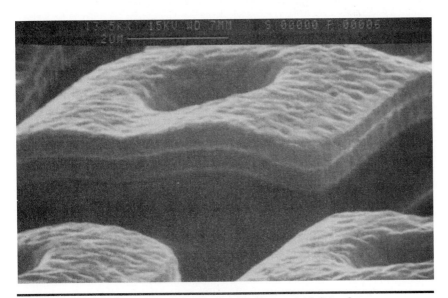

Figure 11.14   1-μm aluminum etch patterns with photoresist left in place.

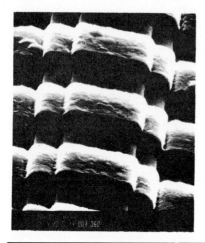

Figure   11.15 Plasma-etched   metallization
(Al/Si). (*Courtesy of Tegal Corp.*)

can be 60-sec oxygen plasma passivation, which removes reactive molecules from the system and wafers.

Prior to etching aluminum-patterned wafers, the resists must be properly treated. Positive photoresists should be postbaked at 110 to 130°C to prevent attack by the plasma. Such an attack causes low CCl_4 concentrations and low etch rates. The need for hardbaking depends on the system conditions and the gas used. In a typical cleaning cycle the operating parameters are water coolant temperature, 18°F; vacuum pump rate, 2000 L/min; reactor pressure, 0.3 to 0.4 torr; rf power, 800 W; and clean time, 60 sec. This is followed by the same parameters for etching the aluminum, except the rf power is reduced later in the process to 300 W and the etch time is 2 to 3 min for etching the aluminum oxide and the balance for etching the aluminum (pure, Al-Si, or Al-Si-Cu) at a rate of about 1500 to 2000 Å/min. The power reduction occurs *after* the aluminum oxide is etched. The final passivation step is identical to the initial cleaning step. The final purging of the reactor with argon or another noble gas removes any residual free radicals on the wafer surface.

The etch depth versus time curve for a typical aluminum layer is shown in Fig. 11.16, where CCl_4 was the etching species. Other cycles are used for aluminum etching, depending on equipment and device process. Another aluminum etch cycle, using CCl_4 (carbon tetrachloride) and a positive novolak-type resist is as follows:

1. CF_4 plasma to preclean the wafer and harden the resist

2. CCl_4 plasma for aluminum etching

3. H_2 plasma to remove chlorides from the parts and the chamber

4. CF_4-O_2 plasma to remove residual silicon precipitates from the Al-Si alloy (2% Si)

The etch rate does not vary significantly (10%) when silicon up to 2% or copper up to 1% is alloyed with the aluminum. There is a problem, however, with water vapor, in that a thin film that prevents further etching forms over the aluminum (aluminum oxide), especially when the humidity ambient is over 50%. This can be eliminated by using a longer pumpdown cycle before etching. After etching with the cycle just mentioned, a residue is occasionally left on the wafer; it is confirmed by Auger analysis to be Al_2O_3 and SiO_2. This film can be removed by partial dry etching followed by immersion in a wet aluminum etchant.

After aluminum etching with positive resist, the wafers should be stripped as soon as possible, because some free radicals may be absorbed and will form hydrochloric acid, which will attack the aluminum. Negative resists are even more absorptive and should be stripped. Air exposure will make the resist more difficult to remove. A 2-min dip in water is one technique used to simplify removal. A good example of plasma-etched aluminum is shown in Fig. 11.17.

Silicon dioxide etching. Dry etching of silicon dioxides is performed in fluorocarbon-based plasma gases. A typical etch is $CF_4 + H_2$, a gas that etches at about 500 Å/min depending upon system pressure, power level, density of the oxide, and gas flow rate and composition.

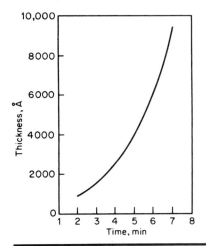

Figure 11.16 Plasma aluminum etch depth versus time.[7]

Figure 11.17   Plasma-etched aluminum (5000 ×). (*Courtesy of ETE Equipments Corp.*)

Doped oxides etch more rapidly than undoped oxides, and deposited oxide also etches faster than thermally grown SiO_2.

A key aspect of etching SiO_2 is obtaining good selectivity over silicon. The selectivity ratio of 20:1 is typical, and that of PSG phosphosilicate glass is 32:1 in $CF_4 + H_2$ at a higher etch rate of 700 to 900 Å/min. To minimize the formation of fluorine atoms which would attack silicon, mixtures of hydrocarbons (CH_4, C_3H_4, C_2H_2) are added to fluorocarbons (CF_4, C_2F_6, C_3F_8) to increase selectivity over silicon. Adding such materials as silicon and carbon to the etch chamber can tie up the flourine atoms and increase etch selectivity. RF power also is used to increase selectivity as well as etch rate. RIE and parallel-plate plasma etches are used to produce very highly anisotropic etch profiles as shown in Fig. 11.18. Shown are three typical etch applications: (1) 1 μm of PSG etched to polysilicon with 30:1 selectivity, (2) submicron doped polysilicon, and (3) heavily doped polysilicon.

Silicon dioxide etching is usually done on a contact mask level. Forming holes in the oxide to make room for aluminum or other metallization requires etching with high anisotropy and high resolution. Small contact windows are needed to conserve silicon real estate and maintain high circuit density. Since metallization follows contact etching, the oxide sidewalls must be somewhat sloped to ensure good continuous metal coverage.

Slope control is accomplished by several methods. One is to add oxygen to the fluorocarbon gas and thereby increase the etch rate of the resist and lower selectivity between resist and oxide. This increases

4 μm deep silicon etched with oxide mask,
demonstrating ultra-clean substrate

(a)

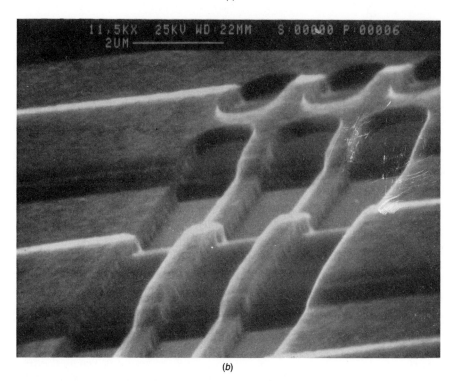

(b)

Figure 11.18 (a) 1-μm-thick PSG etched to polysilicon, 30:1 selectivity. (*Courtesy of Drytek.*) (b) Submicron doped polysilicon etch using Drytek process. Resist is 0.48 μm; polysilicon is 2.11 μm. (*Courtesy of Electrotech.*) (c) Heavily doped polysilicon etched by Drytek process. (*Courtesy of Drytek.*)

(c)

Figure 11.18 (*Continued*)

lateral etching and adds a sloping angle to the wall as the etch species approaches the bottom. Oxides are also heated and reflowed to round off sharp corners and permit better metal coverage.

Polysilicon etching.   Polysilicon etching is used to form gates in MOS devices. A gate determines the channel length of the device and requires high resolution and highly anisotropic etching. After the etch, an ion implant step is used, and thus vertical sidewalls are needed to ensure sufficient polysilicon thickness to mask the implant. Poor control of sidewall taper could result in uncontrollable channel lengths and out-of-spec devices.

Polysilicon etching occurs over thin (250- to 500-Å) gate oxides that protect source-drain silicon junctions. Therefore, selectivity must be high to allow for overetching, which is used to remove residual material along the sidewalls of steps. Since the polysilicon etching process is largely anistropic, it becomes difficult to remove any material that lies vertically alongside a previously etched step.

Combinations of flourine- and chlorine-based plasmas are used to etch polysilicon, since flourine atoms etch silicon isotropically and chlorine by itself etches too slowly. An example of the gas combination is SF_6 and Cl_2. Chlorine is noted for attacking novolak resist chemistry; in an SF_6 gas, it will strip resist and redeposit the polymer on polysilicon sidewalls, where it acts as a banking etch agent to inhibit lateral etching. Some studies show as little as 0.25 μm undercut with this approach. Chlorine plasma also etches polysilicon anisotropically and with good selectivity over SiO_2. It would be a useful gas for etching polysilicon over single-crystal silicon, since the etch rate is slow and would have a controllable end point.

In reactive ion etching, Cl_2-Ar plasmas are used. Highly anisotropic etching may require gas additives containing a recombinant. One example is CCl_4 for reactive ion etching and C_3F_6 in plasma etching. Two examples of polysilicon etching are shown in Fig. 11.18. One exhibits high resolution (0.5 µm), and the other shows nearly vertical sidewall angles. Polysilicon etched in Cl_2 has an etch rate of 500 to 800 Å/min with a 25:1 selectivity over oxide and a 5:1 selectivity over resist.

A key problem is an excess of radicals that will polymerize and form dense polymer films on the wafer surface. This occurs primarily with hydrogen-containing plasmas and thick (greater than 5000 Å) silicon dioxide layers. Polymer deposits also are caused by water vapor entering the etch chamber. The absorption of water by the wafers at ambient is tied directly to polymer formation, and all care should be taken to minimize wafer exposure to the air by using dry-nitrogen purging or storing in a nitrogen cabinet before etching. Very rapid pump-down times are important for this reason. One solution is to use an in-line system whereby the entire process cycle is kept under vacuum, thereby eliminating environmental contamination. Another solution is to add oxygen to the gas mixture.

Planar etching of polysilicon is more difficult than etching silicon dioxide because the etchants tend to be less anisotropic for silicon. Carbon tetrachloride is one etching species, and etch rates of 2500 Å/min at 100 kHz or 250 Å/min at 13 MHz are used. At a 27-MHz frequency, polymer formation has been more frequently noted and suggests a frequency dependence for CCl_4 etching. The undercutting of polysilicon is particularly bad when a silicon dioxide layer underlies the silicon, and the etch rate increases significantly. Uniform oxide deposition is important to minimize overetching.

Silicon nitride etching. Silicon nitride is etched isotropically with CF_4-O_2. It is used as an intermetal dielectric material at a 1.0 to 1.5 µm thickness. In that application, the nitride is plasma-deposited, and the etch chemistry and process conditions are similar to those used for SiO_2 window etching. Silicon nitride is also used as an oxidation and diffusion mask in 1000-Å-thick films over 300 Å of oxide on silicon. In an application of that type, isotropic etching is acceptable because of the thinness of the nitride and the relatively large pattern geometries. Low pressure and atmospheric CVD of nitride films are also made for passivation layers. Selectivity of silicon nitride over silicon is about 1:8 in flourine atom plasmas.

Refractory metal silicides and polycides. As with polysilicon etching, combinations of fluorine and chlorine plasma gases are used in etching to provide the necessary vertical wall profiles and high selectivity (10:1) over silicon dioxide. Refractory metal silicides are deposited over polycide. The structures are used as interconnection paths and as gate materials. Because of the need for close etch control and high anisotropy, two-step etching processes are often used on a polycide structure. CF_4-O_2 and SF_6 plasmas are used to produce anisotropic etching in an RIE system. The SF_6 etch, however, has only 4:1 selectivity over oxide, an unacceptable difference when patterns pass over field oxides. The two-step etching processes overcome selectivity problems by RIE etching the silicide and part of the polysilicon and following with a second anisotropic gas to clear the material remaining along the vertical sidewalls. Since polysilicon etches faster than silicide materials, two-step etches are required to provide the necessary selectivity, anisotropy, and dimensional control.

Refractory metal silicides. TiW-Al structures have long been used to prevent electromigration of aluminum to silicon. More recently, TiW-Al-TiW structures have been used. The top layer of TiW provides two additional benefits: (1) TiW is an antireflection coating which reduces many of the problems encountered during exposure and development and (2) TiW reduces "hillock" formation, which can cause shorts in double-layer metal structures.

There are several important factors that must be considered when etching multilayer structures: (1) cross-contamination of process, (2) etch profile (i.e., no undercutting causing trapping sites, and (3) end-point detection between steps. When TiW-Al-TiW is etched, the processes are selected to provide an anisotropic profile and maximize throughput.

The first step is a very thin (approximately 500-Å) TiW etch. The use of SF_6 results in a very fast, clean etch with no undercut seen. The second step is a BCl_3, Cl_2, $CHCl_3$, standard aluminum etch. It is done in a different chamber than the TiW etch to avoid cross-contamination and memory effect from one process to the next.

The third step is an anisotropic TiW etch using CCl_2F_2 and Cl_2. Sulfur hexafluoride is not used in this step because it has isotropic characteristics. An undesirable amount of undercut results when SF_6 is used for etching films greater than 800 Å in thickness. This step is taken in a third chamber to eliminate cross-contamination effects. A fourth step using SF_6 with very low rf power provides a postcorrosion treatment that eliminates the formation of HCl on the wafer surface. With low power, no measurable etch rate of the TiW films can be seen.

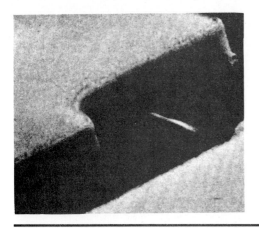

Figure 11.19 SEM of Al-TiW structure after dry
etch. (*Courtesy of Drytek.*)

As can be seen on the SEM in Fig. 11.19, the process provides an
anisotropic etch of the TiW-Al-TiW sandwich structure without any
undercutting. By using a multichambered etching system, it is possible to optimize each step without concern for cross-contamination between processes.

Undercutting, overetching, and control

Undercutting in etching occurs at and below the resist level of the materials being etched. The dry etch gases will erode resists away at various rates, depending on the amount of energy in the dry-etch system, the
condition of the resist, and other system parameters. One example of an
extreme case of resist attack and undercut is physical sputtering, in
which resist is removed at the same rate as the underlying oxide or
metal. Undercutting takes on all sorts of shapes. There is no standard by
which to measure good or bad undercut; it is a function of the needs of the
IC process. At some etch levels, zero undercut is desired in order to maintain line width; at other levels in which a metal or oxide layer is to be
applied over etched steps, a sloped sidewall is needed on the etched layer
to ensure good coverage of the next layer. Undercutting then becomes so
varied by process that specifications are created to cover very specific
process conditions and design rule requirements.

One parameter that deserves special attention with respect to undercutting is overetching, which is defined as the amount of etching
used in a process that purposely extends beyond the point at which a
film has been etched down to the substrate. Overetching is used by
design in many processes to:

1. Overcome nonuniformities in the layer being etched
2. Even out nonuniformities in the etch process (gas, environment, equipment) that cause varying etch rates across the wafer surface
3. "Size" or etch out a specific window dimension
4. Change the etch profile or sidewall angle of the final etched structure

Overetching varies along with the parameters or process needs as described above; as a rule of thumb, the amount increases as the amount of process control decreases. For example, planar plasma electrode etch processes generally require only 10 to 15% overetching since the wall profile is steep because of high anisotrophy. More isotropic etch techniques, such as dry plasma, require more overetch to both size a critical dimension at the substrate interface and to improve wall angle profiles. In anisotropic etching, overetching is used to clean up "stringers" or material remaining along sidewalls.

The primary need in VLSI etching is control. There are numerous types of films to be etched and many different types of etching devices, each with its own special characteristics. As a result, etch technology becomes a complex set of widely divergent process parameters which have at least one common denominator: the need for control. Dimensional control is needed on all sides of the structures to be etched.

Figure 11.6 illustrates the challenges in etching openings (windows) and islands (doors). In the case of the window, the need is to prevent the etch from eroding away the resist in the lateral direction and thereby changing the window width. In the case of the door, the problem is basically the same, i.e., to keep the resist dimension at the wafer interface as constant as possible so that the final etched structure very closely matches the mask dimension.

Organic film etching

Resists used as etch masks in dry etching are often part of the etch process. Positive novolak resists are the most common type of polymer used, and in many etch processes the etch rate of this class of polymer is calculated as part of the total etch sequence. One example, shown in Fig. 11.20, is a polymer layer 2.5 μm thick etched over a previously etched layer of oxide. In many cases, thick organic films are used as part of a sacrificial etch process, wherein the resist etches at the same rate as an underlying film. Controlled sacrificial etching allows removal of very thin layers of topography at the tops of steps while the thick resist protects the substrate below the tops. This type of planar etching is useful when the etch rate of the polymer is known.

In etching sloped contact sidewalls—overetching, which involves

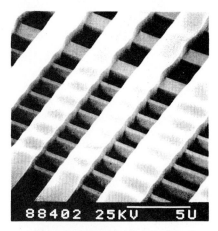

2.5 μm organic material etched over step
topography of thermal oxide.

Figure 11.20   2.5-μm organic material etched
over step topography of thermal oxide.

etching the resist at the sides of contact openings—the polymer etch
rate also is critical in determining the final etch slope angle. After
etching, the resist is removed by oxygen or similar reactive plasma.

Organic film etching is especially common in multilayer imaging
processes in which there are planarizing layers and the plasma polymer etch transfers a pattern to the substrate. One example is placing
an entire resist-covered wafer, after resist exposure, in an HMDS vapor. The HMDS reacts with the exposed resist to form a thin silicon
mask, and the unexposed resist does not react chemically. The process
is called "silylation." The wafer is now polymer-etched to leave a
silicon-coated resist mask behind as the final etch mask. The outline
in Fig. 11.21 shows a typical sequence.

After the oxide is etched, the resist-silicon mask is stripped. Organic
layers are usually etched in plasmas containing pure oxygen at moderate pressures. The addition of small amounts of fluorine to oxygen
plasmas increases the etch rate significantly. Isotropic etching occurs
at low pressures, but reactive ion etching can be used to achieve
anisotropy. The by-products of oxygen plasma etching of polymers and
organic layers are CO_2, CO, and water. Resists used in VLSI and
ULSI have a wide range of plasma resistance. Increasing etch resistance is accomplished by high-temperature baking in oxygen, which
causes oxidative cross-linking.

Trench etching.   Increasing device density has accelerated the need for
both device isolation and efficient use of the silicon area within the
die. Trench etching is one method that has been used for device isolation; it allows designers to lay out more device per square centimeter

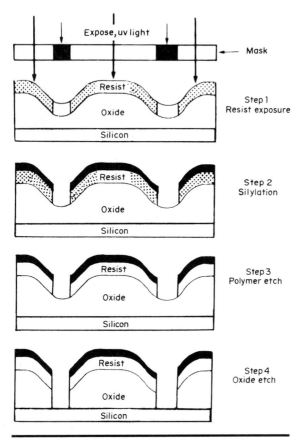

Figure 11.21   Silylation process sequence.

of silicon. In addition, it results in low leakage currents and low-capacitance effects at high frequencies.

Another application of trench etch is in the formation of buried capacitors used for storage in memory devices. Unlike the trench isolation, which is usually filled with silicon dioxide, the buried capacitor utilizes a thin oxidation on the sidewalls followed by a polysilicon fill of the trench. This results in a vertical capacitor which requires far less area than capacitors produced by conventional techniques.

The critical parameters in trench etching are (1) removal of surface contaminants to prevent micromasking, (2) vertical sidewalls, and (3) smooth, rounded bottoms. The need for vertical sidewalls follows from the requirement of minimizing the device area. A film, such as silicon dioxide, is deposited in the trenches after they are etched. Smooth, rounded bottoms are necessary to minimize the effect of thermal stresses on the silicon crystal during film deposition. The rounded bottoms also minimize the effects of mechanical stresses on the film. The parameters used are as follows.

Parameter	Step 1	Step 2
Chemistry	BC13, C12 (Drytek)	BC12, C12 (Drytek)
Pressure	100 mtorr	275 mtorr
RF	750 W	500 W
Time	1:00 μ	25 min/8 μ

In the first step of the process, a high dc bias is used to sputter away surface contaminants. A high BC13 flow rate is used to getter surface moisture.

In the second step of the process, chlorine chemistry is used for sidewall passivation to get a vertical profile. The rate of polymer formation is regulated by the BC13 flow rate so that the polymer is constantly removed from the trench bottom by the ionic bombardment of the etch. Figure 11.22 shows an example of trench etching. As can be seen from the SEM, this etch produces a trench with vertical sidewalls and a smooth round bottom.

Dry-etching equipment summary

Dry-etching systems come in a variety of configurations, and the over-riding factor in structuring the internal chamber where etching occurs is the electrode arrangement. In cylindrical barrel etchers, electrodes are on the edges of the quartz vessel, and no electrical connection is made to the wafers. The wafers are surrounded by a metal slave that keeps energetic ions and electrons away from the wafer surface, and the result is a chemical, isotropic etch.

Parallel-electrode or planar reactors use parallel electrodes, which

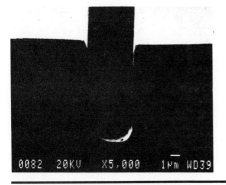

Figure 11.22 Example of trench etch. (*Courtesy of Drytek.*)

let energetic ions accelerate across the potential difference between the plasma and the electrode surfaces. This permits anisotropic or quasi-isotropic etching, because physical etching combines with chemical etching reactions. For an all-plasma reaction, wafers are placed in the grounded electrode; when wafers are placed on the powered electrode, reactive ion etching reactions are used. Gas is pumped in and out of the chamber while rf power is connected to one of the two circular electrodes.

Multiple-electrode systems, called "stacked parallel-electrode reactors," are used to provide the advantages of batch etching (throughput is higher) and parallel-plate (uniformity is higher) dry etching. Stacked electrode etchers have up to six single-wafer chambers with pressure and power densities in between the high pressure and power of planar reactors and the low pressure and power of hexode reactors.

Hexode reactors are cylindrical bath-type etchers in which an internal hexagonal electrode is used and the outside chamber walls serve as the opposite electrode. Wafers are placed on the powered hexode, which has about one-half the surface area of the outside wall ground. This configuration makes possible the high ionic bombardment of the hexode, when the wafers are vertically placed, with minimal attack on the rest of the system. The asymmetry of electrode area is the reason for this selective etching, and it makes hexode reactors well suited for reactive ion etching. Figure 11.23 shows the configuration of a single-wafer RIE plasma system with four individually controlled chambers, each with laser end-point detection. A pick-and-place robot transfers wafers between chambers under vacuum for up to 200-mm wafers.

Etching data. The etching data for planar plasma etching of various semiconductor materials are given in Table 11.7.

Plasma etching summary. Although plasma etching (barrel or planar) has reduced the wetting, disposal, safety, and some of the isotropy problems associated with wet etching, there are problems distinctive to dry etching. Etch uniformity is still not achieved without single-wafer etching or wafer movement. Polymer particle formation in reactors still occurs, and vertical wafer orientation is not the ultimate solution. Photoresists that are more resistant to the plasma gases are required, as well as improved thermal stability in resist systems up to 200°C. Resists have plastic flow problems in the 110 to 140°C range, temperatures not uncommon to plasma reaction chambers. Many equipment variations are utilized to avoid the entrapment of water

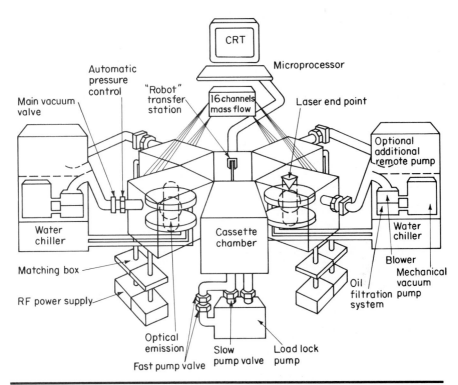

Figure 11.23 Typical hexode reactor configuration. (*Courtesy Drytek.*)

vapor during and after plasma treatment. Moreover, detecting the end point of an etch cycle is important when the etch does not have good selectivity.

Sputter etching

Sputtering or sputter etching is used as a method for dry removal of a variety of metals, dielectics, and silicides. The process is typically run in a vacuum chamber in which a plasma is created by the breakdown of a heavy noble gas, usually argon. A PCF field provides the energy to sustain the glow discharge across two water- or gas-cooled electrodes. As electrons move between the electrodes, collisions occur with the gas molecules and cause ionization. Positively charged argon ions and negatively charged electrons accelerate toward opposite electrodes and create ion bombardment normal to the electrode surface. More collisions occur on the way and create avalanche effects. The wafer, mounted on the cathode, is thus sputtered, and surface removal by

TABLE 11.7 Typical Plasma Etching Parameters

Etch process	Etch rate, Å/min	Selectivity	Uniformity, %	Undercut, %
Al or Al + 2% Si	1500	6:1 over oxide	±7	<2
Polysilicon (doped) (anisotropic)	600	3:1 over oxide	±5	<3
Polysilicon (undoped) (anisotropic)	1200	5:1 over oxide	±5	<5
Polysilicon	2000	8:1 over oxide	±5	<2
Si_3N_4 (over oxide)	800–2000	2:1 over oxide	±5	<2
Si_3N_4 (over silicon)	400–800	7:1 over silicon	±5	<2
SiO_2 (thermal)	500	7:1 over silicon	±5	<2
Silox	800	7:1 over silicon	±5	<2

SOURCE: ETE Equipments Corporation, Jerico Turnpike, New Jersey.

momentum transfer occurs. By-products are scattered toward the anode, and some of them deposit on the chamber wall.

Sputter etch rates in the 100 to 500 Å/min range are fairly constant for all materials. Since removal rates for resist and oxides or metals is similar, very thick resist layers are used as sputter etch masks. The result is a sacrificial etch in which the portion of resist removed equals the thickness of the film being etched. Resist temperatures are kept below 150°C by cooling the cathode, and a high degree of anistropy is made possible by the arrival of ions at normal incidence to the wafer. The relatively low selectivity between etch mask and the film being removed limits sputter etching to a small number of applications, including refractory metal etching.

Sputter or ion etching becomes reactive ion etching when chemical species are introduced to the chamber. Commonly used species are CF_4, C_2F_6, and CCl_4. Two primary types of reactions then occur in the chamber: surface desorption (chemical) and momentum transfer (physical). The sputter etching environment is outlined in Fig. 11.24.

The ion milling system differs from sputtering by separating the discharge from the wafer with an extraction grid to suppress electrons and extract argon ions and using a separate ion source such as a hot filament. This permits separate control over ion density and ion energy. Momentum transfer is still the basic mechanism for material removal. Ion milling is a much more controllable process, but it still lacks the etch selectivity ratios needed in many VLSI and ULSI processes. Consequently, its use is limited.

Ion beam etching

The principal mechanism of ion beam etching is one of directing high-energy ions toward a solid. The capability of anisotropic etching is

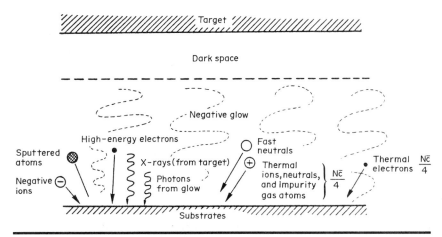

Figure 11.24   Sputter etching environment.

very good, inasmuch as control of the etching species and its effect on the incident target is maintained. Since the particle sizes are in the ion and atom category, the resolution potential is high, as is the aspect ratio (etch width to etch depth).

The major limitations of this technology in the past have been the high temperatures generated in the ion milling or ion bombardment process. This causes resists to flow and also makes their removal extremely difficult. Production ion beam etching has evolved slowly as the need for finer geometries has increased. The major obstacles to utilizing this new etching technology are high substrate temperatures and physical damage to the substrate. High temperature changes junction depths and warps the wafer.

Ion beam etching, also called "ion milling, sputter etching, sputtering, and ion beam sputtering," differs from chemical removal and thermal vaporization removal processes in that the removal mechanism is one of momentum transfer between an ion and a target atom. The only division of ion etching of concern here is physical ion etching (physical sputtering), wherein the bombarding ion is from an *inert* gas, versus chemical ion or reactive etching (chemical sputtering), wherein the bombarding ion is a reactive gas. Chemical sputtering is treated separately under "Reactive Ion Etching."

Sputtering, a relatively old technology, is, then, simply bombarding a solid with ions to remove material from the solid. As the high-energy ions collide with the surface ions of a solid, the energy is transferred from the incident ion to the surface atoms of the solid. The surface atoms of the solid are displaced or removed by the incident ion *if* the energy force that binds the atoms of the solid is less than the energy force of the incoming ion and the direction of the incoming ion force is away from the solid itself. The amount of energy necessary to dislodge an atom of a solid is referred to as "threshold" or "onset" energy; it is typically 5 to 40 eV.

The range of threshold energies is mainly determined by the mass of the incident ion and the crystal orientation of the solid. The sputtering or ion-etching yield is a function of the number of atoms displaced by a given incident ion. The threshold yield increases as the incident ion energy increases to a saturation point; at several thousand electronvolts, damage to the solid and other channel effects negatively affect the sputter efficiency. Sputter yields are most efficient at between several hundred and several thousand electronvolts.

The angle of incidence of the incoming high-energy ion is, of course, a major parameter. The sputter yield is usually best for a 45° angle of incidence, and it is decreased by changing to smaller (normal to the surface) or greater (more glancing) angles.

Ion beam etching equipment requirements. The ion beam that dislodges the atoms of the incident solid must have uniform density and be of the same wavelength of energy to provide uniform etching. The beam must be controlled within the system to prevent beam spreading, and it must be highly collimated to provide good angle control. Wafer cooling must be provided to keep wafers below 120°C. In the etching chamber, pressure must be low enough to prevent beam scattering and back sputtering. Figure 11.25 shows an ion-milling system schematic. The ion-etching process begins in the discharge chamber, where the heated tungsten cathode directs electrons toward the chamber walls. The anode is set at 30 to 60 V—low to maximize the efficiency of the electrons in the ionizing gas. A magnetic field is provided (to further increase the efficiency of ionization of the electrons), and it gives the electrons cycloidal trajectories to increase the probability of an ionization collision. The argon gas, introduced at 10^{-3} torr pressure, is ionized by the electrons. The anode chamber then contains the dense argon plasma of Ar^+ ions and electrons. The precision grids extract and focus the ions from the argon plasma. The first grid prevents electron movement in the grid system and extracts and focuses the argon ions into numerous collimated beams, and the potential applied between the grids determines the acceleration energy of the ions. The ion beam intensity is a function of the rate of arrival of thermal ions at the grid sheath, which in turn is a function of the source-generated plasma density. Thus the ion beam is extracted by applying voltage between the cathode and anode, which produces the ionized gas. The beam is neutralized by a hot filament, which emits electrons on the

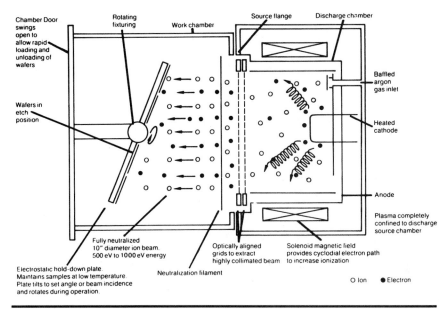

Figure 11.25 Schematic of production ion milling system.[8]

target side of the grid and thereby balances the arrival rates of ions and electrons at the substrates to prevent surface charging.

Etch rate versus beam angle.   The etch rate of various materials in ion milling varies considerably and is a major function of the beam angle incidence, as shown in Fig. 11.26. Note that an angle of 40° or less is required if the AZ-1350J resist is to be used as a mask for removing the other materials. The etch rates, in angstroms per minute, are given in Table 11.8 for many of the materials typically ion-beam-etched.

A typical etch cycle for silicon dioxide consists of loading the wafers, "roughing" or pulling a rough vacuum in the chamber, high-vacuum pulldown, etching at 500 Å/min to remove 1 μm of silicon dioxide, and venting and unloading. The wafers are held in place by an electrostatic device and kept below 100°C temperature by a water cooler.

The results obtained with an ion-etching system are impressive, because 0.2-μm dimensions can be achieved in material thicknesses of 1.0 μm. Figure 11.27 is a micrograph of a chevron bubble memory pattern with 1-μm images and etched structures, straight sidewalls, and no redeposition or trenching. The 5:1 depth-to-width etch ratio is obtainable with virtually any material. The low beam-operating energy eliminates the possibility of radiation damage to the device, and yet

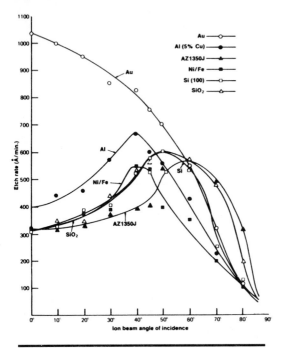

Figure 11.26   Etch rate versus angle of incidence (600-eV argon ion beam).[8] (*Courtesy of Rockwell International.*)

high etch rates are provided by beam angle and energy control. The control of etch depth is very exact, and etching can be stopped precisely at any point by changing the beam energy. Another application of the ion beam etcher is in pretreating or cleaning prior to material deposition. Bubble memories and high-frequency transistors are among the applications of ion beam etching. A major remaining need is to increase the wafer throughput of these systems by either increasing the overall size of the system or increasing beam size.

New masking materials with lower etch rates also are desirable. Stainless-steel and molybdenum masks are used, but they leave a film on the substrate. Positive photoresist also is used, but it has a rather high etching rate relative to the materials being etched. New resist materials are desired for that reason. The materials sputtered from resist include hydrocarbons and carbon clusters, which indicate the polymer fragmentation that occurs. Increased resist resistance to sputter or ion-etching processes is obtained by:

1. Baking the resist in an inert atmosphere. (Baking in air is worse than no bake at all.)

2. Limiting the maximum bombarding voltage to the following: AZ-1350J, 700 V; AZ-111, 600V; and negative resists, 900 to 1000 V.

3. Postbaking resists at 100 to 125 °C.

Reactive ion etching

Reactive ion etching differs from ion beam etching in that the latter relies strictly on physical sputtering, whereas the former utilizes both chemical reaction and physical sputtering. The reactivity of the ions that are formed is very high, especially when chemical sputtering occurs. Using carbon tetrafluoride as an example, reactive ions of fluorine (negative charge) and trifluoromethane (positive charge) are formed in an electric field at low pressure, much as in ion beam etching. The ions are accelerated and directed to the substrate, where they displace atoms through two processes: (1) physical displacement and (2) chemical reactions forming volatile by-products. RIE etching is performed in a planar plasma reactor and is highly anisotropic. Sidewall etch angle control is another benefit of RIE.

The use of CF_4 reactive ion etching techniques has been reported for the production of holographic gratings in silicon dioxide using AZ-

TABLE 11.8 Typical Etch Rates for Ion Beam Etching

Target material	Etch rate, Å/min
Chromium	200
Silicon	360
Gallium arsenide	2600
Thermal silicon dioxide	420
Deposited silicon dioxide	380
KTFR Photo Resist	390
AZ-1350J Photo Resist	325
PMMA	840
Silver	2000
Gold	1600
Platinum	1200
Palladium	900
Nickel	540
Aluminum	550
Zirconium	320
Niobium	300
Iron	320
Ferrous oxide	660
Molybdenum	400
Titanium	100
Alumina	130
Soda-lime glass	200
Riston	365
Copper	601
Stainless steel (304)	375

Figure 11.27   Micrograph of resist-oxide structure after ion milling.[8]

1350 photoresist and AZ developer. The images were made with an He-Cd laser at 325 nm, and the resist thickness was 0.13 μm. Structures of 0.3 μm can be produced in silicon dioxide by using this technique.

Reactive ion etching technology provides improvement of the silicon-to-silicon dioxide etch ratio. Ratios of up to 10:1 are possible by using planar plasma etching, but reactive ion etching yields etch ratios of up to 35:1 by using a CF_4-H_2 gas mixture (with no HF pollution problems). The dependence of silicon dioxide, silicon, and resist etch rates on the percentages of H_2 and CF_4 is calculable. As the H_2 percentage increases, the etch rates of materials shown will change, but silicon dioxide remains fairly high despite the lower etch rates of silicon, PMMA, and AZ-1350B.

This technique permits the etching of contact holes with minimal attack on the photoresist or the underlying silicon (after oxide removal). Thus if the silicon dioxide layer is of nonuniform thickness, as it generally is, overetching to completely remove the oxide does not result in attack on the silicon *or* degradation of the resist. The vertically etched sidewalls in the oxide permit conservation of IC device area. Reactive ion etching is a key technology used to solve the problem of isotropy in etching IC layers. Vertically etched silicon dioxide layers are essential to providing increased device density, and the amount of etch control and selectivity will add to the quality and yield of ICs in a production line.

Reactive ion etching, which relies heavily on chemical reactions at

less than ⅒-torr pressures, can use lower-energy beams because potential energy (energy that is used to break other chemical bonds) is
tied up in the ions themselves. The lower-energy beams used in reactive ion etching result in lower processing temperatures. Figure 11.28
shows a typical reactive-ion-etching apparatus.

Reactive ion beam etching

Reactive ion beam etching or ion-assisted etching works on the principle of an ionized beam of energy created by the introduction of gas
that will generate a reactive beam. The gas is injected into a broad-
beam ion source as shown in Fig. 11.28. The plasma that is generated
with the gas and ion source is the point of origin for the reactive ions
that will combine with the wafer films to be etched away. These ions
are removed from the plasma by using separately controlled power
grids operating at 500 to 1000 V. The highly energetic and reactive
ions are then directed, as a uniform beam, perpendicular to the wafer
surface to provide anisotropic etching. In addition to the specific ions
chosen to etch the wafer, there are also active neutral etch species
that emanate from the ion sources that participate in the etch reaction. The example shown in Fig. 11.28 is for etching a metallization
layer alloy of aluminum, copper, and silicon. Note the wide variety of
etch species removed from the metal film during etching.

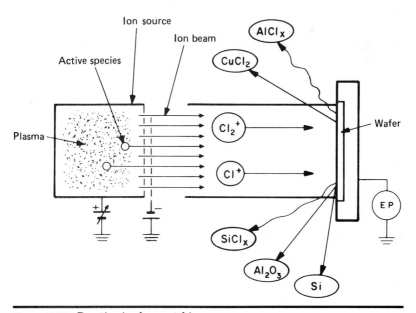

Figure 11.28   Reactive ion beam etching.

Reactive ion beam etching is based on a combination of chemical and physical reactions, analogous to combining reactive plasma and sputtering reactions in a single chamber. The first consideration is selecting a gas formulation suitable for removing the film or layer to be etched. The gas composition is responsible for volatilizing the semiconductor or metal layer.

Along with etching the desired film at a rate that permits good wafer throughput, the gas used in an RIBE system must not react with the substrate or the layer under the layer being etched. The gas must also be inert to the resist layer; it must keep the resist image intact so that veils, scums, or any portion of the organic layer, by not being removed, degrades the substrate by redepositing material or contaminates the etch chamber. For example, nonvolatile parts of a resist system, such as trace metal contaminants, could be freed from a resist layer under attack by the etch gas and become entrapped in the IC or wafer matrix after etching. The amount of gas used, with respect to the physical etch component, also is important. An excessive amount of etch gas chemistry will produce nonuniform results in ways similar to those associated with reactive gas plasmas.

The second major component of an RIBE system is the power or energy used to excite the gas. Optimizing the power requires setting the level so that a sufficient etch rate is produced at the same time and maintains adequate etch selectivity ratios of resist, the substrate, and the film being etched. Technology developments in VLSI device design have caused etch selectivity ratios to be increased, mainly to preserve increasingly tighter tolerances on critical dimensions that are brought about by smaller geometries and thinner layers. Control of the etch process has emerged as a separate technology requiring its own set of analytical tools, procedures, and parameters that must mesh with the actual etching step. One example of this is the integration of laser end-point detection as a component built into etch equipment.

The power or energy component of an RIBE system must not be excessive with respect to the gas component, or selectivity will be reduced just as it is in an ion beam milling system. Sputtering artifacts degrade the sharpness of the resist pattern and therefore the etch result. In RIBE, an optimum balance of the chemistry and the energy that activates that chemistry must be achieved. The ratio, or balance, is a function of the gas composition, mixing of gas components, vacuum level, number of active neutral ions, spatial configuration (wafer-to-source distance, etc.), gas flow rate, and power level, to name the major parameters.

RIBE equipment. Many configurations of RIBE equipment are commercially available, but conceptually they do not differ significantly. The major system components are:

1. Ion source
2. Gas source
3. Etch chamber
4. Wafer load-unload mechanism
5. Etch control panel (rate of gas and ratio energy)
6. Vacuum system and controls
7. Parameter setup, control system, and access panel
8. Automatic end-point detection system

A diagram of a typical system is shown in Fig. 11.29. The ion source will generate highly uniform beams of reactive ions. The overall beam width must be sufficient to deliver extremely uniform energy and reactive ion intensity across the entire wafer (or other substrate) diameter. As wafer diameters increase, this aspect of RIBE etch systems must change in order to preserve the anisotropy of the etch. Most systems provide for wafer cooling by water, Freon, or air to keep the resist below its thermal distortion point. End-point detection with either automatic optical or electrical sensing permits precision, programma-

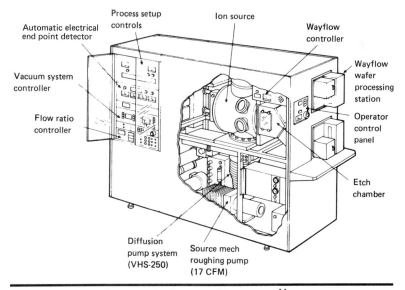

Figure 11.29 Typical reactive ion beam–etching system.[14]

ble overetching for the purpose of changing the slope of the etched film, or simply overetching by 50 to 15% to overcome expected nonuniformities in the etched layer thickness.

The purity of the etch environment must be maintained at a very high level to avoid particulate contamination. In a serial-type processing machine, in which only one wafer at a time is etched, the etch chamber must be protected from the outside environment, and the protection is usually accomplished with a load-lock arrangement in the process chamber. The cost of this physical separation of individual wafers is added process time. In a typical production etch operation, an operator loads a cassette full of wafers onto the end of the etcher and keys the process parameters into the control panel. One wafer at a time (serial-type processer) is then conveyed through a load-lock mechanism (basically a separate chamber that separates the wafer from the outside environment) and directly onto the platen where the etch reaction begins. The automatic end-point detector senses the point of completion (generally including an intentional overetch), and the wafer is sent out while the next one enters the load-lock. Complete cassettes are processed in this fashion and are brought to and from machines by clean-room operators, since this operation is usually performed in a Class 10 to Class 50 clean room.

Etching of dielectrics

One of the desirable features of any etch system is the ability to perform all etch operations in a single system. This feature reduces capital equipment expenditures, saves costly plant floor space, and reduces wafer handling that is correlative to defect density. In wet-etch technology, it is virtually impossible to find a single etch that will provide the necessary characteristics on various films, so multistation etch chambers are required. Each chamber is exposed to the processing environment. In dry etching, the wafer is isolated in a vacuum chamber in which, like CVD processing, it can be subjected to a series of sequential and individually different etch materials (gas mixtures, or plasmas, beams, etc.)

1. Metallization

   - Aluminum (pure)
   - Aluminum alloys (copper, silicon)
   - Metal silicides (several types)

2. Dielectrics

   - Silicon dioxide (doped, undoped)
   - Polysilicon

- Silicon nitride
- Single nitride
- Single-crystal silicon
- Polyimide
- Resists (PMMA, optical positive resists, beam resists)

Other organic films or protective layers (scribe coats)

The dielectric films most commonly used with RIBE methods are silicon dioxide, silicon nitride, and polysilicon. These films are generally etched with carbon tetrachloride (CCl_4), and in some cases CF_4 is used to etch polysilicon. The layer thicknesses generally used vary with the device type and fabrication method. They fall in the range of 1000 to 4000 Å for most applications, but in some processes films of polysilicon as thin as a few hundred angstroms are used. The dielectrics cited above, especially resist and other organic films, are etched in oxygen plasma by using SiO_2 or a similar O-resistant material as a mask. Silicon oxides (doped and thermal) etch at about the same rate in RIBE, and some typical rates are tabulated below.

Film type	Etch rate (Å/min.)	Conditions
Phosphorous doped SiO_2 (7%)	1100–1200	650 eV/CCl_4/~ 0.7 μÅ/cm^2
Silicon nitride	1100–1200	675 eV/CCl_4/~ 0.75 μÅ/cm^2
Thermally grown SiO_2	1100–1200	600 eV/CCl_4/~ 0.6 μÅ/cm^2
Polysilicon	1400	750 eV/CCl_4/~ 0.8 μÅ/cm^2
Positive optical resist	600	750 eV/CCl_4/~ 0.8 μÅ/cm^2

When CCl_4 is used with the tabulated materials, it does not have the sensitivity to oxygen content during etching that occurs in metallization etching. Figure 11.30 shows an example of RIBE results with tantalum silicide over silicon dioxide.

Selectivity ratio

The material that resides under the film being etched is important as to its not being attacked by the etch. The difference in the etch rates of the layer being etched and the underlying layer, called "selectivity," is expressed as a ratio of the etch rates. If phosphorus-doped silicon dioxide, often placed over silicon, etches at 1000 Å/min and the underlying silicon etches at 50 Å/min, the selectivity ratio is 20:1. That is a relatively good and healthy ratio, since a 5 to 10% overetch of the oxide will cause a loss of only about 20 Å of silicon to be etched; almost

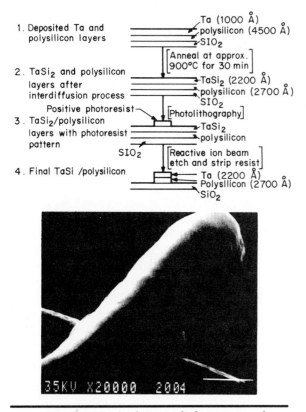

1. Deposited Ta and
   polysilicon layers

Ta (1000 Å)
polysilicon (4500 Å)
SiO$_2$

[Anneal at approx.
900°C for 30 min]

2. TaSi$_2$ and polysilicon
   layers after
   interdiffusion process

TaSi$_2$ (2200 Å)
polysilicon (2700 Å)
SiO$_2$

Positive photoresist

[Photolithography]

3. TaSi$_2$/polysilicon
   layers with photoresist
   pattern

TaSi$_2$
polysilicon

SiO$_2$ [Reactive ion beam
          etch and strip resist]

4. Final TaSi /polysilicon

Ta (2200 Å)
Polysilicon (2700 Å)
SiO$_2$

Figure 11.30 Reactive ion beam–etched structure and process. [TaSi$_2$ (2 kÅ)/polysilicon (2 kÅ) on SiO$_2$].[15]

beyond the limits of measurability. Silicon nitride is often placed over silicon dioxide. In that case, end-point detection is critical, since both materials have similar etch rates. In many cases, the etch rates are fairly close in the area of dielectrics because related atomic numbers and physical configurations are involved. Fortunately, precise end-point detection is possible with electrical and laser systems.

Metallization etching

Metallization systems used on advanced ICs have traditionally posed etch problems. For example, the reflectivity of pure aluminum causes many reflections back into the resist layer during exposure. Since reflections follow an often nonuniform circuit pattern, the change in resist geometries is not uniform throughout the pattern. This requires special line-width compensation programs, software driven, that overcome the problem at the design and exposure stages. In design, the circuit geometries are compensated for by oversizing the traces or conductors that are most severely affected, with reduced compensation on other areas. In the exposure step, reduced overall exposure time and precise exposure control are needed to prevent overexposure.

A second problem with metallization systems has to do with the shape of the patterns, typically long runs of conductors. These circuit lines are closely spaced and narrow in dimension, and they are easily disrupted by a small particle, process artifact, or any physical deviation in the process, such as a scratch, that disrupts the circuit line. Metallization traces often run below 1 μm in width, and some approach the ½-μm level.

Plasma and reactive ion etching of aluminum, which is generally alloyed with copper and silicon to enhance its physical properties (ductility, conductivity with silicon), often causes a third problem: residue formation. The residue is hygroscopic, and it draws from the environment residual moisture which combines with the metals in the etched pattern to form corrosion by-products.

One of the benefits of RIBE over plasma etches is the ability to use both chemical and physical mechanisms in the etch chamber. In the case of aluminum alloy metallization etching, this is very important because the aluminum is reacted chemically, as the silicon is, and produces volatile end products. The copper, a nonvolatile, is taken care of by the physical aspect of the reactive beam. It is sputtered away along with the nonreactive aluminum oxide and any other nonreactive materials, thereby leaving a clean wafer surface.

This physical-chemical action eliminates the hygroscopic residues that create the post-etch corrosion that can short out and destroy a device. A typical reaction process for removing AlCuSi allows for several different operations occurring in sequence. The first step is generally an aluminum oxide sputtering phase, important because the aluminum oxide is chemically inert. As the etch sequence progresses, the sputtering action to remove additional aluminum oxide will continue. The Al_2O_3 etch rate is a function of ion energy, and the aluminum oxide content depends upon the amount of oxygen present in the system or in the sample before etching.

The second reaction to occur, just after Al_2O_3 is cleared away to expose the metal alloy, is the violation of aluminum in CCl_4. The chloride by-products are evacuated into the vacuum pump oil or similar exhaust port. The copper component of the alloy reacts with the CCl_4 to form a nonvolatile chloride, which also is sputtered off. The aluminum alloy corrosion, referred to earlier, that occurs with RIE and plasma systems stems from this nonvolatile copper-containing chloride, a by-product that bonds with atmospheric water to form hydrochloric acid–cuprous chloride. The sputtering energy level that typically removes these chlorides is in the 600- to 650-eV range.

The silicon component of the alloy is removed by a combination of chemical and physical reactions. In most RIBE systems, a threshold ion energy level is required to remove all of the alloy. In the specific case of the system cited previously, the energy level was approximately 675 eV, below which point residue or "speckle" was left behind. The speckle is illustrated in Fig. 11.31; it is a common occurrence in plasma systems in which nonvolatiles cannot be sputtered away because the mechanisms of the etch reaction are chemical and not physical. Residue-free anisotropic etching is the goal of advanced VLSI processes and should be achieved when the process parameters are properly established. Optimization of the source-to-substrate distance, water content, oxygen content, voltage, and other key variables is required to achieve results like those shown in Fig. 11.14. The sample shows 1-μm aluminum-silicon oversteps etched with positive resist by using CCl_4/He. Selectivity is 20:1 with thermal oxide.

The etch rates run up to approximately 5000 Å/min, and the resist etch selectivity ratio is 3.5:1 when a standard positive optical resist such as HPR-204 or AZ-1350J is used. The uniformity of the etch over the wafer diameter is typically excellent with RIBE techniques, running generally less than 4%.

In summary, reactive ion beam etching uses an ionized, broad-beam source from which reactive ions are extracted and delivered to the wafer or other substrate as a highly uniform anisotropic beam. Involved along with the so-called reactive ions (energized ions that combine chemically with the substrate films to be etched away) are neutrals that enter into the overall reaction kinetics. RIBE methods supersede

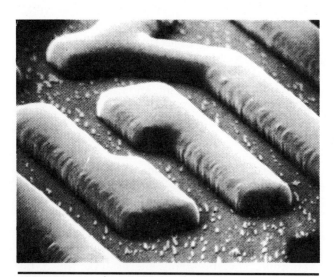

Figure 11.31 Speckle residue from etching.

many of the earlier dry-etch methods by combining the chemical (reactive plasmas, wet etch) and physical (sputtering, RIE) aspects in a single chamber and using each chamber to accomplish part of the total job.

In advanced IC processes, RIBE is popular because of its excellent process control in which each key variable, such as gas flow or power, is managed to deliver good selectivity, stopping power, clean residue-free etching, and high uniformity. RIBE processes are compatible with commonly used resist formulations, a necessary qualification of any etch system, yet the compatibility is not at the expense of wafer throughput. Radiation damage is minimal, and the obtainable resolution meets the submicron requirements of advanced VLSI. All of these factors actually constitute a good checklist for any major etch technology and are summarized for reference purposes:

1. Resolution below 0.5 μm
2. Critical dimension control of 5 to 10% of the etched geometry
3. Low rate of attack on resist or other mask
4. Acceptable differential solubility (etch ratio selectivity) of film being etched and mask or underlying layer
5. Minimal radiation damage from high-energy species
6. Automatic end-point detection
7. Good independent control over key variable
8. Clean, residue-free surface after etching
9. Compatibility with in-line automation
10. High degree of etch uniformity (less than 4% over wafer diameter)
11. Highly anisotropic (little to no lateral etching)
12. Ability to etch a wide variety of dielectrics, metals, and silicides in the same system to contain defects and minimize handling
13. High level of process automation
14. Relatively low process temperatures to prevent wafer distortion and resist flow
15. Short etch times that permit high wafer throughput

Troubleshooting

In IC fabrication, the proof of the pudding is in the etching, because this step can either enhance or degrade the lithography performed on the wafer or mask surface. The etching also verifies the quality of imaging and highlights any weak or substandard areas. For example, a marginally clean area on a wafer will not have the same degree of resist adhesion as other well-cleaned areas and will be undercut more by the etchant. The undercutting or lateral etching not only will be greater but will be uneven and leave a rough or jagged etch edge. Excessive undercutting can result in lower device yield by exceeding cotolerances.

Etch concentration

Transferring resist geometries with high fidelity in the etching step depends on several factors besides good cleaning. The concentration of the etchant species must remain relatively uniform if the etching is to proceed at a uniform rate. The causes of poor etch uniformity are as follows:

1. Gas or ion species flow changes
2. Change in etch species concentration
3. Poor dry-etch chamber configuration
4. Improper wafer spacing in the cassettes in wet or dry etching
5. Incomplete developing or resist monolayers on a portion of the wafer
6. Poor wetting of the wet etchant
7. Nonuniform thickness of the material being etched

Blocking

"Blocking," or the nonetching caused by an interfering layer or poor wetting of the etchant, is a common reject-causing problem. In both metal and oxide etching, blocking occurs both randomly and uniformly across the wafer or mask surface. A major source of blocking in wet oxide etching is contamination of the buffered HF etchant. If the oxide etch bath is in close proximity to a nitric acid–containing etchant, the nitric acid fumes (portions of the etchant) will be absorbed by the oxide etch and cause nitrates to form in the bath as well as oxide to grow on the wafers in the etchant. Oxide etchants are so sensitive to this type of contamination that as little as 10 ppm of nitrates in a bath will change the etching properties. Removal of this contamination is accomplished by loading the bath with pieces of a silicon wafer that will tie up the nitrates. A way to check for the presence of excessive nitrates in an etch bath is to remove the resist and etch the remaining oxide. If a latent image of the resist pattern remains after oxide removal, excess nitrate concentration is probable.

Developer residue, or scum, is the most common cause of etch blocking. Positive-resist developers absorb carbon dioxide from the air, which depletes the developer's active ingredient, usually sodium or potassium hydroxide. A depleted developer bath will not completely remove exposed positive resist and will therefore leave a thin resist layer that is practically undetectable except when etching. Resist developers may become so loaded with dissolved resist that some redeposition of dissolved resin prevents the oxide from etching.

Poor surface wetting can be caused by hydrogen bubbles released during etching. That, coupled with the high surface tension of acid etchants, causes the etchant to bridge the small-geometry areas and leave them unetched. Ultrasonic agitation has been tested to dislodge bubbles and carry the etch into the fine structures, but ultrasonic agitation is not recommended because it also may separate the resist from the substrate. A wetting agent added to the etchant will reduce surface tension and dislodge hydrogen bubbles, thereby permitting the etch to wet into the smallest openings on the wafer.

REFERENCES

1. B. Schwartz and H. Robbins, "Chemical Etching of Silicon," *J. Electrochem. Soc.*, vol. 123, no. 12, 1976, p. 1903.
2. J. Dey, M. Lundgren, and S. Harrell: "Parameters Affecting the Etched Edge of Silicon Dioxide," *Kodak Seminar Proc.*, vol. 2, 1968, p. 4.
3. G. I. Parisi, S. E. Haszko, and G. A. Rozgonyi, "Tapered Windows in Silicon Dioxide: The Effect of NH_4F:HF Dilution and Etching Temperature," *J. Electrochem. Soc.*, vol. 124, no. 6, 1977, p. 917.
4. Technical Manual, Shipley Company, Newton, Mass.
5. S. M. Irving and J. Hayes, "Localized Plasma Etching of Dielectric and Silicon Films," Kodak Interface Seminar, 1977.
6. "Dry Etching" and "Plasma Dry Stripping," LFE Corporation, Process Control Division, Waltham, Mass., 1978.
7. D. K. Ranadive and D. L. Losee, "Plasma Etching of Aluminum Using Carbon Tetrachloride," Abstract No. 116, Electrochemical Society Meeting, St. Louis, 1980.
8. "Microetch" Technical Literature, Veeco Instruments, Inc., Plainview, N.J.
9. Gene Goebel, Shipley Company, private communication.
10. S. Broydo, "Important Considerations in Selecting Anisotropic Plasma Etching Equipment," *Solid State Tech.*, April 1983, p. 159.
11. M. Hutt and W. Class, "Optimization and Specification of Dry Etching Processes," *Solid State Tech.*, March 1980, p. 92.
12. R. Powell, "Gases for Plasma Etching: What's in a Name?," *Solid State Tech.*, April 1984, p. 301.
13. S. Wolf and R. N. Tauber, *Silicon Processing for the VLSI Era*, 1986, p. 346, fig. 13.
14. D. Downey, W. Bottoms, and P. Hanley, "Introduction to Reactive Ion Etching," *Solid State Tech.*, February 1981, p. 121.
15. A. Baudrant, A. Passerat, and D. Bollinger, "Reactive Ion Beam Etching of Tantalum Silicon for VLSI Applications," *Solid State Tech.*, September 1983, p. 183.
16. J. Maa and J. O'Neill, "Reactive Ion Etching of Al and Al-Si Films with CCl_4, N_2, and BCl_3 Mixtures," *J. Vacuum Sci. Tech.*, June 1983, p. 636.

Resist Removal

Resist Removal Criteria

Resist removal from wafers is a fundamental step in IC fabrication, the main objective being complete removal of the resist images without adversely affecting the wafer surface. Resist stripping can pose some significant challenges, including the following:

1. The resist layer or images must be completely removed without leaving any residues, including metals, that may have been present in the resist. Metal content of the resist, especially sodium content, must be low. Residues may be extremely difficult to detect, and a subsequent diffusion or implant step would drive the impurity into and alter the electrical properties of the device.

2. There must be no adverse effects on the underlying surface, including undesirable etching of the metal or oxide surface. Etching of the surface may cause fracturing in doped-oxide layers. Excessive etching may reduce the masking resistance of an oxide layer for implant or etching steps.

3. A cost-efficient and safe process is necessary.

4. Environmental regulations are putting increasing pressure on waste treatment procedures for resist stripper effluents, wet or dry. Resist removers must incorporate improved "ecologic" chemistry or be supplied with simple cost-efficient waste treatment procedures.

5. Operating safety has been an increasing concern in chemical operations involving corrosive acids, bases, or solvents. Proper exhausting of fumes and adequate protective clothing are essential, and the stripper that requires a minimum of these kinds of precautions is desirable.

6. Chemical or gas action of a resist stripper should be dissolution, not lifting and peeling, to prevent redeposition of resist particles.

7. Removal time should be reasonably short to permit good wafer throughput.

8. Stripping solutions or gases should be free of metal ions to prevent contamination of devices.

9. A relatively simple method to detect the completeness of removal is essential.

Table 12.1 indicates a few of the different types of applications that must be handled.

Since there are many types of liquid and dry resist-stripping approaches, we will divide the types into three main categories: (1) solvent-type strippers, (2) oxidizing-type strippers, and (3) dry strippers.

Solvent-Type Resist Strippers

Solvent-type strippers are characterized by their ability to break down the structure of the resist layer. The molecular bonding within the resist makes a significant contribution to the ability of the stripper to remove the resist. Three chemical structures (shown in Fig. 12.1) illustrate different degrees of difficulty in resist removal. Structure (a), in which the polymeric chain backbone incorporates a double bond, is more difficult to remove than structure (b), in which the double bond is external to the chain. Structure (c), with a six-membered ring arrangement, may require longer strip times still.

TABLE 12.1   Applications and Requirements for Resist Stripping

Application	Special resist-removal requirements or problems
Lift-off process	Stripper must penetrate, swell, and lift a metal layer and provide a clean fracture.
Direct ion-implant resist masking	Highly baked resist is "chemically" reacted by the dopant ions at high temperatures, making the resist extremely insoluble in known strippers.
Etching noble and seminoble metals	Oxidizing acids react with resist images to form insoluble by-products. Resist must be lifted off at substrate interface and bypass insoluble surface species.
High postbake temperatures used to provide exceptional adhesion or etch resistance	Cross-linking (negative resists) or extreme hardening to form a bakelite-type product (positive resist) requires rigorous stripping chemistry.
Simple removal of misaligned resist patterns for reimaging after development and inspection	Positive resists and some negative resists leave a "memory," or ghost, image of the resist pattern, and microetching may be required to remove this memory.

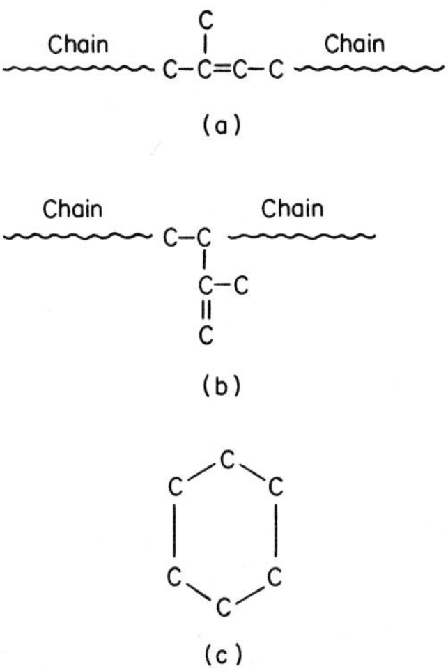

```
              C
 Chain        |        Chain
~~~~~~~~ C-C=C-C ~~~~~~~~
             (a)


 Chain            Chain
~~~~~~~ C-C ~~~~~~~
            |
            C-C
            ‖
            C
           (b)


          C
       C⁄   ⁀C
       |     |
       C     C
        ⌡ C ⌡
          C
         (c)
```

Figure 12.1 (a) Photoresist molecules in which the double bond forms part of the polymeric chain backbone require longer stripping times than (b) those with the bond external to the chain. (c) Hexagonal ring configurations strip even slower.[1]

The simplest of solvent-type stripping is for nonpostbaked positive photoresists. Since such resists have been processed only through an undoped oxide etch, they are readily soluble in acetone, methyl ethyl ketene, simple sodium hydroxide solutions, or one of a variety of proprietary strippers. Proprietary strippers are formulated with special surfactants or wetting agents and often contain active ingredients designed to dissolve a particular resin system. Positive resists not postbaked can generally be reexposed and developed off the wafer surface. In IC fabrication, the number of situations in which postbaking is not required is small. However, the need for lower process temperatures has caused a general reduction of hardbake temperatures.

In considering the various types of solvent or solvating-type strippers (as opposed to oxidizing and dry-etch type), we will discuss each individually according to its attributes, functional properties, and applications.

Acetone

Acetone is widely used as a stripper for many types of resists, and its application is typically limited to nonpostbaked resists. Acetone is used at room temperature, and a two-bath system is preferred to minimize residue formation from dissolved photoresist. Acetone will remove most positive IC resists by simple immersion for 5 min in the first bath and 2 to 3 min in a second bath. The effectiveness of acetone for complete removal (with no residues) is limited to postbake temperatures of 120°C. Above 120°C, residue formation increases as a function of postbake temperature, as shown by Kaplan and Bergin[2] and in Fig. 12.2. The resist is AZ-1350J; and at 130°C, residues are only approximately 1 nm thick or essentially undetectable. At 135°C, the residues are about 40 Å thick; at postbake temperatures above 135°C, residues increase sharply in thickness, especially if an adhesion promoter is used. To improve the ease of stripping with acetone, or with any stripper and a positive resist, a postdevelopment exposure is recommended. Since it is the diazo that will cross-link the polymers at postbake temperatures and the cross-linking increases removal difficulty, converting the diazo *before* postbaking to another form (carboxyl) by exposure of the entire wafer will simplify removal. The exposed resist will then display good solubility in the acetone until temperatures in excess of 160 to 170°C are reached.

Acetone is an effective stripper for thick coatings of positive resist, as in lift-off applications in which the resist is 2 to 3 μm thick. In such applications, the acetone is heated to about 40°C to increase its penetration of the thick positive-resist layer. The swelling of the resist layer causes the evaporated metal film on top of the resist to separate cleanly and leave good, sharp metal edges.

Acetone baths should be covered to prevent evaporation and contamination of other process solutions. Because of its low-cost and low-temperature operation, acetone is an excellent stripper for positive resists. Negative resists are not as easily broken down by acetone, and

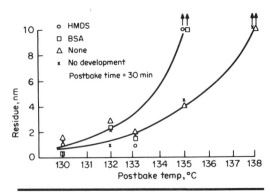

Figure 12.2   Residue thickness after acetone stripping as a function of postbake temperature.[2]

other solvents, such as methyl ethyl ketone (MEK) and methyl isobutyl ketone (MIBK), are often used.

Trichloroethylene

Trichloroethylene (TCE) is used to strip nonpostbaked negative resists, and stripping is done by using the same two-bath operation. Wafers are generally best held just above the solution level in the trichloroethylene vapors. A reagent grade, electronic grade, or "transistor" grade trichloroethylene is recommended, since lower grades may contain high levels of impurities that will contaminate wafer surfaces. TCE is generally always used heated and is mixed with equal parts of isopropyl alcohol to form a good rinsing solution after stripping. A spray rinse is particularly effective to provide a clean wafer surface, but most processes incorporate immersion or dispense-spin rinses. The wafer track handling systems can generally be adapted to handle a spray remove and a spray rinse, provided the resist is not so highly baked as to make removal times too long.

A fresh solvent spray, in which the dissolved resist is not recirculated in the bath, is recommended. Following a vapor phase solvent clean, wafers should be completely free of any residues, provided bake temperatures are below 120°C.

Phenol-based organic strippers

Phenol-based resist strippers are widely used for removing postbaked positive and negative resist from wafers. Phenol-based strippers are used at 90 to 100°C immersion in a two-bath system. Wafers are typically immersed for 3 to 5 min (shorter or longer depending on postbake temperatures and surface) in the first bath, 1 to 2 min in the second, "cleanup" bath, and then finally rinsed first in hot water and then in a solvent that removes both the water and any other residuals. The solvent can be MEK, acetone, methyl alcohol, isopropyl alcohol, or any number of proprietary solvent mixtures used in spray or vapor phase.

The stripper temperature of 100°C may be exceeded up to 150°C for resists postbaked above 150 to 170°C. At high temperatures, phenol-based strippers are often fuming, and adequate safety precautions are important. They include:

1. Use all stripping solutions in well-ventilated areas.
2. Provide good exhaust for complete fume removal.
3. Wear rubber gloves, eye and face protection (goggles or face shield), and protective clothing (acid-resistant laboratory coat).

4. Do not breathe vapors.

5. Do not allow any skin contact with solutions.

6. In the event of contact, remove clothing and flush affected area with water. Wash affected clothing and get medical attention if necessary, especially if eyes are involved.

Phenol-based strippers are used in conjunction with solvent strippers in some negative-resist processes. A removal sequence for a highly baked negative resist would be as follows:

1. Immerse in hot trichloroethylene for 10 min.

2. Immerse in phenol-based stripper at 140°C for 15 min.

3. Immerse in hot trichloroethylene baths (two) for 30 sec each.

4. Dip in acetone for 3 sec.

5. Rinse to resistivity in high-purity water.

6. Dry with filtered dry nitrogen or heated dry filtered air.

Low-phenol and phenol-free organic strippers

A class of strippers has been developed to overcome the objections to phenols. This group has many of the same attributes and functional properties as other organic formulations, but it also has some enhanced performance properties in the areas of substrate compatibility, disposal, and safety of operation, the three major problem areas for the phenol stripper family. These strippers are often based on organic acids, and they are used in two-bath systems followed by water rinsing and spin drying.

A typical bath life, using a two-container system, is several hundred wafers with 5000 to 7000 Å of resist-coating thickness and a 150°C postbake temperature. Bath life is a function of the amount of resist dissolved, the level of polymerization (higher-level postbaking depletes dissolving components more rapidly), and the time during which the bath is used at operating temperature. The heated bath will lose some solvents by evaporation as a function of time.

The compatibility of these removers with metals must be measured, because some removers exhibit slight attack on aluminum. On PROM wafers, fuse values are measured before and after treatment in the stripper to see if the nichrome fuses are attacked.

The most important parameter is wafer cleanliness, since residues not only interfere with subsequent processes but also cause electrical shifts in the devices. The level of cleanliness can be tested by measuring three parameters: (1) capacitance-voltage shift; (2) Auger analysis, and (3) hydrocarbon film. Capacitance-voltage (CV) drift is measured best after aluminum dot evaporation (490° alloy in N_2 for 20 min), checking positive drift under positive bias, and subtracting the drift of a control wafer (no stripping) from the drift of the test sample. Auger analysis is used in this case to check for metallic contamination on the surface of the wafers. Carbon comes from carbon dioxide absorption and is of no concern.

The hydrocarbon test involves processing wafers in a way that is similar to that by which CV test wafers are processed, except that a 30-sec dip in 10:1 buffered HF is used before aluminum evaporation and alloy to remove any oxide. The presence of a hydrocarbon film residue is indicated by a triangular-shaped alloy mark on (111)-oriented wafers. The marks are counted in the contact holes, where the presence of a hydrocarbon film will prevent good ohmic contact (aluminum-silicon alloy).

Another good test for any stripper is a die-sort yield test, in which wafers are processed through a metal mask up to wafer sort. Several strippers can be compared in this test by splitting a lot of wafers *before* resist stripping at the metal mask and measuring the average good die per wafer. Ideally, this test *should* be run by using each stripper at all mask levels and giving it identical processing at all other steps. While other fabrication parameters affect die yield, these will be constant for all the split lots using the various strippers tested.

Remover 1112A, another phenol-free stripper (Shipley), is also free of phosphates, chromates, fluorides, and metal ions. Remover 1112A is a unique product among strippers in that its use and application can be varied considerably, according to the concentration and temperature of the bath. Table 12.2 shows its positive-resist removal properties as well as etch rates for semiconductor materials under a variety of concentrations and temperatures.

Inorganic Strippers

Inorganic resist strippers are typically solutions of sulfuric acid and other components. Often called "oxidizing strippers" (because they oxidize the resist to carbon dioxide and water), these inorganic formulations are generally "home-brew" baths one of which, a sulfuric acid–nitric acid stripper, is made up and used as follows:[4]

Makeup:

Nitric acid	12%
Sulfuric acid	88%

Use:

Temperature	100°C
Time	5 min
Rinse (deionized water)	5 min
Dry	Filtered dry nitrogen

This stripper will remove ion-implanted positive resist *if* the resist is baked at 400°C for 1 hr in air. This oxidizes and cracks the resist film, thereby making stripper penetration easier. Sulfuric acid–nitric acid strippers cannot be used on resist applied over aluminum without causing some attack on the aluminum. Since the use of the stripper is limited to temperatures above 90°C, where aluminum is then etched, other strippers, such as those mentioned earlier, must be used for resist stripping on metal. Another type of inorganic stripper is a mixture of chromic acid (chromium trioxide dissolved in sulfuric acid) in water used at 70 to 100°C. Strippers containing oxidized chromium are suspected of causing electrical defects in ICs and are therefore not widely used, despite their low metal content and free water-rinsing properties. Because their rate of attack on aluminum is low, the cold chromic acid strippers are used occasionally in this application.

A very common inorganic resist stripper is "Caro's acid," or a solution of peroxymonosulfuric acid in concentrated sulfuric acid, first disclosed by Beck et al. in U.S. Patent No. 3,900,337 (1975). Concentrated sulfuric acid is mixed with 85 to 90% hydrogen peroxide, and the following reaction takes place:

$$H_2O_2 + H_2SO_4 \rightarrow HO-(SO_2)-O-OH + H_2O$$

This stripper can be highly exothermic when the water-containing peroxide is added for dilution. The solution will boil when laboratory-grade H_2O_2 is added; therefore, excess water is to be avoided. Caro's acid is a particularly effective stripper because both oxidation and dehydration mechanisms are involved, and up to 3 weeks or longer pot life is achievable because the acid converts resist into carbon dioxide and water. That makes disposal relatively simple.

Another advantage of Caro's acid is room-temperature operation, which eliminates the fume and hot-solution dangers of phenol-based removers. Caro's acid is a fast remover for positive resists and will strip positive novolak-type resists in 5 min.

Higher degrees of cross-linking pose real challenges for strippers.

TABLE 12.2 Etch Rates in Angstroms per Minute of Semiconductor Materials in Shipley Remover 1112A

	(Undiluted) 1X Temp., °C				(1pt. H₂O) (1pt. 1112A) 2X Temp., °C				(1pt. 1112A) (2pts.H₂O) 3X Temp., °C				(1pt.1112A) (3pts.H₂O) 4X Temp., °C			
	20	35	50	65	20	35	50	65	20	35	50	65	20	35	50	65
(100) Si	<1.3	<2	<7	<14	4.99	15.4	50.8	127	2.87	11.0	41.2	107	2.00	6.18	24.7	94.7
(111) Si	—	—	<0.2	<3.4	—	—	6.40	15.5	—	—	3.75	13.0	—	—	3.15	9.61
99.999% Al	<4.7	<18	<52	<112	19.9	50.8	92.6	156	51.7	132	249	366	72.1	161	204	415
Al, 4% Cu, 2%Si	<9.7	<18	<64	<128	23.3	54.8	122	165	57.9	162	275	272	75.6	191	256	506
Poly Si	—	—	—	<4.5	—	—	—	77.8	—	—	—	51.6	—	—	—	43.0
SiO₂	—	—	<0.12	—	—	—	—	—	—	—	<0.12	—	—	—	—	—
AZ-1370(150°C)*	—	No	Yes	Yes	—	No	Yes	Yes	—	—	No	Yes	—	—	No	Yes
AZ-1370(120°C)	Yes	Yes	—	—	Yes	Yes	—	—	Yes	Yes	Yes	—	Yes	Yes	Yes	—
AZ-111S(90°C)	Yes	—	—	—	Yes	—	—	—	Yes	—	—	—	Yes	—	—	—

*AZ-1370 postbaked at 150°C for 30 minutes. If photoresist cannot be removed at 65°C bath temperature, increase 1112 temperature up to 100°C for complete removal.

SOURCE: Shipley Company, Newton, Mass.

Three examples of cross-linking, tested by Kaplan and Bergin,[2] are as follows:

1. CF_4 plasma etch over thermal oxide

2. Arsenic ion implant

3. 15-kV electron bombardment (SEM)

After the plasma treatment, using a 130°C postbake before and after the plasma (separately), a 10-min immersion in Caro's acid removed the positive novolak resist with no residues remaining. In the ion bombardment case, a 400°C air bake was required prior to the Caro's acid to completely remove a 150 Å residual film caused by the ions, and even then a 75-min immersion was required in Caro's acid to remove the resist with no residue remaining. The SEM-exposed resist is equally difficult to remove in Caro's acid because even small electron dosages render positive novolak resist insoluble in Caro's acid.

An inorganic stripper that is used for removing highly cross-linked resist is made by mixing $K_2S_2O_8$ (potassium salt of persulfuric acid) with sulfuric acid. This solution results in the formation of persulfuric acid. The preparation is as follows: Add 700 g of $K_2S_2O_8$ to 2000 mL of sulfuric acid while mixing with a stirplate (magnetic stirrer) for several hours or until dissolved. The solution is then used, like other strippers, in two baths with a 5-min immersion in each bath. It is relatively easy to make up and provides results equivalent to those with Caro's acid. In using either Caro's acid or the potassium persulfate, it is clear that simple oxidation alone is not sufficient to remove highly cross-linked resist and that the carbon atoms must be more available to the stripper than they are in a highly cross-linked state resulting from high-energy exposure.

In summary, inorganic chemical resist strippers are required for the residue-free removal of highly postbaked resists, and such removers take over where the organic strippers leave off in terms of degree of difficulty. The use of an adhesion promoter adds to the degree of difficulty. Moreover, the longer the immersion time, the cleaner and more residue-free the wafer surface will be.

A simple remover for positive optical resists, reported in *IBM Technical Bulletin* **22**(12), 1980, is formulated as follows:

7.5 parts 0.1 N sodium silicate (Na_2SiO_3)

1.5 parts 0.1 N sodium hydroxide (NaOH)

1.0 part 0.1 N tribasic sodium phosphate (Na_3PO_4)

This solution is used at 90°C with a 5-min immersion followed by a hot-water rinse. Its advantages are the elimination of phenol strippers, shorter strip time, and a less hazardous bath to operate compared with conventional positive-resist strippers.

Detection of Stripper Residues

Residues left after resists are stripped from wafers are very difficult to see or detect with the more conventional inspection techniques. Residual monolayers and submonolayers are often less than 200 Å thick and in many cases are completely undetectable with optical and even scanning electron microscopy.

Ellipsometry is very useful for measuring the apparent thickness of an oxide. The surface is measured before resist application and then again after stripping. When the optical constants of the surface and one optical constant of the residue (by measuring on the transparent silicon dioxide layer) are known, the residue thickness can be calculated. Since the ellipsometer measurement is made on an oxide surface (in exactly the same place before and after stripping), the thickness of the residue is true for the residue "in oxide form" only. However, because the refractive index correlates closely with computed thickness, the measurement is reasonably accurate.

Plasma chromatography is one of the most sensitive analytical techniques for detecting residual layers. High-energy spectroscopy (ESCA and ISS) can be used to detect the presence of carbonaceous material and provide a quantitative figure. An old qualitative test is the "breath test," which is useful for showing the presence of a nonuniform residue only. It is administered by simply breathing onto the wafer surface and observing for the random patterns on a residual layer on the surface. It is a crude but simple and practical test for detecting monolayers.

Plasma Resist Stripping

Plasma removal of resist from silicon wafers offers several advantages compared with wet chemical stripping, namely:

- Safer operating conditions for operators
- No metal ion contamination
- No polluting problems
- More cost-effective than wet chemical strippers
- No attack on underlying surfaces
- Fewer processing steps
- Controlled reaction
- Higher level of automation possible

Plasma removal takes place by using an oxidation reaction wherein the resist (C_xH_y) combines with active oxygen as follows:

$$C_xH_y + \text{active oxygen} = CO\,a + CO_2\,a + H_2O$$

The gas products produced by this reaction are pulled out of the plasma chamber by the vacuum pump. In production, wafers are loaded into the reaction chamber and a vacuum is pulled. The oxygen is then bled into the chamber at a rate of several hundred cubic centimeters per minute and is followed by rf energy, which is inductively coupled to the gas at a fixed frequency. The rf energy excites the oxygen and creates the active species (most of which is atomic oxygen) that reduce the resists' polymeric chains to simpler and lower-molecular-weight groups. These then volatilize and leave the system. A diagram of a typical system is given in Fig. 12.3.

If in breaking down the resist the oxygen plasma seeks to interact at the double-bond (unsaturation) part of the polymeric chain "backbone" of the resist, longer plasma-stripping time will result. Conversely, double bonds external to the polymeric chain will result in shorter strip times. Some negative resists containing cyclized polyisoprene structures pose more difficult situations for the plasma in that, being closed rings, they exhibit reduced tendencies toward oxidation and fragmentation.

The attack of the active, or atomic, oxygen in the plasma on the unsaturation(s) in the resist structure is a result of the electron-attraction (electrophilic) nature of atomic oxygen. A number of species of oxygen exist in an oxygen plasma; they include O^+, O_2^+, O_2, O_2^-, $O(^3p)$, and $O_2(^1\Delta)$. Resist removal by oxidation is believed to be caused by the $O(^3p)$ species. The oxygen plasma contains between 10 and 60%

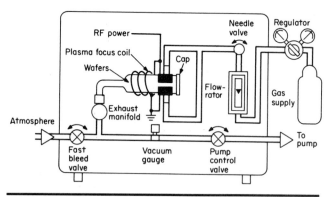

Figure 12.3 Plasma system diagrammatic cross section. (*Courtesy of LFE Corp.*)

atomic oxygen, depending on the interior chamber wall state and the additives in the gas.

Dissociation of the molecular oxygen determines the rate of resist removal. The excitation reactions that cause the dissociation are in turn a function of the initial electron impact. Proprietary gas mixtures or systems may contain additives (gaseous species) that combine before entering the discharge zone; as a result, they may increase the oxygen dissociation. In addition, treatment of the interior chamber wall to make it more passive to atom recombination would help maintain a high density of active oxygen species.

Special gas mixtures

Plasma gases for specific applications have been developed and used in IC fabrication. One example is an organohalide vapor mixed with oxygen. This mixture will simultaneously decompose and volatilize resist and reduce inorganic contamination, such as tin, iron, or magnesium, that might exist in the resist. At the same time, such a process inhibits oxide formation, which causes undesirable ohmic contact resistance prior to aluminum metallization. Wafers can be taken directly from the process into diffusion. Reactive ion etchers are used to remove baked resist. A small percentage of chlorine is used in the final stage of removal to remove silicon left from silylated resist.

Dry chemical resist removal

Plasma removal is a simple oxidation reaction and is an essentially self-contained process. The organic resist C_xH_y combines with active oxygen very readily, as follows:

$$C_xH_y + \text{active } O_2 \rightarrow CO, CO_2, H_2O$$

The stable by-products, in a gaseous state, are drawn off by the vacuum pump.

Masks are loaded into their holders or cassettes and, in a batch operation, placed in a vacuum environment of about 0.2 torr. The O_2 gas then is pumped in at a rate of several hundred milliliters per minute, followed by inductive coupling of rf energy to the gas at a specific frequency (13 MHz). This so-called "excitation energy" generates several active species, but the primary one is atomic oxygen.

Atomic oxygen essentially takes apart the resist and breaks it into its simpler components which, by themselves, are volatile gases. The generation of active oxygen is mainly the result of dissociation of molecular oxygen. This may occur by excitation reactions due to initial electron impact:

$$e^- + O_2 \rightarrow O_2^* \qquad \text{(in more than one state } +e^-; O_2^* = \text{excited oxygen)}$$
$$\rightarrow O + O + e^-$$
$$\rightarrow O^- + O$$

Various energetic states of the resultant atomic oxygen are also possible.

Loss reactions of atomic oxygen within the discharge proceed either homogeneously (in the gas phase) or heterogeneously (on a solid wall). Of the reactions that involve neutral species, the following are the most important:

$$O + O + O_2 \rightarrow 2O_2$$
$$O + \text{wall} \rightarrow \tfrac{1}{2}O_2$$
$$O + O_2 + O_2 \rightarrow O_3 + O_2 \qquad \text{(usually on a cold wall)}$$
$$O + O_3 \rightarrow 2O_2$$

Theoretically, the initial attack of oxygen on resists results in abstraction of hydrogen to form carbon- and oxygen-containing free radicals and water. Subsequent attack of oxygen atoms on the free radicals yields CO, CO_2, OH, and hydrogen atoms. A possible sequence may be the following:

$$\text{Photoresist (e.g., } \sim\!\!CH_2\!-\!CH\!\!\sim) + O^* \rightarrow CH_3CHO, OH$$
$$|$$
$$C\!-\!CH_3$$
$$\|$$
$$CH_2$$

1. $-CH_3 + O^* \rightarrow H_2CO + H$
2. $CHO + O^* \rightarrow OH + CO$
3. $CHO + O^* \rightarrow H + CO_2$
4. $OH + O^* \rightarrow O_2 + H$
5. $O^* + CO + M \rightarrow CO_2 + M$
6. $OH + H_2CO \rightarrow H_2CO + H_2O + HCO$
7. $O^* + H_2CO \rightarrow CO + H_2O$

Plasma reactions have many applications beyond simple removal of photoresist. They include

1. Photomask residue or contamination removal
2. Wafer preclean before epitaxy

3. Hybrid circuit cleaning before bonding

4. Cleaning raw, glazed, and metallized ceramics

5. Removal of bonding wax

Certain plasma-etching conditions cause resist removal to be extremely difficult and call for special treatment. Below are two examples of these cases and the recommended removal techniques.

Resist stripping after aluminum plasma etching[5]

A residue from the etching process, usually an Al-Si alloy, remains on resist surfaces after etching and makes removal in oxygen plasma difficult. A better gas mixture in this case is 70% oxygen and 30% Freon 14 (carbon tetrafluoride). If the aluminum alloy contains copper, ultrasonic cleaning in deionized water or a copper etchant is better.

Negative-resist removal after planar plasma etching

In high-power planar plasma etching, considerable resist hardening takes place. A 30-min removal cycle in oxygen plasma at 500 W will strip negative resist that has been in the etch (C_4F_8) for 5 min at 7 A. Wet acid strippers are not entirely effective.

Future chemical strippers must meet a variety of functional and environmental criteria. Table 12.3 is a listing of some guidelines for a future stripper that would hopefully serve the interests of both IC manufacturers and the environment.

TABLE 12.3 Future Resist Stripper: Guidelines for Functional Performance

Property	Comments
Substrate compatibility	No attack on silicon (single and polycrystalline), silicon nitride, silicides, gallium arsenide, aluminum, iron oxide, gold, nichrome, chrome, or copper.
Resist removal	Complete residue-free removal of positive and negative resists: 1. After postbakes of up to 200°C. 2. After medium- to high-dose ion implantation. 3. After ion milling. 4. After dry (RIE, plasma) etching in various commonly used gases.
Toxicity	Nontoxic and nonfuming to meet OSHA and EPA regulations.
Disposal	Nonpolluting; to conform to Federal Water Pollution Act Amendment of 1972, 33 U.S. Code, 1251–1356.

TABLE 12.3   Future Resist Stripper: Guidelines for Functional Performance
(Continued)

Property	Comments
Operating temperature	Less than 100 °C.
Resist-removal time	Less than 30 sec.
Metal content	Less than 0.05 ppm of sodium; less than 3 ppm of total metals.
Wafer throughput	Capable of maintaining the same wafer throughput as the developing step or, if slower, fast enough so as to not be the time-limiting step.
Wafer handling	Compatible with a cassette-to-cassette processing with no operator interference.
Removal cost per wafer	Less than resist cost per wafer.
Rinsibility (if wet stripper)	Ability to be rinsed by in-line pressure-dispense or aspirated spray. Rinse time compatible with balance of process.
CV shift	Less than 0.05 V.

REFERENCES

1. "Plasma Dry Stripping," Publication of LFE Corporation, Process Control Division, Waltham, Mass., 1979.
2. L. H. Kaplan and B. K. Bergin, "Residues from Wet Processing of Positive Resists," *J. Electrochem. Soc.,* vol. 127, no. 2, 1980, p. 386.
3. Technical Bulletin on Burmar 712D Resist Stripper, EKC Technology, Hayward, Calif., 1978.
4. F. W. Karasek, *Ann. Chem.,* vol. 46, 1974, p. 710A.
5. "Plasma Etch Process for Glass Passivated Solid State Devices," Technical Paper T7213, International Plasma Corporation, Hayward, Calif., 1980.

Resist Quality Control

Controlling the physical, chemical, and functional properties of photoresists, batch to batch, has become more important as IC imaging criteria have become more critical. The resist materials supplier must ensure increasingly higher levels of functional performance to meet the more stringent needs of lithography on the IC line. It is the compatibility of each new lot of resist with the current working parameters of the IC fabrication equipment that must be maintained so that resist line and space sizes will remain within specifications.

Quality control includes resist companion products as well, especially the resist developers that play a large role in the functional behavior of the resist. Other imaging-related solutions are resist thinners, adhesion promoters, strippers, and etchants. The most critical by far, however, is the resist material.

Specifications

All suppliers of resist and related imaging materials provide specifications of their respective products. These specifications often are published in the data sheets or are available through the technical division of the supplier company. Typical supplier specifications of imaging materials are given in this chapter.

Metal content

Metal content is specified because of the concern over sodium or other metals interfering with the electrical properties of a device. Increased device complexity means increased sensitivity to metals in resists. Metal analysis is performed by emission spectrographic analysis, which gives the ppm ranges of metals present. Metal content accuracy below 1 ppm (ppb, or parts-per-billion range) can be tested by an atomic absorption spectrophotometer. Most positive photoresists have a total metal content less than 10 ppm and a sodium content less than 1 ppb. The other metals of concern are tin, gold, lithium, and potassium. Metals exist in photoresists as trace impurity contamination and are not part of the formulations.

Solids content

The solids content of photoresists is a key property because it has a major bearing on coating thickness and flow properties. It is determined by measuring the mass of given specimen of resist in liquid form and baking the sample until no weight changes are detected. Then the mass of the dried sample is recalculated. The solids content is expressed as a percent (dried mass) of the original mass (liquid mass). The procedure is described in ASTM-F-66-77T.[1]

Positive and negative resists may undergo a change in solids content as a function of time, depending on storage conditions and internal chemical reactions in the packaged solutions. In negative resists, this may be caused by "autopolymerization," a chemical reaction wherein internal chemically initiated polymerization causes the formation of aggregates or gels that may be filtered out (or deformed through a filter membrane) and thereby change the measured solids content. In positive photoresists, the decomposition of the sensitizer will result in the formation of material that also may be filtered out to bring about a change in solids content. The main differences between these "self-initiated" chemical reactions is that negative resists are more reactive in this sense and will undergo autopolymerization more readily than positive resists will undergo sensitizer decomposition for any given set of storage conditions.

Viscosity

Resist viscosity is important to control, since it is a major determinant of spun-on, sprayed-on, or otherwise coated resist film thickness. Since resist viscosity is a function of solution temperature, close temperature control is imperative to obtaining reproducible results. Viscosity should be measured with a Canon-Fenske reverse-flow viscometer according to ASTM 0445-65.[1]

Most resist suppliers specify viscosity levels, and tighter control can be provided, if necessary, by either negotiating a specification (between supplier and user) or adjusting the resist slightly (preferably downward by thinning, since evaporation to raise the viscosity may cause dust or airborne particle contamination) in the materials control section of an IC fabrication facility.

The amount of viscosity increase is usually only 2 to 4 cSt for several months of storage at room temperature, attributable to intra-molecular reactions such as polymer-chain uncoiling in negative re-sists or molecular expansion of solvated molecules in positive resist.

Another condition that will change the viscosity of photoresists is filtration. Most resists are non-newtonian fluids (positive resists), and when they are subjected to the sheer-stress forces of absolute mem-brane filtration, they will experience a drop in viscosity, again by only 1 to 3 cSt. The viscosity will return to its original value within 2 to 3 days, depending on the amount of sheer stress.

Specific gravity

The specific gravity of resists can be measured in several ways, but the preferred and most commonly used techniques are with a Westphal balance or pycnometer. ASTM D891-59[1] provides the proce-dures for these and other methods.

Water content

The water content of resist materials is a relatively important prop-erty, because major variations will cause resist flow and exposure problems. Determining the water content of photoresists is detailed in ASTM D-1364-78.[1]

The water content of negative resists is generally below 0.2% by weight; that of positive resists is below 0.6%. The higher percentage for most positive (diazo) resists is in part explained by the necessary role of water in the photolytic reaction, wherein the conversion of sen-sitizer intermediate molecules to carboxylic acid molecules has a wa-ter molecule dependency. The reduction of the water content below a certain level in these resists therefore brings about a reduction in the number of converted species (species that should become developer-soluble) and thereby causes an apparent loss of photoresist sensitivity. For that reason, wafers coated with diazo photoresist should not be ex-posed directly from dry nitrogen or dessicator storage; they should be allowed to "water equilibrate" by standing (covered) in the clean-room ambient for at least 20 to 30 min at the normal relative-humidity level (40 to 50% at approximately 20°C). In addition, since resist softbaking brings about the removal of water (and solvents) from the resist film, moisture equilibrium at the resist-air interface and within the film must be achieved for proper exposure response. The presence of excessive humidity in the clean room, for both negative and positive photoresists (especially for positive resists), can cause resist-adhesion failure. The adhesion failure may take the form of increases in under-cutting or complete lifting and sliding around of resist images on the wafer surface.

Flash point, TLV rating, and ash content

The test methods covered thus far represent the most commonly published specifications from resist suppliers that have direct functional realities in the imaging process. Other specifications include flash point, threshold limit value (TLV) rating, and ash content. The flash point is provided mainly for safety and shipping reasons. It is determined by the "tag open-cup technique," as specified in ASTM D-1310, or by the "tag closed-cup technique," as specified in ASTM D-56. The Department of Transportation and OSHA require the use of the closed-cup technique. The "TLV rating" is simply a toxicity rating, generally given in parts per million and defined in detail in a book on the toxological properties of industrial materials by Sax.[6]

The "ash content" is specified as a percent by weight of material left after low-temperature oxidation or burning of the resist in an oxygen plasma. Ash content is a concern because of the use of plasma ashing for resist removal; remaining "ash" from such a removal operation may contaminate the devices during ion implantation, CVD, or other subsequent processing. The ash content of resists is typically specified as less than 0.03% by weight.

Refractive index

The refractive index of a material is the difference between the speed of light through that material and the speed of light in a vacuum. The refractive index of resists is needed in calculating the optical effects of exposure reactions and their impact on resolution. Matching the refractive index of the resist with that of the substrate is desirable to minimize optical distortion and maximize resolution. Lithography simulation models for resist imaging require refractive index data. The refractive index is derived by measuring liquid resist with a refractometer or similar device. The refractive index data for AZ-1300 series and AZ-111 positive resists are shown in Fig. 13.1. Dry film refractive index values are derived by ellipsometer measurement.

Accelerated-aging tests

The shelf life of many imaging products can be approximated by accelerated-aging tests in which samples of resist, for example, are subjected to conditions that simulate a longer period of time (1 year) in a relatively short period (hours). Since resist shelf life and aging are time- and temperature-related phenomena, exposure to environments of extreme temperatures for short time periods will approximate the same amount of chemical aging as temperatures of 21°C over a much longer time. These tests simply ensure the long-term chemical stability of a product and are generally part of the standard

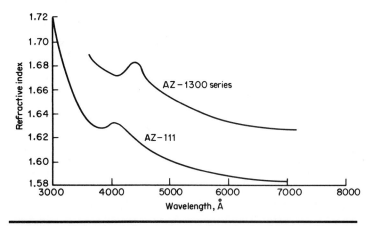

Figure 13.1   Refractive index versus wavelength of AZ resists. (*Courtesy of Shipley Company*).

batch-to-batch testing performed by resist suppliers. Since most resist users consume batches of photoresist well within the published shelf life of the product, it is usually redundant for users to run this type of test.

Quality control of resist developers is important because of the increasingly critical process controls needed to produce high-quality, nanometer-level lithography. Developers are tied directly to resists in the exposure-development reactions, and the interdependence of these reactions dictates the need for quality control of both solutions. Positive-photoresist (diazo-type) developers are typically mild aqueous alkaline solutions whose "active" ingredient is a hydroxide. Since the concentration of the hydroxide (in conjunction with the exposure dosage) largely determines the dissolution rate of the photoresist, variations in hydroxide level will cause apparent changes in resist photosensitivity. Tests have shown that the resist sensitivity change caused by a given developer normality change varies considerably with the products used, the type of equipment used, and a spectrum of process conditions that influence what is commonly referred to as "exposure development latitude." The level of concern, from a quality control point of view, should lie in specifying the developer normality, since the other factors are part of the resist system characterization discussed in Chap. 12.

Normality

Developer "normality" is determined by manual or automatic titration, and typical published procedures are available from most resist suppliers.

Overdevelopment tolerance

"Overdevelopment tolerance" is a simple, functional test for positive and negative resists. In this test, the resist coating is simply given 2 to 3 times its normal time in the developer after normal processing through exposure. This treatment will highlight any weaknesses in developer resistance and may be used to check changes in this property from batch to batch.

There is no specified procedure for the overdevelopment tolerance test except to follow a process sequence in which the parameters are those most likely to be encountered on the production line. In production, wafers are typically overdeveloped by a minimum of 20 to 40%, so 100% overdevelopment is appropriate. The test will indicate important latitude parameters such as line width change, pinhole incidence, and loss of unexposed resist. Developer attack rates on unexposed positive resist above 150 Å/min generally indicate a process problem and may not be a resist product problem.

Turbidity

"Turbidity" is the relative amount of light scattering caused by a liquid, and it is tested when there is concern over precipitation or incomplete dissolution in a given material. It can be measured in resist and developers with a Brice-Phoenix light-scattering photometer, a device that generates data via either a galvanometer or a strip chart recorder. Turbidity testing, although not commonly performed by resist material users, may prove useful when materials handling involves extreme temperature variations and the effect on the functionality of the product is questionable. For example, positive-resist developers shipped when they are exposed to freezing temperatures may experience precipitation, and even after storage at room temperatures, they may not appear to have completely redissolved. Since the turbidity test is relatively simple and quick to perform, it should be considered when there is any question about a product's chemical stability.

Solvent system analysis

The analysis of a resist solvent system is generally a supplier quality control procedure. A gas chromatograph (GC) is used to analyze for solvent composition; a small sample from the solvent system is injected into the GC unit. The instrument will show a peak area for each solvent component. The weight percent of a solvent is determined by multiplying the height of the peak by the width at 50% peak height (equal peak area). The GC system automatically integrates and calculates test values.

Since the balance of solvents in a resist solvent system determines the drying and coating results, improper solvent balance can cause serious processing problems.

Infrared absorption test

The way in which a resist system absorbs infrared energy tells much about system composition. "Infrared absorption analysis" is performed on an infrared spectrophotometer. The location and presence of chemical bonds in the resist system are obtained in the spectrum output.

Ultraviolet visible sensitizer absorption test

This is a quantitative test measuring the concentration of the resist sensitizer (photoactive compound) by diluting a known weight of the resist with its thinner and running a double-beam absorption spectrum in a spectrophotometer at approximately 350 to 450 nm (uv visible) or a region that is appropriate for the particular sensitizer being tested. An equation derived from Beer's law (absorbance = absorptivity × path length × concentration) is used to calculate the sensitizer concentration. Again, this test is one typically run by resist suppliers and some large-volume users of resist materials (who use several batches in a relatively short period and want to compare batch-to-batch control). In positive resists, the sensitizer concentration, relative to other ingredients, plays a key role in the exposure properties, the developing latitude, and therefore the practical resolution obtainable in production. Quality control specifications are generally not published externally but are often made available by the resist supplier when there is a need for them.

Resist purity tests

Resist filtration is used to ensure a given level of solution purity. Flow-rate decay is one type of purity test, but several methods are used to monitor the purity of resist solutions:

1. *Particle counting.* Particle counters or filter membrane grids can be used to determine the level of objects of a given size in a volume of resist. Particles may come from nondissolved solids in the resist, containers that are dirty before resist packaging, contamination in manufacturing, or chemical reactions of the resist (autopolymerization, precipitation) and nonsoluble gel slugs. Specifications for cleanness based in part on particle counts can be established by the user, but they are usually part of the supplier internal specifications. Figure 13.2 shows examples of particle contamination, most of which

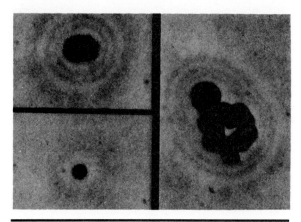

Figure 13.2 Particles filtered from a resist solution.

can be removed by a 0.2-μm absolute membrane filter. Some particles, such as insoluble species in the form of gel "slugs," may deform their way through a filter and still show up on a wafer surface.

2. *Void analysis.* "Voids" (Fig. 13.3) are micrometer-size circular dewetting spots that leave the substrate exposed in the center and will break down in etching. As in particle counting, they should be identified and specified in terms of a maximum level. If they are caused by some form of impurity, passing the resist through, or letting it stand for 2 days in, a molecular sieve or diatomaceous earth may result in absorption of the contaminant. If the voids are a result of incomplete solubilization of the chemical species, for example, they may not be

Figure 13.3 Void in a resist coating.

removable. Carbon filtration is also used to remove organic contamination.

3. *Residue or haze analysis.*   This test, usually run with the Gautner Haze Meter, detects the tendency for residuals to form in imaging.

Thermal flow test

Many resists have a point or narrow temperature range in which thermal flow occurs. Some processes depend on the amount of flow either for line sizing, to close micropinholes, or to maintain or maximize CD control. Most positive resists have a thermal flow point between 90 and 140°C. Thermal flow behavior should be periodically (each lot) checked by the quality control or process people. Checking each lot of resist for thermal flow is generally a resist supplier function and need not be done by the IC manufacturer. If there is some question about the amount of variability in this parameter, the customer should check with the supplier and review the test history.

Characteristic curve tests

As explained in Chap. 8, the "characteristic curve test" measures photosensitivity by depth of penetration at discrete wavelengths. Figure 13.4 shows the results of this type of test for AZ-1350J. Testing each lot of resist for sensitivity variation (performed by all resist manufacturers) by this method is especially valid for users who expose with aligners using only one or two wavelengths. Since the only sensitivity of concern is at those wavelengths, the user-related test would be to plot two such characteristic curves.

Molecular weight testing

Weight-average molecular weight testing of resist systems is done by using gel permeation chromatography (GPC). It is more sensitive to high molecular weight species, whereas number-average molecular weight testing is sensitive to low molecular weight species. Testing for number-average molecular weight is accomplished by using membrane osmometry.

Since these molecular parameters may change from batch to batch, resist manufacturers must "adjust" batches within limits that will keep the functional properties as constant as possible. The resist user does not generally run these tests.

Resist Filterability Test

The relative purity or cleanness of photoresists can be determined by filtering a known volume of the material through an absolute mem-

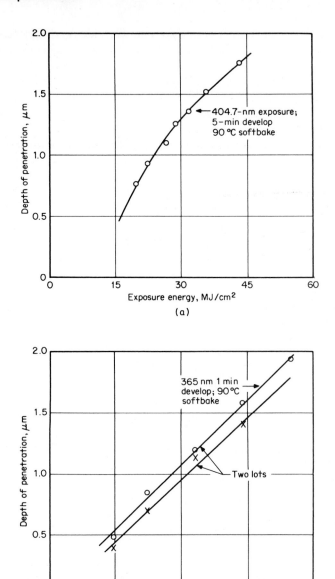

Figure 13.4   Characteristic curve for AZ 1350J at 404.7- and 365-nm wavelengths.[2]

brane filter of known pore size. The time required for the material to flow through the filter is its "filterability" parameter or constant. This "flow-rate decay" (rate slows as solid material in the resist blocks the filter) is calculated by a weight-versus-time parameter, and the filterability constant mentioned is expressed as N/N_0 (the ratio of the number of particles present capable of blocking the filter to the number required to totally block the filter). Thus a lower filterability constant indicates a cleaner resist product.

Filter types

1. *Absolute filters.*   Absolute filters retain contaminant particles on their surfaces and will produce the cleanest material. Particle elimination can be to a level of 0.1 µm, which is essential for many semiconductor and microelectronic applications. However, absolute filters usually have slower flow rates and less retention capacity and are more expensive than other filters. Submicron absolute filtration should be conducted in a Class 100 clean-air environment. Moreover, this type of filtration requires special cleaning for packaging materials and all equipment downstream of the last absolute filter.

2. *Depth filters.*   Depth filters entrap contaminate particles within the filter media. They will exhibit a faster flow rate and a greater retention capacity, and the filters themselves are less expensive. Particle elimination can be achieved to a level of 0.3 to 0.5 µm. Large batches of resists are filtered through filters of these types.

Testing on the IC Line

A series of test methods that cover most aspects of IC fabrication, aside from those already covered, is published by the National Aeronautics and Space Administration (NASA) under the title "Quality Control during IC Processing."[5] The tests covered in this document are listed by method number in Table 13.1. The document itself may be obtained by writing to the Superintendent of Documents, U.S. Government Printing Office, Washington, D.C. 20407; request TSP78, Vol. 4, No. 2, MFS-25112.

Summary

The concern for near-absolute purity and consistency increases as requirements for IC processes tighten. The use of relevant tests in quality control of resist and other imaging solutions is highly recommended. We have outlined a wide variety of physical, chemical,

TABLE 13.1   IC Quality Control Tests

Method 1	Water resistivity
Method 2	Measuring particulate matter in water
Method 3	Total dissolved and suspended solids in water
Method 4	Measuring oxidizable organic impurities in water
Method 5	Determining bacteria count in high-purity water
Method 6	Lifetime measurement of semiconductor materials
Method 7	Resistivity measurements of semiconductor materials
Method 8	Orientation of substrate with respect to crystal plane
Method 9	Substrate orientation
Method 10	Measurement of oxygen content of silicon
Method 11	Silicon crystal perfection
Method 12	Crystal perfection using bent x-ray topography
Method 13	Substrate conductivity type
Method 14	Substrate thickness, taper, and flatness
Method 15	Substrate width or diameter
Method 16	Substrate surface finish
Method 17	Thickness of epitaxial layer (infrared spectrophotometer)
Method 18	Thickness of epitaxial layer
Method 19	Thickness of epitaxial layer (stacking fault)
Method 20	Defects in epitaxial layer
Method 21	Temperature monitoring during epitaxial deposition
Method 22	Measurement of film thickness
Method 23	Capacitance-voltage measurements to determine oxide quality
Method 24	Measurement of line width and spacing
Method 25	Viscosity measurement of photoresist
Method 26	Measurement of specific gravity of photoresist
Method 27	Photoresist exposure control
Method 28	Detection of pinhole defects in insulating layers
Method 29	Gross pinhole detection in photoresist films
Method 30	Oxide and photoresist pinhole determination
Method 31	Photoresist spinner calibration
Method 32	Junction depth and base-width measurement
Method 33	Furnace profiling
Method 34	Determination of ion-implanted dose and uniformity
Method 35	Diode and dielectric isolation
Method 36	Isolation leakage
Method 37	Metallization adherence
Method 38	Quality of ohmic contacts
Method 39	Grain boundary inspection
Method 40	Electromigration
Method 41	Sizing and counting of airborne particles (0.5 to 5.0 μm) in dust-controlled areas
Method 42	Cross-sectional uniformity of metal
Method 43	Integrity of glassivation

functional, and analytical tests each of which has a degree of merit to be weighed against the needs of the process. Certain basic tests should be run by both supplier and user; others are more strictly supplier-oriented. Good communication with the supplier will often obviate the need to run many of the latter.

REFERENCES

1. American Society for Testing and Materials, Published Methods and Procedures: F25-68, F-66-77-0445-65, D891-59, and D-1364-78.
2. J. Bohland, Shipley Company, private communication relating to exposure-determination methods.
3. "Optoline Resolution-Sensitivity Test Pattern," Optoline Corporation, North Andover, Mass., 1980.
4. Technical Data Sheet, "Filtering Photoresists," Shipley Company, Newton, Mass., 1980.
5. Technical Support Package, "Quality Control during IC Processing," George C. Marshall Space Flight Center, Alabama, 1980.
6. N. I. Sax, *Dangerous Properties of Industrial Materials,* Reinhold, New York, 1980.

Mask Fabrication

Introduction

Substrates containing IC patterns that are transferred to wafers are called masks or reticles. Masks are 1× plates; reticles are generally 5× or 10× the final image size on the wafer and are optically reduced in a wafer stepper. A mask covers the entire wafer; a reticle contains a master pattern that is stepped and reimaged many times on a single wafer. Thus, 1× reticles are to be distinguished from 1× masks.

Masks are produced by a variety of manufacturing "routes," all of which begin with the mask designer. Figure 14.1 traces alternative routes, moving from the designer to the computer, which optimizes the layout of each mask level so that the entire mask set needed to fabricate an IC utilizes surface area optimally. The computer-aided design (CAD) information, generally on magnetic tape, is fed into an e-beam direct-writing system and an e-beam mask generator or pattern generator. Only in the case of the e-beam direct-writing system is the use of a metallized glass mask avoided and the pattern information transferred directly to the wafer. All other pathways to finally delineate the images onto the wafer involve the use of an electron beam or optically generated photomask.

We will trace the steps necessary to generate a high-quality photomask beginning with the elements of the mask: glass or quartz and metal, metal oxide, or emulsion. The overall requirements for these mask materials is to produce a mask that is consistent in quality, provides high resolution, has low defect levels, is optically compatible with exposure equipment, resists, and cleaning and etching solutions, and is cost-effective. Technology changes that impact mask fabrication are increases in wafer diameters, reduced tolerable defect levels, shorter resist exposure wavelengths, and smaller IC geome-

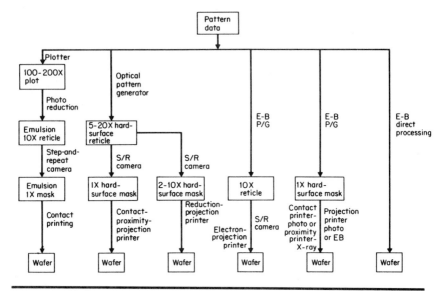

Figure 14.1 Pathways to mask fabrication.

tries. Dimensional control for every aspect of the lithography process has increased to meet these trends.

Pattern Design and Data Production

The use of CAD in VLSI design is widespread and is considered by many experts to be one of the most developed applications of CAD. There are several aspects of VLSI CAD, including:

1. Current VLSI CAD capabilities and trends
2. Device modeling
3. Logic simulation, device analysis, and time studies
4. Interconnection strategy
5. CAD relationship to lithographic processes
6. CAD relationship to testing chips and designs

The level of integration seems to be a common measure of the rise in IC complexity. Early integrated circuits had up to 10 logic gates; the next generation, medium-scale integration (MSI) circuits, typically had between 10 and 100 logic gates. After MSI came large-scale integration (LSI) with between 100 and 1000 gates. The current VLSI devices have over 1000 gates each. The next layer of integration is ultralarge-scale integration (ULSI).

The basic steps used to conceive, design, test, fabricate, and chip test are essentially the same ones as were used years ago. In summary form, they are as follows:

1. Define and optimize the IC fabrication and interconnection processes.

2. Electrically define circuit elements.

3. Design logic schematic.

4. Convert logic design to mask patterns (geometries).

5. Electrically test simulation of design to detect flaws.

These steps originally involved many hours of engineering time, since all were performed manually. The pattern dimensions were arrived at by evaluating several different geometries and measuring the yield of each test chip. Final-pattern geometries were heuristically derived. The end result of this largely trial-and-error approach was a set of specifications for the electrical parameters and design rules (physical parameters) to match. Tolerances for each geometry were specified, as they are today, to keep yield at a maximum level.

In the next decade, as device complexity increased, the manual steps became so cumbersome and labor-intensive that shorter, more effective methods had to be devised to preserve overall process economies. The problem was solved by digitally encoding the IC pattern geometries. The machine that accomplished this task became known as the "optical pattern generator." The actual method involved converting the final drawings into data tapes by placing an electromechanical digitizer over a drawing and tracing the circuit pattern.

The major advance made possible by digitizing the artwork was computer handling of the IC pattern data. Now the computerized artwork could be run through a design-rule-check (DRC) program to test for open or short circuits or other design flaws. Once a flaw in the IC pattern artwork was detected, it could be easily changed via the digitizing system. In earlier times, where Rubylith was used, an entirely new piece of artwork would have to be scribed to correct an error. The electronic manipulation of interactive graphics saved hundreds of hours of artwork generation time.

The next innovation was software to simulate the circuit functions. Many different programs were developed to perform IC simulation, and they resulted in good verification of the layout designer's work. In addition, the interactive graphics equipment could print out a drawing of the final digitized tape after modifications. An important benefit of this software was the ability to automatically verify the match between design layout metallization or interconnections and the balance of the design. An example of an interactive CAD system is shown in Fig. 14.2.

The increased use of automatic placement computer programs was helped by newer and simpler design styles. For example, the gate arrays and other designs with standard cells in a regular pattern were more easily designed because of the repeatability of the units. The gate array, or standard slice, shortened the total design time and did not sacrifice precious silicon real estate in the process. Thus, free-form layout began to give way to automatic design and test programs. The overall job of producing first a design and then the data from which masks are made is complex and involves many individual steps. The basic building blocks used to reach the point of reticle production are shown in Fig. 14.3.

Design rules

Design rules are defined as the physical constraints on the circuit geometries imposed by device physics and pattern lithography processes. The rules give the designer the minimum geometrical boundaries for line widths, overlay tolerances through the entire mask set, and the rules for extensions of given circuit elements on the chip.

Figure 14.2   CAD system with interactive graphics. (*Courtesy of VIA Systems, Inc.*)

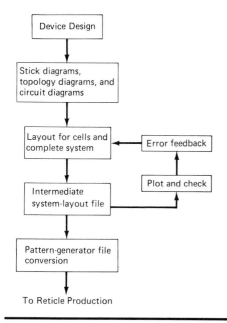

Figure 14.3 Primary steps in design and data production.

Device physics changes only as the substrate material and overlaid materials (deposited oxides, silicides, metals, dopants) change. The electrical parameters for silicon and silicon oxides are fairly well mapped, and so also are those for the semiconductor materials commonly used, including polysilicon, silicon nitride, and aluminum and its alloys. New metal silicides, gallium arsenide, and new polymer dielectrics are examples of changes in material technology that affect the designer's activities. They possess different electrical properties and react differently in the fabrication process than the other materials mentioned. Also, the scaling rules will most likely change for new materials, and the limits must be defined.

Pattern elements and their electrical results

In considering the design rules to follow for a given circuit, the designer first looks to the basic mask levels. The relations among pattern sizes and electrical working parameters essentially determine what a chip does. Diffused regions, polysilicon layers, ion implant regions, contact sizes, metal spacing, and pad area are examples of sections that have distinct electrical requirements that in turn are directly related to the pattern size.

Diffused regions. Diffused regions always have a minimum spacing that is determined by both lithography and electrical requirements. In the design stage, placing two diffused regions too close together could cause serious problems. For example, depletion layers are associated with the junctions of the diffused regions. When spacing is not sufficient, the layers will overlap and the result will be current flow where none was desired.

Ion implant regions. Design rules for ion implant regions depend on the type of device: If the implant is to become a transistor gate, its spacing parameters will be different from those in a device in which the ion implant region is placed close to but is separate from an enhancement mode transistor gate region. In the first case, the design should provide for extension of the ion-implanted area so that it overlaps the entire gate area. The actual distance it extends beyond the gate is also specified in the design rules.

Polysilicon regions. Line widths of polysilicon are not associated with depletion layers and generally can be placed closer together than diffused regions can be. The reflectiveness of the polysilicon is a factor in spacing widths in design rules; reflections from polysilicon during resist patterning affect the dimensions formed in the resist. Design considerations for polysilicon lines overlapping on other areas must be stated; for example, the overlapping of a polysilicon line and a diffused line can cause an unwanted capacitor. Polysilicon-gate dimensions are also very critical to specify. For example, polysilicon lines running over a diffused area will form a transistor. If the polysilicon-gate area is too small and does not extend a sufficient distance beyond the diffused area, a short circuit will occur.

Contacts. The contact mask establishes ohmic contact between a metal layer and polysilicon or a diffused layer. The design rules for contacts will specify the diameter of the contact. If the contact area is too small, there may be problems getting good electrical connection to the bottom layer. Contacts that are too large may make electrical connection to another, unintended layer. Other rules for contacts apply to placement inside the IC pattern. For example, contacts cannot occur over certain gate regions if the gate regions are too thin. When the gate or polysilicon thickness is sufficient, contacts that are placed only partway into the polysilicon layer can be designed.

Metal layers. Interconnection level metal or metallization is the covering that goes over all previously etched oxides, nitride, polysilicon-doped glass, and other materials. The design rules for metal are given somewhat favorable treatment, since they are expected to cover steps higher than any previous level.

Aluminum metallization is typically required to completely cover the contacts. Metal or silicide sizing for both contact coverage and current-carrying capability throughout the chip to the pad areas is critical. In some circuits, small nicks or edge defects will occur, and since the design rule for the metal level was wide enough or sufficiently large, the circuit will continue to function.

Design rule limitations. Overall electrical parameters are derived through careful study of the design rules and the establishment of electrical values that fit the chip specifications. The smaller the designs for IC devices become, the closer the devices approach the physical, electrical, and lithographic limitations of the technologies used to produce the circuits. Many parameters do not scale in a linear fashion. New semiconductor materials (silicides, gallium arsenide, etc.) are then used to extend the technology.

Silicon compilers. The concept of a compiler is one of automating the definition of the macro- and microfunctions of an integrated circuit from inception to final layout. The description of the circuit given to the compiler is more abstract than that used for manually designed circuits. The silicon compiler starts with the functional circuit description and essentially synthesizes the pattern, including descriptions of the major blocks provided by automatic definition.

The input for a silicon compiler can be in the form of a flow diagram (Fig. 14.4). In this case, the example used is a programmable logic array (PLA) moved from the functional level to the layout level by using a silicon compiler. The flowchart is converted into a circuit diagram (Fig. 14.5) and then into the actual pattern (Fig. 14.6). The placement and routing of the circuit are performed automatically with a chip assembler, which also draws the final layout shown in Fig. 14.6.

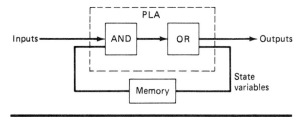

Figure 14.4   Flowchart for a PLA.

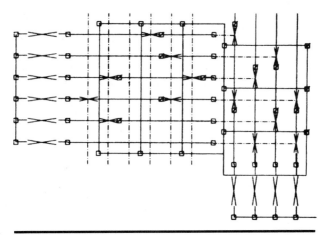

Figure 14.5 Circuit diagram of a silicon compiler.

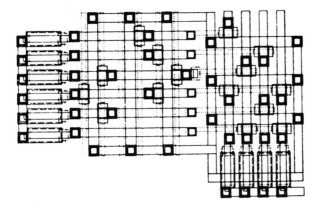

Figure 14.6 Pattern data for a silicon compiler.[8]

Device modeling

Device modeling helps circuit designers to better predict the behavior of their designs; it also serves to show how a given IC will operate before the cost of producing a mask set is assumed. Semiconductor device modeling is essentially a numerical simulation of a given IC topology, an example of which is shown in Fig. 14.7.

Several mathematical equations (electromagnetic field equations) are used to model IC device behavior; they match the behavior of electrons in the working device and hence are considered valid in modeling. Since some of them are nonlinear and some of the behavior in the IC is linear, additional models that incorporate approximations of de-

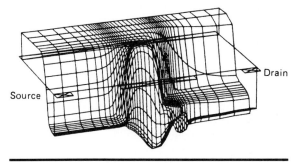

Figure 14.7 Three-dimensional diagram from a device-modeling study.[9]

vice behavior are used. The resulting three-dimensional model allows designers to study the internal workings of the device they have tentatively structured. The example given in Fig. 14.7 is log-scaled hole density resulting from a simulation including avalanche generation.

In the future, additional computational power and more sophisticated programs that include circuit, process, and device parameters in a single program will make modeling an even more powerful tool in the hands of design engineers. In summary, the IC design process and functions have evolved rapidly, as shown in Fig. 1.4.

Photomask Quality Glass

The glass used to fabricate high-quality photomasks must meet several important criteria:

1. It must be nearly defect-free both on the glass surface and in the internal structure.
2. It must have high optical transmission in the wavelength range used to expose photoresists.
3. It must be chemically resistant to glass- and mask-cleaning solutions and environments.
4. It must have low thermal expansion.

Glass types

A number of different glass types, each with properties suited to a specific use, are used in mask fabrication. Green soda-lime glass is relatively easy to draw into flat, parallel sheets. It is the most common type used for mask manufacturing, and its supply is abundant. The quality of green soda-lime glass is good, being relatively free of major surface defects or volume inclusions. The large sheets originally pulled are inspected, and selected areas are taken out to be used for photomasks. Green soda-lime glass is the most economical of glasses used in mask making.

In electron beam resist imaging, the mask substrate under the resist can reach temperatures up to 180°C or higher. On chrome-coated glass, the heat can cause a sodium-chrome reaction that results in resist undercutting (called "mouse nipping"). The lower sodium content of white soda-lime glass compared with green soda-lime glass is desirable because it minimizes this problem. In applications in which the substrates reach high temperatures, low-sodium white soda-lime glass or quartz should be used in place of green. The main factors in the use of low-sodium glass are the much greater polishing costs and a lack of suppliers, again compared with green soda-lime glass.

Borosilicate glass is considered a low temperature expansion (LTE) material; the green and white soda-lime glasses are classified as high temperature expansion glasses. Borosilicate glasses are even better suited to the high-temperature situations encountered in either mask making or mask use. Thermal stability makes this material well suited for reticle manufacturing and e-beam mask making, in which dimensional stability is more critical. In projection printing, ambient temperatures can change up to 5°C and cause shifts in the image dimensions on the wafers. For example, a 2°C change in temperature can cause 2 μm of run-out on a 5-in plate with soda-lime glass. Alumina-borosilicate glasses are very low in alkali content, many being below 2% and some below 0.15%. Borosilicate glasses are, unfortunately, very difficult to grind and polish and typically have a greater incidence of opaque inclusions that can cause printing defects. The availability of borosilicate glass is even lower than that of white soda-lime glass, and the price of substrates is several times that of green soda-lime glass.

Quartz is classified as an ultralow thermal expansion (ULTE) material; it has less than 0.2 μm of run-out over a 6-in area for several degrees Celsius temperature change. Figure 14.8 compares the run-out of quartz with that of the borosilicate and green and white soda-lime glasses. Quartz is nearly pure silicon dioxide with only small amounts of other oxides; it is desirable mainly for reason of its very low thermal expansion properties, which are due mainly to a very low sodium content. The transmission of quartz is also very high in the near- and deep-uv portions of the spectrum.

The disadvantages of quartz are greater incidence of inclusions that may act as opaque spots in printing and the very high cost—at least 7

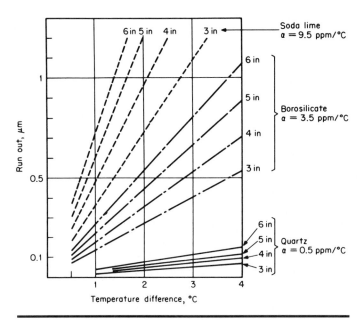

Figure 14.8 Run-out as a function of wafer size, substrate material, and mask temperature.[2]

times the cost of soda-lime glasses. However, the cost of quartz for deep-uv applications in which few masks are required in projection printing is small relative to the gains in resolution and dimensional stability. Individual materials costs must always be weighed against their cost contribution to the total process. In applications outside deep uv, the added price of quartz for its added thermal stability could not be justified. Quartz also fractures easier than the other glasses.

Glass composition and properties

The approximate chemical compositions of the major types of mask glass are shown in Table 14.1. Silicon dioxide (SiO_2) is the major ingredient in all glasses, and quartz is almost pure SiO_2, with a combination of other oxides comprising the balance. Very small changes in the percent composition of glass ingredients can cause major changes in the functional properties. For example, small changes in the alkali content, on the order of tenths of 1%, will dramatically alter the ease of grinding and polishing the alumina glasses.

There are only a few materials (LiF, CaF_2, MgF_2, sapphire, and quartz) which are fully transparent in the deep uv and which can be used to make optical elements like lenses or prisms. For masks, only quartz is of interest because sapphire is more expensive and has a 10

TABLE 14.1   Characteristics of Common Types of Glass

Characteristic	Type of Glass			
	SLW (Soda lime)	SL (White crown)	LE-30 (Low-expansion)	Qz (Fused quartz)
Chemical composition, %				
SiO_2	73	70	60	100
B_2O_3	1		5	
Al_2O_3			15	
Na_2O	15	8		
K_2O	1	9		
CaO, MgO	10	12	20	
Others		1		
Physical properties				
Thermal expansion (10^{-7})	94	93	37	5
Transformation temperature, °C	542	533	686	1120
Refractive index, N_D	1.52	1.52	1.53	1.46
Young modulus, kg/mm^2	7200	7340	7540	7410
Knoop hardness, kg/mm^2	540	530	657	615
Electric resistivity $(\Omega \cdot cm)$	10^{12}	10^{15}	10^{15}	10^{18}
Specific gravity	2.50	2.56	2.58	2.20

times higher thermal expansion. The fluorides have a still higher thermal expansion, are less hard, and chemically are not sufficiently stable against the cleaning processes necessary for masks.

Transmission.   Pure quartz is fully transparent from 200 to 1300 nm. Metallic impurities like Al, Ca, Fe, and Ti (1 ppm in synthetic and 10 ppm in good natural quartz) reduce the uv transmission, while the H_2O content (approximately 600 ppm for mask-grade quartz and 5 ppm for special IR grades) produces absorption bands in the IR at 1.4, 2.2, and 2.7 μm. Usually only one kind of impurity can be brought to a minimum; quartz for mask substrates will therefore have a minimum of metallic impurities in order to obtain the highest transmission down to 185 nm. Shorter wavelengths are strongly absorbed by O_2, which requires the light path to be held under vacuum. In general, very clean handling (fingerprints to be avoided!) is necessary to maintain the full transmission for wavelengths below 200 nm.

Chemical properties.   The outstanding chemical durability of SiO_2 is well known. Noticeable attack occurs only by hydrofluoric acid (HF), hot alkali, and hot phosphoric acid and its compounds. The attack rate at room temperature is 1 $\mu m/min$ for 40% and 0.07 $\mu m/min$ for 5% HF.

Fluorescence.   All glasses will fluoresce when illuminated with uv light below 300 nm. Most grades of quartz will also. Synthetic quartz is an exception; it shows no or only extremely weak fluorescence.

Devitrification.   Amorphous (vitreous) SiO_2 will crystallize if heated above 1000°C. During cool-down, quartz parts can crack because of partial crystallization. Devitrification is no problem in mask technology, in which such high temperatures are never attained.

Thermal hysteresis.   It is often not taken into account that glass will not follow rapid temperature changes immediately. Quenching usually leads to a shrinkage of the mask (the outer, more rapidly cooled glass compresses the interior), and annealing can help to elongate plates. Dimensional changes of 0.3 μm over 2 in within 3 days and 0.6 μm over 6 in within several months have been observed. The effect is less pronounced in borosilicate glass and can nearly be neglected in quartz. Rapid temperature changes can occur due to resist baking, plasma etching, or cleaning operations. Insensitivity to such temperature changes is a further advantage of quartz often not taken into consideration.

Mask structure

Mask plates used today in production are typically coated with a thin layer of chrome or antireflective chrome ranging in thickness from 600 to 1000 Å. In the case of antireflective chrome, a thin layer of chrome is deposited onto the glass, followed by a 200-Å coating of chrome oxide. The illustration in Fig. 14.9 shows the cross-sectional structures of chrome and an antireflective chrome plate.

The transmission properties of all these materials is very important, since any significant loss of usable resist exposure energy will reduce wafer exposure throughput. The transmission properties of the materials cited in Table 14.1 are shown in Fig. 14.10. Note the high transmission of quartz at the deep-uv wavelengths and the high transmission of all the glasses relative to the primary mercury lines indicated on the chart. Transmission specifications have been prepared by the Semiconductor Equipment and Materials Institute for all glasses

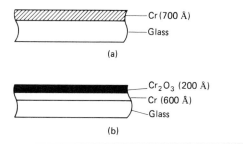

Figure 14.9 Cross section of (a) chrome and (b) antireflective chrome substrates.

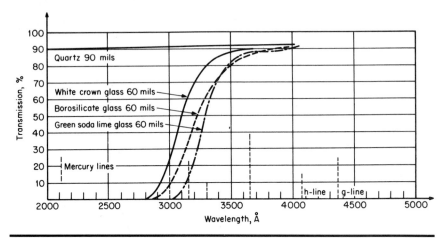

Figure 14.10   Transmission versus wavelength of different substrate materials.[2]

mentioned. The specifications call for the minimum transmission percentages at 1.5 mm of substrate thickness.

Glass melting and forming

Sand and silica are reduced to a molten mass and refined, and impurities are removed until an acceptable level of quality is reached. Frequently, raw glass sheet is made at this point. Borosilicate glasses, for example, are produced in quantity and then melted. The processes that lead up to the production of a raw glass plate are shown in Fig. 14.11. The melting of different types of glass is a complex and highly specialized operation. Borosilicates are melted in special platinum-lined furnaces that were originally used to produce high-purity optical glass.

By means of computers, melting furnaces are able to maintain close control over the physical and chemical properties of the glass. One example of composition control is the reduction or even elimination of alkali in mask-quality glass. Alkali increases the dimensional instability of glass in which it is present. Reduced alkali reduces the blank thermal expansion coefficient. The chemical and physical resistances of the glass also are increased as alkali content is decreased, and ion migration is reduced.

The following are the typical steps taken to produce the final mask blank: melting, raw glass sheet, cutting, smoothing, polishing, cleaning, surface inspection, flatness check, and final glass substrate.

Glass Preparation

The glass selected for the photomask fabrication process is cut from larger sheets of drawn glass. The demand for inclusion-free and sur-

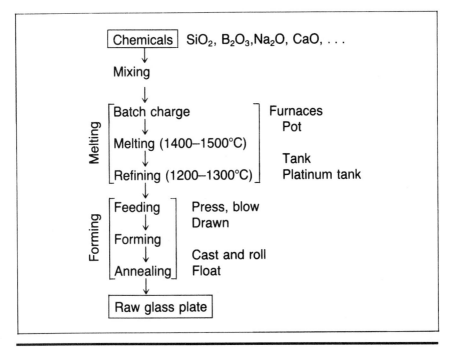

Figure 14.11 Glass melting and forming.[10]

face defect–free material calls for careful custom selection. After being cut to size from the larger sheets, the glass plates are cleaned to remove chips along the edges and are then graded for flatness. A cleaning procedure for soda-lime glass prior to metallization is given in Table 14.2.

Mask Flatness

Mask flatness is checked with a variety of equipment types including grating incidence interferometers, laser interferometers, capacitive gauges, air gauges, and other optical interference devices. Laser interferometers are widely used for their resolution capability and accuracy. They have resolution and accuracy down to below 0.3 μm with precision of 0.6 μm. Laser interferometry can provide full flatness profiles in an off-contact mode and glass positioning at 20°. Off-contact measurements are needed to prevent distortions that would cause erroneous readings.

Warpage

Warpage is a deviation of a mask surface that results in a concave or convex shape that equals the mask diameter. It can be caused

TABLE 14.2   Soda-Lime Glass–Cleaning Procedure (Prior to Metallization)

Glass grade	Wavelength transmission	Percentage
High thermal expansion (HTE)	3660 Å	80%
	4050 Å	85%
	4360 Å	90%
Low thermal expansion (LTE)	3660 Å	80%
	4050 Å	85%
	4360 Å	90%
Ultralow thermal expansion (ULTE)	4050 Å	90%
	3660 Å	85%
	2540 Å	70%

by stresses in the glass that were not properly relieved or by having the mask under unequal pressure for a protracted period. High temperature also can cause warp in a mask blank. Warpage is closely correlated to substrate thickness, and mask blanks that are too thin are easily warped.

Plastic flow

More subtle factors that affect microlithography are internal changes in the material itself. Plastic deformation is one example of such a change. Mask blanks in process are subjected to various temperatures, followed by much cooler rinses or air drying. The temperature itself can cause internal changes in the glass structure. Nonuniformities in glass composition, for example, will give rise to nonuniform stresses, warpage, or plastic deformation. These process-induced distortions are generally small; they cause 0.1- to 0.3-μm changes.

Vacuum chuck distortion

Nonflatness in masks can be caused by pressing masks under vacuum against various types of vacuum chucks that will stress and bend the glass. Chucks have many different configurations, including concentric grooves, exhaust ports, and other machined areas that cause nonuniform mask pressure. Generally, the smaller the size of the groove or open area in a chuck, the more uniform the pressure. Pinhole ports are thus the best for minimizing this contact-printing problem.

Gravitational sag

The force of gravity on a mask blank is sufficient to cause a sag that will in turn distort the patterning of images. Gravitational sag has a predictable pattern and varies with the type of fixture used. The plots in Fig. 14.12 indicate the relations between the types of fixturing used

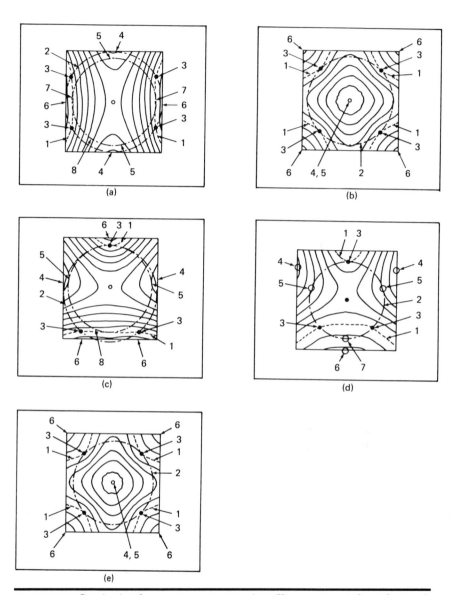

Figure 14.12 Gravitational sag versus support points. Key to points indicated on each contour plot: 1. Contour line of zero deflection. 2. Support pin circle. 3. Support pin. 4. Point of maximum negative deflection on substrates. 5. Point of maximum negative deflection within flatness quality circle. 6. Point of maximum positive deflection on substrate. 7. Point of maximum positive deflection within flatness quality circle. 8. Flatness quality circle. (*Courtesy of Corning Glass* Technical Bulletin.)

to hold the mask blanks and the surface contours that result. The information is based on computer calculations and plots that started by assuming a perfectly flat and parallel plate. The variables that are included in this test, run at Corning Glass Works in Corning, New York, are as follows: substrate size (length, width, and thickness), support mode (number of support pins, location of support pins relative to the substrate length and width, and the plane of support relative to the horizontal), and glass composition (as composition relates to the glass density, modulus of elasticity, and Poisson's ratio).

The five contour plots in Fig. 14.12 indicate the deflections due to gravitational sag, which is inversely proportional to the square of the glass thickness. A good horizontal support fixture for a glass plate is a continuous vacuum chuck, which has a grooved channel around the edge of the mask and generally on its top or working side. Many new types of vacuum fixtures have been developed to reduce the problem of applying improper pressure to mask blanks. A key parameter in all areas of concern for mask flatness is glass thickness. In short, the thicker the glass, the greater the possibility of maintaining overall flatness.

Flatness classifications are given in terms of the amount of run-out, in micrometers, over a specified plate size. Additional specifications for the number of visual defects also are used.

The capacitive gauge works by comparing capacitance values between the internal structure and external values, measured from a probe tip to the glass surface at various points. Single-probe testers move across the glass surface and take multiple readings; a multiprobe tester allows the glass to be stationary. The capacitance signals are converted to a digital display of flatness. Single-probe testers yield a typical resolution of 0.25 µm, accuracy of 1 µm, and precision of 1.1 µm. A multiprobe tester yields typical accuracy of 1 µm, resolution of 0.03 µm, and precision of 0.6 µm. The main advantages of capacitance gauges are high resolution and rapid digital flatness readout. The potential disadvantages are substrate movement in the single-probe system and lack of a full contour display to evaluate the substrate visually.

Stabilization time

Masks heated in the production line require various times to stabilize. The stabilization time versus run-in and run-out is given in Fig. 14.13 for three types of glass blanks. The SL type is white crown glass (Hoya), and the LE-30 is a borosilicate glass, also from Hoya. The Q-18 and Q-25 glasses are Hoya quartz.

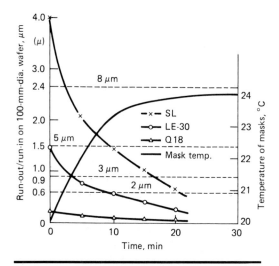

Figure 14.13 Run-in and run-out versus stabilization time for three types of mask glass.

Cleaning, Polishing, and Inspecting

Glass plates cut from larger sheets must be polished and thoroughly cleaned and inspected before being coated with the mask-forming material. Polishing is used to increase the flatness of the glass and remove surface irregularities and defects. It is a multistep process involving several grades of abrasive compounds working toward a fine rouge that puts an optical quality surface on the plate. Both the front and the back sides must be polished, because backside defects will cause reflections that may result in printed defects. As a rule of thumb, any defect visible at 1000× magnification should be removed. After polishing, plates are given a multistep wet-chemical cleaning in which glass-cleaning solutions and ultrasonic agitation are used to dislodge loose chips on the edges as well as particles embedded in the glass surfaces.

Following multiple rinsing in deionized water and/or an alcohol-water drying, the plates are ready for inspection. Filtered dry nitrogen also is used for drying prior to inspection to completely remove water without staining.

Properly cleaned photomask blanks are then inspected for degree of cleanliness and defects. Plates should be free of surface stains, residues, or other contaminants as well as such bulk or surface defects as pits, scratches, bubbles, and other irregularities visible with an optical microscope at 1000× magnification.

Photomask Glass Coating or Metallization

The inspected glass blanks are coated with one of several materials to form the mask layer for subsequent imaging. Chrome (reflective and antireflective), iron oxide, silicon, chrome oxide, and oxides of other metals are used to form the thin masking layer. Mask coatings are deposited by using one of three basic technologies: (1) sputtering, (2) vacuum deposition, or (3) chemical vapor deposition.

Sputtering

Sputtering works on the principle of producing an electric discharge between two electrodes under a few millitorrs of pressure that causes the cathode to disintegrate and coat the anode (plate). The advantages of sputtering include high-speed deposition of material, simple fixturing of the substrate, high energy of the particles reaching the substrate for better adhesion (10 times vacuum deposition), and high homogeneity. Moreover, by using bias sputtering, the substrate can be etched during the deposition process to ensure a clean and well-prepared surface. Ion bombardment *before* the deposition is used for the same purpose. This technique gives better adhesion than glow discharge vacuum deposition.

Sputtering does, however, have some disadvantages, including more complex equipment, the need for operators who are more highly trained, high-energy particle damage to the substrate, and contamination problems. Radio frequency and dc sputtering are used to deposit iron and chrome oxides, and heating of the substrate is sometimes considered a disadvantage that overrides the simplicity of this deposition technique.

Triode, tetrode, and similar sputtering discharge techniques are applied at low pressures, which help provide high-purity materials at high efficiencies. The cost and complexity of these techniques are, however, higher than those for rf and dc sputtering.

Chromium applied by sputtering is more reflective than vacuum-deposited chromium (60 to 65% compared with approximately 50%). The higher reflectivity indicates a more dense or compact layer. The same statements apply to iron oxide, which also is generally deposited by sputtering techniques.

Vacuum deposition

Chromium is coated onto photomask blanks by vacuum deposition: heating in a vacuum system at 10^{-5} to 10^{-6} torr pressure. The heated chromium, which has a melt temperature of 1900°C and a sublimation temperature of 2200°C (only 1500 to 1600°C at 10^{-3} to 10^{-2} torr), is reduced to a vapor and coats the substrate, which lies above the chromium crucible or heated source. As the chromium cools, high surface tension (approximately 10,000 kg/cm^2 for a 1000-Å-thick layer) forms between the glass and metal layers. A drawback of vacuum evaporation of chromium is the relatively low energy of the chromium particles reaching the glass, which results in relatively low metal adhesion or bonding energy. Subsequent cleaning of chrome blanks or fabricated masks causes chrome lifting or separation. In vacuum deposition, the glass is often heated to provide better chromium adhesion and increase the density of the film. Another technique used to improve the quality of the chrome layer is rotation of the glass during deposition, which permits layer uniformity better than 2%.

The older crucible technique for chrome evaporation is slow and expensive but makes efficient use of the chromium. Electron beam evaporation is now widely used; it provides rapid and relatively clean films at good production rates.

Chemical vapor deposition

Chemical vapor is widely used in both IC and photomask fabrication for the deposition of dielectric layers such as silicon dioxide and iron oxide. CVD processing occurs at low pressure in a reactor at about 200°C. The low temperature of the gases and the glass minimizes thermal distortion of the blanks. The oxide is formed by the reactions of gases in the chamber, and the by-products of the reaction are rapidly removed by the gases flowing through the system. Bonding of the deposited oxide is aided by the energy of the deposited material forming on the glass.

The major variables that will affect the quality of deposition and overall reaction parameters are:

1. Types of gases used
2. Gas concentration
3. Gas flow
4. Reactor design
5. Process temperature and control

Reflectivity

Another property to monitor in exposure through a photomask is the amount of reflective interference. The formula for determining interference minima is

$$2ne + z = \lambda k$$

where e is the layer thickness, k is any number, n is the refractive index, λ is the wavelength, and z is the phase variation at air-oxide and oxide-glass interfaces. Thus any optical thickness that is a multiple of $\lambda/2$ will provide maximum transmittance and minimum interference.

The proper deposition of these various materials is critical to the subsequent imaging and etching processes. For example, contamination in a vacuum system, from either poor cleaning or another source, may codeposit with the film and cause resist adhesion problems in etching, the result being poor edge sharpness. Quality control of these deposition steps and maintenance of the equipment are essential.

Mask materials

Chrome is the most commonly used hard-surface mask material; it has high film hardness for durability and cleaning resistance, high resolution, and long mask life. Its main disadvantages are high reflectivity and visible opacity. The high reflectivity causes a greater variability in printed line sizes in photoresist (Fig. 14.14), and the visible opacity presents alignment problems on some mask levels. Antireflective chrome coatings, which are chrome layers with 0.02 μm of chrome oxide on the surface, solve the reflectivity problem yet retain all the good properties of chromium. The reflectivity of various

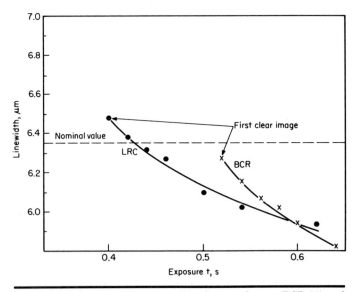

Figure 14.14 Stepping of masters directly onto chrome (BCR 20) and low-reflectivity chrome (LRC 30) versus line width.[5]

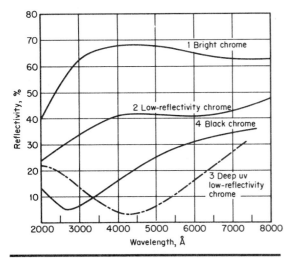

Figure 14.15 Chrome coating reflectivity.

chrome coatings is shown in Fig. 14.15. Note the low reflectivity of the
deep-uv chrome at approximately 2700 Å.

Emulsion masks are inexpensive and can be autoprocessed in special plate-handling equipment. The emulsion offers an advantage
by its thickness, which can support or contain an embedded particle
or epitaxial spike without adversely affecting the printing results.
However, the thick silver halide emulsion also limits the resolution
of the mask and, because of its softness, limits mask life to a fraction of that of a hard-surface mask. Emulsion masks are still used
mainly for larger-geometry applications or as master intermediates
to produce hard-surface working plates.

Inspection and Cleaning

Plates are inspected and, if stored in an environment in which some
particle contamination is possible, are cleaned prior to imaging. The
inspection can be made by using automatic or manual techniques,
whereby the plate is checked for pinholes, defects, or discontinuities of
any kind by using transmitted light. The inspection should include a
measurement of the optical density of the plate with a densitometer.
Reflectivity also is measured against a predetermined process specification, and parts are graded and separated according to the results of
these tests. In addition to the analytical tests, plates should be func-
tionally tested for etch resistance by first baking the chrome at 150 to
250°C for 1 hr and imaging. The plate is then etched and checked for

good edge acuity of the etched layer, an indication of how well the deposition was accomplished. Poor chrome or deposited layer adhesion to the glass will show up as "mouse nips"—undercut areas that make the edges look as though a mouse had chewed them. Other tests for chrome layer adhesion can be used; they include mechanical abrading, ultrasonic agitation, or a combination of those tests.

Cleaning is necessary both before a plate is imaged and periodically as the plate is used. Even if plates are freshly deposited and exposed to clean-room conditions for several hours, contamination can occur. Particle adhesion to photomask plates is caused by molecular forces (van der Waals–London forces), capillary forces, and electric forces. The van der Waals forces can become extreme, particularly as the interaction between atoms and molecules is increased by closer proximity. As separation occurs, the forces decrease very rapidly.

Electric forces play a large role in mask contamination and cleaning. Most (greater than 85%) airborne particles are electrically charged. Because of different work functions of the electrons in different materials (glass and dust), a potential is generated between the mask surface and the particle. For example, an electric force of 1 dyn has been calculated for a 50-μm glass particle contacting a nonconducting surface. Coulomb interaction, which is more difficult to measure than other electric forces, also is responsible for particle adhesion to surfaces.

The electric forces of attraction are very difficult to measure because they are functions of numerous physical and environmental factors. Particle contaminants are removed by simply overcoming the adhesion forces. Most studies indicate the need to create, on the surface, bursts of energy that introduce a quasi-steady updraft. The burst of air or fluid must overcome the surface-adhesion energy, and then the aerodynamic lift force of air bursts (or hydrodynamic lift force of a liquid) will carry away the contaminants. The use of a mechanical brush with bristles, along with airflow, has proved to be very effective in removing particles in both photomask and silicon wafer cleaning. Adding vibration increases the removal efficiency of the brushes. Thus the process for mask cleaning can be a completely dry one, thereby eliminating potential staining with liquid solutions.

A special type of air device using a vacuum nozzle is very effective in removing surface particle contamination. Placing the vacuum nozzle 7 mils from the surface, the maximum distance recommended, generates a 5-in vacuum. A gap of 2 mils generates a 22- to 25-in vacuum, wherein the rapidly moving air molecules grab particles and remove them swiftly. This type of system, with either a series of nozzles or a vacuum air knife under which plates are conveyed, permits automatic mask cleaning with good throughput time and high reliability. The system could also incorporate a vibrating, rotating bristle brush to first remove layer particles.

The dry cleaning of new or reprocessed masks is effective for particles down to about 50 μm in size. Particles about 100 μm and larger weigh more than the adhesive bonding them to the mask surface. However, a 10-μm particle will have an adhesive force (van der Waal, electric, and capillary combined) over 35 times its weight, and the bonding force increases significantly as particles become smaller.

Chemical and mechanical methods both are used to clean new and reprocessed masks that are contaminated from printing operations. Chemical methods range widely from solvent vapor (or solution) to alkaline soak cleaners, oxygen plasma–acid cleaning, and water or alcohol rinsing.

Chemical mask cleaning by wet processing in which solutions are used only in an immersion mode is lengthy but is a low-cost and practical method. The recommended method for new and reprocessed chrome masks is given in Table 14.3.

An air dry will permit a small amount of chrome oxide growth, which will aid in bonding positive resist to the surface. A shortened procedure for plates with little to no soil or contamination is simply an ultrasonic cleaning in Freon TF for 10 min followed by dry nitrogen purging. These procedures are used for new masks that inevitably receive some form of particle contamination the most common particle size range, 0.5 to 10.0 μm. Masks that have been used for exposure and contain not only dust particles but chunks of positive or negative resist must first be processed through a proprietary resist stripper according to the type of resist contamination. Most positive-resist-contaminated masks can be immersed in Shipley Remover 1112A at 120 to 160°F for 2 to 5 min. This solution is highly compatible with

TABLE 14.3 Chemical Chrome Mask Cleaning

Solution	Time	Temperature
1. Hot-soak cleaner	5–10 min	Ambient to 175°F, as required
2. Deionized-water rinse	5–10 min	Ambient, not below 21°C, preferably 25 to 30°C
3. Isopropyl alcohol	15-sec	Ambient to 21°C
4. Dry forced nitrogen	30 sec	Ambient
5. Chromic-sulfuric acid	2 min	Ambient to 120°F, as required
6. Deionized-water rinse	5 min	100°F
7. Triple-cascade (overflow) deionized-water rinses	3 min each	Ambient
8. Oven-dry (nitrogen or air ambient)	20 min	100°C

TABLE 14.4   Particle Size versus Water Velocity Required for Particle Removal

Particle size, μm	Water velocity, m/sec
500	0.108
250	0.08
80–100	0.07
30–40	6.17
5–10	0.41

chrome and iron oxide mask surfaces and is free of metal ions, phenols, phosphates, chlorides, and fluorides. In addition, it is a clear aqueous solution that permits observation of the cleaning process. Proprietary strippers are recommended over acetone for reasons of flammability, operator safety, and waste disposal.

Although the chemical immersion process described earlier is very effective for removal of all organic particles and films, it will not remove the small (< 10 μm) inorganic particle contamination that is a major source of mask rejects. The solutions are, however, more effective than gaseous removal. For example, airflow of 100 m/sec will not overcome the adhesive force of a 1-μm particle, whereas a water flow of only several meters per second will remove the same particle. Particle size versus water velocity for removal is shown in Table 14.4.

High-velocity streams of water are especially effective in overcoming the adhesive forces between small particles and mask surfaces. Average jet velocities of 100 to 170 m/sec were generated at pressures ranging from 100 to 3000 lb/in^2, as reported by Kroeck of Bell Laboratories. The mask-cleaning system reported in this work uses pulsing high-pressure ($\cong$ 2500 lb/in^2) deionized water streams through oscillating nozzles positioned about 2 in above a spinning mask. The parameters with this system are as follows:

Water pressure	2500 lb/in^2
Mask spin speed	3000 rpm
Nozzle pivot angle	90°
Nozzle orifice	0.008 in
Nozzle-to-mask distance	1.5 to 2 in
H$_2$O improvement angle	30 to 45 in
Nozzle traverse	7 per minute
Drying	Infrared spin
Elapsed time per mask	1.5 min per side

A special feature of this system is a sheet of magnesium metal that allows magnesium ions to enter the water stream from the pump and neutralize static charges caused by water moving across the mask surface. The metal ions prevent an erosion of chromium edges by an electric discharge.

Another mask-cleaning technique is mechanical scrubbing with a nylon brush. Designed by Oswald of Bell Laboratories (U.S. Patent No. 3,585,668), a nylon retractable rotating brush scrubs the spinning mask surface while deionized water is fed to the mask. Water or a mask-cleaning solution is dispersed at a rate of 0.2 to 0.4 gal/min. The top and bottom surfaces may be cleaned in this fashion. Operating parameters are as follows:

Deionized water flow (chrome side)	0.4 gal/min
Deionized water flow (glass side)	0.2 gal/min
Brush speed	260 rpm
Turntable spin speed	200 rpm (clean)
	1800 rpm (dry)
Filtered nitrogen flow	10 L/min
Elapsed time	1.5 min per side

The filtered dry nitrogen is dispensed after the water or mask cleaner rinse cycles for drying. The mask-cleaning applications of these methods differ. The high-pressure water technique is recommended for stored masters or working plates just before they are put into aligners. The rotating brush method should be used on new mask products or after an acid soak on all plates that are being recleaned after printing operations. The main advantage of high-pressure gas and water cleaning, as well as mechanical brush cleaning (with deionized water flush simultaneously), is a dramatic increase in throughput. The old immersion techniques may have an advantage in removing stubborn organic contamination, but the slow cleaning cycle and disposal and handling problems are severe compared with the newer methods.

Masks should always be stored in containers that prevent glass-to-glass contact and are covered to prevent environmental contamination. Masks should be oriented vertically to prevent sag in storage or transit, and all packing and unpacking should be done under Class 100 minimum clean-air conditions. The outside containers should indicate "This side up" and carry an instruction regarding careful handling and opening only in a clean room.

Mask Imaging and Etching

Many different paths can be taken between the digitized pattern files and the wafer. Figure 14.16 shows the variety of process options avail-

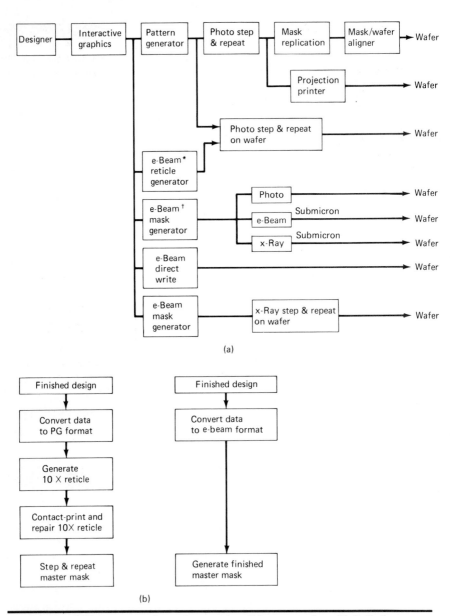

Figure 14.16 (*a*) Process steps for various mask-making approaches. (*b*) Optical versus e-beam processing for master-mask production. *At 2:1 for geometries between 1 and 2.5 μm. †1:1 for geometries > 2.5 μm.

able to arrive at a mask. The trends in mask imaging have generally been toward a shorter process between the pattern data in the digitizer and the wafer-imaging operation. Optical and e-beam imaging techniques are used to produce reticle and mask patterns. Examples of the finished mask are shown in Figs. 14.17 and 14.18.

Figure 14.17 is a single-image 10× reticle used in a wide-field step-and-repeat machine. More than just one image on a reticle fills up the imaging lens and limits movement of the stage during imaging. Several images at 10× on the reticle allow the operator to select and pellicize a "perfect" pattern. A pellicized multiple-image 1× master is shown in Fig. 14.18.

Imaging the coated mask and reticle substrates is the most critical step in the mask fabrication sequence. Much that has been done prior to the imaging step was to prepare for the rigors of high-resolution image formation. The imaging process takes advantage of the high-quality glass and its flatness features. The uniformity of the layer to be imaged and the layer to be etched are held to tight tolerances so the

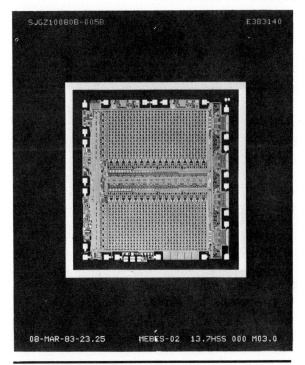

Figure 14.17 A single-image 10× reticle for a wide-field step-and-repeat camera. (*Courtesy of RCA.*)

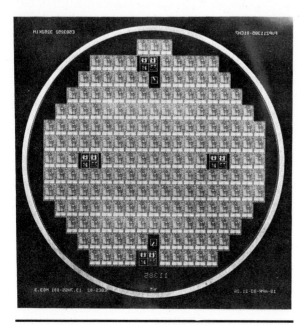

Figure 14.18 A pellicized multiple-image 1× master. (*Courtesy of RCA.*)

imaging step will result in patterns of uniform size across the entire active field.

Electron beam imaging

Electron beam exposure of resists is widely used for mask imaging. The parallel development of resists sensitive to electron beam energy and the exposure equipment to deliver the energy doses needed has resulted in systems and processes for volume production. Also, electron beam imaging systems can be software-driven to write any sequence and configuration of pattern elements needed to create a given set of masks. The smaller dimensions required for mask making brought the need for closer layer-to-layer registration tolerances. Larger chips created time pressure on the optical equipment used to pattern 4× and 10× reticles, so electron beam writing was used to generate reticles with better turnaround. The new, complex designs for these large chips also went through a series of timely design and mask set evaluations before being reduced to production. The electron beam writing approach allowed for quick test design, thereby saving precious time in the commercialization of a new semiconductor device.

A popular way to increase profitability of a given IC design is to shrink the overall dimensions, a process called "scaling." The reduction of all the dimensions sometimes causes lithography problems on one or two areas of the device, perhaps at only one mask level. This lack of good scaling behavior is corrected by changing only part of the design, a debugging process made easier by e-beam generation of test masks. Rapid tooling for scaled down designs is a major e-beam imaging application.

In summary, the benefits of e-beam technology for mask fabrication and eventual device writing are:

1. Rapid turnaround on new designs and scaled designs
2. Unlimited device size (larger chips)
3. Reduced defects (vacuum-processed)
4. Software capabilities (changing pattern and feature sizes easily)
5. Compensation for distortions
6. Beam writing directly onto surface
7. Submicron resolution
8. Good critical dimension control
9. Good overlay and registration accuracy

Electron optics versus light optics

The wavelength of an electron optics system is less than 1 Å; that of an optical system is 4360 Å. Some mask-making equipment uses 3650-Å light, and recent exposure at mid- and deep-uv brings the shortest optical wavelength down to about 2000 Å. In this area of comparison, electron sources have a strong resolution potential advantage.

Uniformity of exposure energy is always a primary concern, and the dose profile for a typical electron beam is shown in Fig. 14.19. Energy uniformity for electrons is about 3% over a 6-in area, compared to 4 to 5% for the same area with light optics. Energy uniformity of both types of systems has improved as refinements in the equipment and sources have been incorporated into commercial systems. Depth of focus, 30 to 100 μm, gives the e-beam system an advantage over light optics, which have a 1- to 3-μm depth of focus. These figures are based on an f-number rating of approximately 100 for electron optics and approximately 3 for light optics. Resolution for mask making is less than 1 μm for electron optics and also for light optics with recently developed shorter-wavelength lenses and laser direct writing.

Finally, the speed of flashing of electron optics is greater than or equal to 20×10^6/sec compared to the 100/sec of light optics. Overall, the e-beam system unquestionably rises above the optical system in

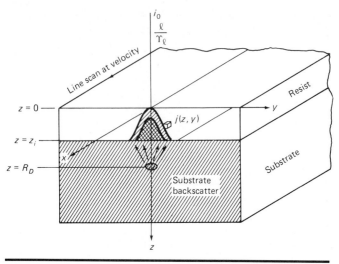

Figure 14.19   Electron dose profile.

total resolution potential. That at least gives it a niche in the applications in which extreme resolution, combined with the other key attributes of very low defects and rapid throughput of 10× reticles, allows it to surpass optical alternatives. An electron-optical column is shown in Chap. 15.

Pattern generation

Pattern generators are optical flash exposure systems used to produce the pattern data for a mask or reticle from digital input. The optical pattern generator converts the material software result of the CAD digitized tapes to a series of exposures on either an emulsion- or photoresist-coated chrome plate at 10×. Typical pattern generation time on one level can exceed 30 to 40 hr/level, with over 600,000 flashes required to print all the pattern information. The large amount of time required has been the main reason for e-beam writing as a method for reticle generation.

High-resolution pattern generators for imaging emulsion and chrome are widely used in reticle production, both arrayed and nonarrayed. Figure 14.20 shows a photograph of the GCA/Mann Type 3600 optical pattern generator. This system features interchangeable light sources for various light-sensitive materials and interchangeable lenses for patterning at different sizes and formats. Highly linear stage movements are controlled by a laser interferometric metering system over the active area. The 3600 pattern generator has image-

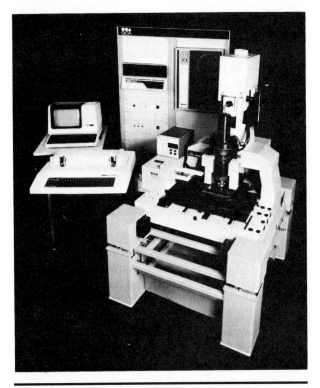

Figure 14.20 GCA optical pattern generator. (*Courtesy of GCA Corp.*)

placement accuracy controlled by directly referencing the laser interferometer system to the optical column.

Plate stability

The optical flashing of the pattern generator can be very precisely controlled in terms of exposure dose, exposure position, and mechanical stability. However, the plate being imaged is a supercooled liquid structure that moves enough to upset the critical dimension tolerances of VLSI patterns. For example, gravitational sag alone can cause a given mask level to be run out of specification while the pattern generator is running over the 20- to 40-hr imaging time. Plate flatness may be poor even when the unexposed plate is brought into the production area for exposure. The temperature of the plate, the imaging environment, and even the temperature maximum and minimum levels the plate has experienced up to 24 hr before reaching the pattern generator are important. Hysteresis effect, caused by the glass relieving itself in the clean-room (pattern generator) area after experiencing very warm or cool ambients, should be measured and accommodated so they do not influence the patterning process.

The stability of the glass or quartz substrate is first controlled by verifying the thermal history of the glass plate 24 hr before reaching the pattern generator. Next, the plate should be quickly checked for real-time flatness. Then it can be loaded, assuming cleanliness and other data needed for processing are complete. The variety of features possible in the imaging process, all software-delivered, include device scaling, aperture incrementing, origin offset, mirror imaging, and rotation.

Photorepeaters

When the pattern generator produces the VLSI pattern on a reticle or mask from digital data, the photorepeater generates arrays of the IC images. Photorepeaters typically image directly onto resist-coated chromium.

Control of most photorepeaters is by a computer terminal with video or disk input. The computer is used to establish major parameters of the images, including array size, stepping intervals, measurement units, array origin, drop in test patterns, orthogonality scaling, and exposure values. Typically, the software will allow arrays in a variety of shapes including squares, rectangles, and circles.

Photorepeaters use mercury arc lamps as the exposure energy source. Lenses have different reduction capabilities and various field sizes, and although most are used at 436-nm wavelengths, optical variations result in different resolution capabilities. For example, numerical apertures run from 0.20 to 0.35 for $5\times$ lenses with production resolutions of 1.75 and 0.8 μm, respectively. Other lens properties tested and supplied as part of a customer specification include tested field distortion, rotation error, tested field reduction error, and tested field trapezoid, all at specified magnification, wavelength, field size, and numerical aperture. Lenses can, of course, be custom-built to provide a combination of specific physical parameters to match a given application.

Aligning the reticle in a photorepeater calls for several key mechanical features to ensure accuracy, precision, and good parallelism between the two stages. Reticles used in the industry range widely in thickness and dimension, so platens or reticle holders must be able to accommodate a wide range of plate dimensions.

Reticle production

Reticles for wafer steppers can be produced in one of several ways, all beginning with CAD digital tape information. One process begins with CAD tapes and uses a pattern generator to generate a reticle that is then placed directly into the wafer stepper. The pattern is generated at either 5× or 10×, depending on the reduction ratio of the stepper lens.

Pattern generation is most efficient for IC patterns that do not have large sections of repetitive elements, which can be transferred more efficiently with image repeaters (Fig. 14.21).

The additional step of image repeating, by using frequently repeated pattern information, fits well into designs, such as a calibration reticle or memory array, in which a few basic types of structures are used.

A hard-surface master mask can be dropped from the pattern-generated images. This variation accommodates the reversal of mask polarity and permits an all-positive-resist process. The process uses a digitized circuit pattern, and the use of negative (normal) or reversal emulsion pattern-generated plates provides any field orientation desired. The final chrome reticle is then contact-printed from the emulsion original while care is taken to maintain reference edge to reticle

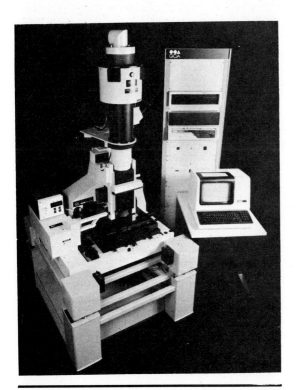

Figure 14.21 Optical photorepeater. (*Courtesy of GCA Corp.*)

alignment pattern spacing. This process is probably the least preferred of the three because of the contact-printing step, which introduces defects.

The resist application and imaging parameters for hard-surface masks, reticles, or e-beam masks basically follow the same course as those for wafers, the major exception being the difference in exposure equipment and a much thinner resist coating. Refer to each of the imaging chapters for detailed guidelines for patterning a microelectronic surface, since only the specific mask-related imaging parameters will be covered here. (The reader has presumably seen the more detailed areas such as resist characterization and surface chemistry.)

Prior to resist coating, all hard-surface mask substrates should be thoroughly dried and cleaned. Dryness on hard-surface chrome is especially critical because of the absence of mechanical resist adhesion sites that do exist on wafers and to a lesser degree on ion oxide surfaces. A "flash" infrared bake or short burst of high-intensity light (to provide heat) will raise the plate temperature to 100 to 150°C for 30 to 90 sec, constituting a prebake. After a spin cool with nitrogen purge, blanks are ready for coating. All resists used for mask fabrication should be filtered to a level of 0.1 to 0.2 µm with an absolute filter.

Mask blanks are then spin-coated to a uniformity of ± 2%, and the functional area of the mask should be free of fringes or color changes as well as particles interrupting the flow of resist toward the edges of the plate in spin coating.

Resist thickness plays a very important role in obtaining resolution and good line control latitude. Thicker resist coatings (5000 Å) give more line control latitude in exposure and better protection against pinholes than thin (2000- to 3000-Å) resist coatings. There is a difference in exposure latitude for different thicknesses of photoresist. The thinner films do have the advantage of reducing the line width error because the modulation transfer function (MTF) is smaller. A 2000-Å-thick film yields a 0.05-µm error in line width; under the same conditions, a 5000-Å-thick film yielded a 0.2-µm error. The process engineer must consider the inherent tradeoff in resolution and line control with resist thickness changes.

Softbaking is rapid because of the high reflectivity of mask surfaces and the thin coatings used. The guidelines in Chap. 7 for wafer softbaking apply to mask softbaking.

Exposure is more critical in mask fabrication than in wafer fabrication because of increased optical reflection and tight line tolerances. The ideal exposure mechanics would involve only forward-traveling light waves into the resist and through the mask blank, where backside mask reflection would be absorbed in the glass before it reached the resist-blank interface.

At the same time, resist-air interface reflection would be about 7% given the refractive index of Shipley AZ-1350 at 1.68 μm (in the 3800- to 4500-Å range). This surface reflectance would remain constant despite changes in thickness or wavelength.

In practice, surface reflectance as well as resist exposure can vary considerably with resist thickness change, wavelength change, and other variables, all because the refractive indices of the resist and mask-layer material are mismatched. The optical-coupling parameters, detailed by Penfold,[3] will vary considerably between chrome and iron oxide. For example, exposing light strikes the resist, where it achieves its maximum amplitude, and then attenuates as it penetrates the resist coating and loses intensity. The light then reaches the resist-blank interface. Reflected light has its maximum amplitude at the resist-blank interface and travels back to the resist-air interface, where it reaches a minimum amplitude. On its way back, however, the interference of standing light waves produces several maximums and minimums of intensity (five in a 6000-Å film in the 3800- to 4500-Å wavelength range). Since the interfering light waves reduce both exposure latitude and exposure time, the ideal exposure is one with *no* interfering light waves. This is possible only with the ideal optical coupling described earlier, a situation wherein the ratio of maximum intensity to a neighboring minimum intensity at the resist-blank interface (called the "standing-wave ratio") is equal to 1, thereby providing the most uniform exposure possible.

One must consider the variations in intensity caused by reflectance that give rise to standing-wave ratios much larger than 1. The argument for low-reflectivity chrome blanks is strong; the actual reflectivity of chrome is about 33% at 405 nm. Reduced exposure and, more important, increased exposure latitude is obtained by reducing the mask-blank reflectance by using "black" or antireflective chrome, essentially an oxidized chrome layer over shiny chrome. Even more dramatic results are seen by comparing line widths over a large area, using both "shiny" and black chrome. Exposure time control should be ±2%.

After exposure, blanks are developed by using methods described in detail in Chap. 9. As with exposure, the main difference between wafer and mask fabrication lies in the more critical nature of the mask process. The developer temperature should be controlled to ±0.1°C, and the developer concentration should be optimized to provide good resolution, good throughput, and maximum contrast. Using AZ-1350B Resist and AZ Developer, a higher concentration seems to provide this balance. Development should be controlled to ±3%, so a 30-sec immersion cycle would need ±1-sec control.

The resist postbake step for hard-surface masks is really more of a drying step and is generally kept below 100°C. Postbaking, as described for wafers in Chap. 10, is not required. Postdrying is, however, advantageous for positive resists on chrome and, to a lesser degree, on iron oxide when wet etching follows. A simple nitrogen dry is used prior to plasma etching.

Mask Quality Checks

The overall quality of a mask and its constituent elements must be measured in terms of array centering and rotation. Title, sizing marks, and orientation are other examples of overall quality checks.

In general, a defect will be an unintended opaque spot in a clear field or a clear spot in an opaque field. Figure 14.22 shows the most common types of defects and their general appearances.

Mask quality and yield

The increase in the die size is the single largest factor requiring strong attention to mask quality. A simple mathematical yield equation shows why this is so:

$$Y = \frac{1}{1 + D_0 A}$$

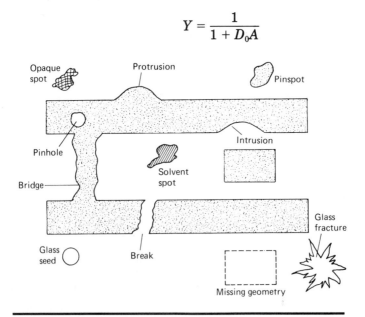

Figure 14.22 Typical photomask defects.

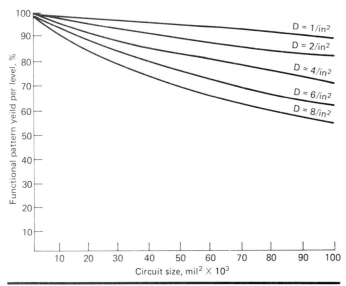

Figure 14.23 Functional pattern yield versus circuit area for several defect densities.

where D_0 = density of random defects on wafer
A = die area on wafer
Y = yield

Yield decreases rather quickly as die area increases and defect density increases.

In Fig. 14.23 the relation between yield required per masking step and final yield is plotted for several mask level sizes. Note that, in order to get even a 50% final yield for any of the multilevel examples, the yield of any single level must be greater than approximately 88%.

Mask Etching

Chromium is by far the most common material used in mask fabrication. A wide variety of wet etchants are used for chromium. The most common ones are composed of ceric ammonium nitrate with an oxidizing agent, which is typically acetic, perchloric, or nitric acid. Another type of etchant used for chromium is potassium permanganate, but the most popular of the etchants is ceric ammonium nitrate. Although it is difficult to monitor the strength of ceric ammonium nitrate etches, a variety of process techniques are used to identify the relative activity of an etching bath. The etching time used is typically determined by calculating the thickness of the chromium and the etching time of the chrome, along with density or thickness variations in the chrome layer. Chrome etchants should always be filtered to 0.5 μm with an absolute filter to eliminate particles that may interfere with the etching process. Typical wet etchants are listed in Table 14.5.

Wet versus dry etching

Primary problems associated with wet-etching technology in mass manufacturing are the corrosion caused by wet etchants in clean-room areas, disposal difficulties, toxic fumes, and the splattering of acids, which creates a human safety problem. In addition, chemical costs are high, as are costs for the equipment and etch hoods required for wet etching. Special drains, special tooling, and administrative headaches caused by government safety (OSHA) and pollution (EPA) regulations are additional disadvantages. The idea of dry etching is appealing for those as well as for some serious technological reasons.

Plasma dry etching not only simplifies the process by reducing the number of steps required but also improves the resolution and yield of parts in a production process. Further, it can be used as a precleaner for the glass prior to deposition, a selective remover of chromium

TABLE 14.5   Wet Etchants for Chrome and Other Mask Materials

Etchants for iron oxide:
1. H_2PO_4 heated to 60 to 90°C
2. 1 part 97% HF and 3% HPO_3
   1 part HCl (conc.)
3. 575 cm^3 HCl
   285 cm^3 H_2O
   150 g $FeCl_2$
4. 530 g $AlCl_3 \cdot 6H_2O$
   630 g CaBr
   110 g $SnCl_2 \cdot 2H_2O$
   850 mL H_2O
   260 g HBr
5. 100 g $MgCl_2 \cdot 6H_2O$
   200 g $CaCl_2 \cdot 2H_2O$
   25 g $SnCl_2 \cdot 1H_2O$
   165 g H_2O
   60 g HCl 32%

Etchant for silicon:
1. 70 g $FeCl_3 \cdot 6H_2O$
   9 cm^3 HF (conc.) (38 to 40%)
   220 cm^3 H_2O (room temperature, 1 μm/min)

Etchants for chrome:
1. Activation: HCl + Al, Fe, Zn, etc.
2. 50 g Na OH
   100 mL H_2O } 1 part
   100 g K_3 (Fe(CN)$_6$
   300 mL H_2O } 3 parts
3. 9 parts saturated Ce (SO$_4$)$_2$ solution
   1 part HNO_3 (conc.)
4. 100 g Ce (SO$_4$)$_2 \cdot 4H_2O$
   50 mL H_2SO_4 (conc.)
   1000 mL H_2O
If milky deposit is formed on mask during rinse, dip mask after etching first in 10% vol H_2SO_4 and rinse afterwards.

without undercutting, and a cleaning environment for masks after photoimaging.

Dry etching

Innovations in plasma equipment and the refinement of plasma techniques to improve uniformity have made dry plasma etching extremely competitive with wet chemical methods for the fabrication of photomasks. Nonuniformities in etched profiles, caused by perturbations of the electric discharge (typically $\pm$ 10 to 15%), resulted in a severe limitation of the use of dry plasma in mask processes. An "Equi-Etch" system (LFE Corp.) decreased the temperature of the substrate during etching and permitted the use of the system on thermally sensitive devices. The uniformity provided by the system was made possible by employing an inductive rf power coupling and a perforated metallic material-handling zone within which the masks are exposed to evenly disbursed electrically neutral active species. By allowing the preferentially electrically neutral active species within the perforated metallic envelope, the temperature during etching is reduced and extremely uniform etching throughout the batch of materials in the presence of the ever-present electric field nonuniformities is possible.

Etch duration versus temperatures.   One of the important parameters in dry etching of photomask substrates is the reaction temperature. Since photoresists and electron resists deform at different temperatures depending on the resins used, both the thermal flow temperature of the resist and the maximum reaction temperature during the dry etching should be established. Several types of plasma environments are used; they include capacitive plasma, inductive plasma, and an inductive medium. The capacitively excited plasmas stabilize at about 430°C after 6 to 8 min of etch time. Inductively generated plasma reaches an equilibrium temperature of about 320°C within the same time period. The Equi-Etch medium reaches a temperature of about 215°C and stabilizes at about 250°C after 15 min of 350-W rf power.

One of the inherent advantages of dry etching is the absence of impurities and dust particles that are typically found in liquid etchants. Also, the solvents used in wet etching can cause mask defects, whereas gases are filtered and cleaned to a much higher degree.

Etch rate versus pressure.   The rates of etching chrome and chrome oxide are functions of the gas pressure, among other parameters. Figure 14.24 shows the relation between etching rate, in angstroms per sec-

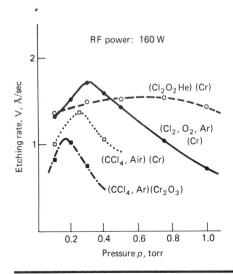

RF power: 160 W

Figure 14.24   Etch rate plotted against pressure.[11]

ond, and pressure, in torr. These data are for an rf power level of 160 W. As they indicate, the etch rate reaches a maximum at a particular gas pressure, and the maximum is different for each type of plasma. Note that separate plasmas are indicated in the figure. Of particular interest is the mixed gas plasma containing helium, which provides a weak dependence of etch rate on pressure. The greatest etch rates of chrome as well as antireflective chrome films were found to be at about 0.3 and 0.15 torr, respectively. Note also that the etching rate of the chrome is greater than that of the chrome oxide films at the same power level even though different gases are used.

Etch rate versus rf power.   Figure 14.25 shows the relations between etching rates of chrome and chrome oxide, in angstroms per second, and the rf power level, in watts. Gas mixture 3 (CCl_4, air) etches chrome with a saturation point of about 300 W. The same peak for the other plasmas is reached at as low as 160 W for the argon-chlorine-oxygen mixture and the helium-chlorine-oxygen plasma (gas mixture 2). The antireflective chrome has a maximum of etching rate at 200 W.

Etch rate versus oxygen gas concentration.   Another important parameter in determining the process for chrome plasma etching is the percent of oxygen in the gas mixture. In test data, the etching rates of the chrome films applied by sputtering and the antireflective chrome films are increased as the oxygen concentration increases in the gas mixture. When the oxygen concentration reaches approximately 40%, a maximum sputtering rate is achieved.

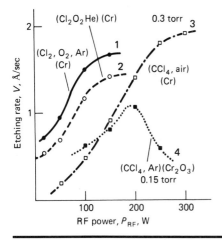

Figure 14.25 Etch rate plotted against rf power.[11]

Plasma etching characteristics are extremely sensitive to the stoichiometry of the deposited chrome oxide layer as well as the structure on the surface and beneath the surface of the metallic film.

Pattern width of master mask versus copy mask. One good way to compare the etching results of wet- and dry-etching techniques is to plot the line widths and variations that occur between the readings on positive- and negative-working master masks. The dry-etched masks result in a more faithful reproduction of the masters than the wet-etched masks do. There is very little difference between original and copy of dry-etched masks because of the elimination of undercutting, a factor that works to the advantage of both chrome and antireflective chrome films.

Undercut profiles

There is generally a predictable pattern of resist and chromium undercut with wet and dry etching, as long as the process variables are held reasonably constant. In Figure 14.26 we can see the cross section of the resist and chrome layers and their dimensions. This example permits us to know the percent transmittance along an etched chrome edge. At $T = T_0$ the chrome is just completely etched, with zero overetch. At $T = 2T_0$, the etch time is doubled. Note that, at $T - T_0$, the undercut is a negative value (larger feature size than the mask).

One problem associated with microimaging is proximity effects. As the pattern widths get smaller and more closely spaced, the shadowing of light between various pattern elements affects imaging param-

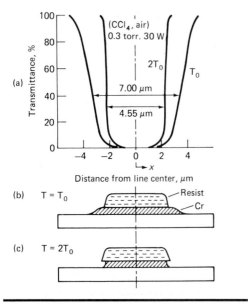

Figure 14.26 Cross-sectional profiles after etching.[11]

eters. Since the relation between the pattern size and shape and the resultant optical effect (shadow, reflection) vary, so does the optimum resist exposure dosage. Optimum exposure doses are therefore given for a specified image size and configuration.

Pattern shift

Pattern shift, or the difference between the size of the mask original (master) and the copy (submaster), is precisely controlled in mask manufacturing. In the etching step, where imaged masks become finished masks, measurement for pattern shift is essential. Pattern shift can be expressed as

$$\Delta S = W_c - W_m$$

where ΔS = pattern shift
W_c = pattern width of copy mask
W_m = pattern width of master mask

Reverse etching

Reverse etching is removal of the chrome and chrome oxide areas under the resist images before the nonimaged mask areas are etched. In Fig. 14.27 conventional and reverse etching are compared. The pri-

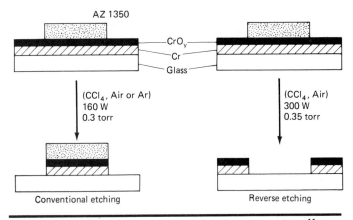

Figure 14.27 Comparison of conventional and reverse etching.[11]

mary difference is a doubling of the power level. The breakdown of the resist in the plasma environment accelerates the etching of the areas under the resist. The gas by-products of resist breakdown produce a new gas plasma species. This "mixed" plasma is in high concentration immediately near the resist images; hence, the rapid etching and reversing of the expected result. Different resists will produce changes in the nature of this etch phenomenon.

Resist Removal

Resist removal after etching can be accomplished in wet-stripping solutions, in dry plasmas, or in a combination of wet- and dry-stripping environments.

There are several guidelines that all wet- or dry-resist strippers must adhere to in order to facilitate high yield. They include:

1. Complete, residue-free removal of the resist coating
2. A removal process that is completely inert with respect to the mask surface
3. Removal process economics that do not adversely affect the overall process economics
4. A removal process that is nonpolluting, to meet environmental regulations, and nontoxic, to meet industry-approved safety guidelines

The most common technique for removing thin positive-resist coatings is immersion in liquid stripper baths.

Wet removal of resist coatings does pose some problems of its own.

For example, chips or small flakes of resist will stick to a chrome or glass surface and bond strongly enough to remain through rinse operations. Fibers and other airborne contaminants also may remain through etch operations and alter the IC pattern slightly. Problems related to surface tension and static charges in wet chemical operations are not often overcome by even high-pressure jet spray rinsing.

Remover 1112A (Shipley Company) is widely used to remove resists from mask surfaces, especially iron oxide. Iron oxide is particularly sensitive to attack by other chemical resist strippers, but Remover 1112A does not react with it. Further, Remover 1112A may be heated without causing safety problems and will then readily strip or dissolve postbaked resist. Since some chemical etchants react with the surface of the resist to form high molecular weight species or structures more insoluble in strippers, a more aggressive chemical is needed. See Chap. 12 for dry resist stripping.

Pattern Measurement

The main challenge in mask pattern measurement lies in edge-sensing. Aberrations, diffraction, and other optical variables cause changes in the aerial images transmitted through microscopes. Diffraction will cause a very sharp edge on a chromium pattern element to have a gradation from dark to light at the line edge. This transition region, similar to the Airy disk in an optical system (its diffraction-limited spot diameter), must then be measured by using an optical threshold. In an optics system, the Airy disk has a 1.0-μm diameter in a 0.65-NA lens. Figure 14.28 shows the material or substrate profile, along with its corresponding optical image profile and the optical threshold. The threshold will vary considerably with the equipment used to perform the measurements. A 50% threshold will produce either a dark band or a line between the two images. In image-shearing microscopes, the profiles of the split image allow for the formation of a third "displaced" image and profile shown in Fig. 14.29.

Depending on the equipment used and the method of illumination, other image profiles can be derived. A theoretical model of the relation between a line object and its coherent spectrum is a useful aspect to study. It is essential to know the profile of image intensity and its correlation to the object and the influence of the measuring instrument in order to derive a pattern measurement that is meaningful. Figure 14.30 is a schematic of a metrology tool capable of 20-nm accuracy over 9-in^2 reticles.

Laser measurement

Lasers offer another accurate and semiautomated pattern measurement technology. They are relatively low-cost, high-quality point

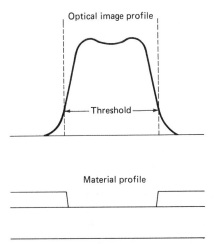

Figure 14.28 Comparison of optical image profile and substrate dimensional profile.[11]

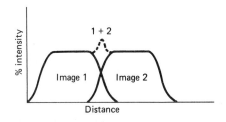

Figure 14.29 Overlapping images, in an image-shearing microscope, creating a third, displaced image.[11]

sources whose beams are diffracted on the edge of the image to be measured. The scattered light is sensed by detectors that record and transmit the data. Unlike principally simpler optical microscope systems, the laser is insensitive to variations in the density of the image being measured because diffraction occurs at a physical edge or step. Optical systems also are likely to vary on measurement according to the intensity variation in the primary pattern illumination system. Laser systems operate independently of that parameter.

Scanning microscope measurement

The highest level of precision in measurement technology is obtained with the scanning electron microscope. SEM analysis is needed for the high-magnification analysis of microelectronic structures; it provides

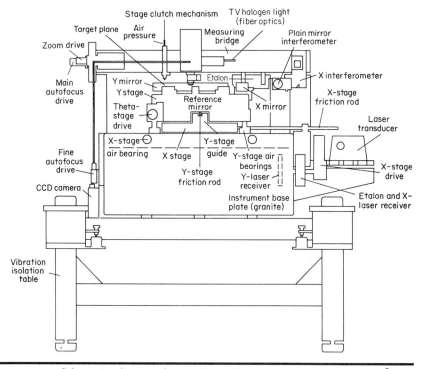

Figure 14.30 Schematic of a metrology tool capable of 40-nm accuracy over 9-in² reticles. (*Courtesy of Image Micro Systems.*)

resolution on the order of 50 to 100 Å and will detail shapes of microstructures as well as their dimensions. The manufacturers of scanning electron microscopes have responded to the needs of the IC industry by providing the following capabilities:

1. Insertion of entire mask or wafer into vacuum chamber for observation

2. Provision for avoidance of damage to the substrate by the electron beam

3. Ability to see several specimens per hour

4. Ease of operation so that process personnel can be readily trained

In a typical system the sample can be moved 100 mm in both the x and y planes, tilted up to 60°, and rotated 360° continuously. The working distance in the z plane is 10 to 48 mm. An option is to automatically correct the image to a predetermined magnification. This means that, with focus adjustment, an image of specified magnification is obtainable regardless of the substrate thickness.

The sample is protected from the potentially damaging beam by blanking (electromagnetic shuttering) the beam and using an optical microscope to search for areas of interest.

Mask inspection: adjacent die comparison method

The adjacent die comparison method for mask inspection has several inherent advantages for automatic inspection. The conclusion has already been reached that individual inspection by using manual techniques is simply too laborious and that automation of the task is essential. The adjacent die comparison method is automatable because

1. All dies are identical.

2. Reference masks are not required (one mask used).

3. Focus on one mask is simple.

4. Mechanical tolerances for mask scanning can be relatively loose.

5. Automatic focus capability is used to compensate for mask sag and bow.

Several systems have been developed to provide automatic inspection by this method. Some use laser-scanning systems with computerized information storage and printout; others use a color TV camera with which defects are found by illuminating three die patterns simultaneously, each with a primary color. After recombination of the three die images, a black-and-white image of the patterns is formed. However, a defect on one or two or all three dies will not have the fully recombined color scheme and will appear as one or two of the primary colors, thereby signaling a defect. Some systems compare mask pattern data with the fabricated master.

The general requirements for any of the adjacent die comparison methods are stage precision and accurate placement of the two sets of optics over the dies. Figure 14.31 is a schematic of an automatic mask inspection system. Measurement applications in a MOSFET process are shown in Fig. 14.32.

Pellicles for IC Mask Protection

The most yield-sensitive area of IC device manufacturing is lithography, and airborne particles comprise one of the largest sources of lithography-related defects. A pellicle is a very thin membrane that is used on top of IC masks to insulate the masks from airborne particles and other environmental contaminants. Pellicles are made of Mylar or transparent nitrocellulose material so they do not interfere with the

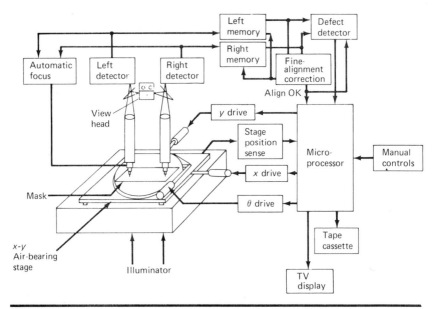

Figure 14.31   Block diagram of an automatic mask inspection system. (*Courtesy of KLA Instruments.*)

exposing energy on its way through the mask onto the resist coating. The pellicle membrane is stretched over a frame and placed directly onto the imaged side of a mask; in thickness it ranges from as thin as 0.7 µm to over 12 µm. High transmission is at wavelengths above 350 nm at thicknesses between 2.9 and 12 µm.

Pellicle particle protection

The reason for pellicle existence is to "catch" the particles and other airborne contaminants that would ordinarily land on a reticle or mask surface. Pellicles thereby reduce the need to clean a reticle or mask except when a pellicle is damaged and must be taken off the mask and replaced.

The pellicle is attached to the mask and remains with it during all standard exposure operations. The question of pellicle cleaning and insulating the mask from particles raises the issue of clean attachment. Pellicles can be blown off with filtered dry nitrogen or rinsed with ultrapure deionized water to remove dust, dirt, or any relatively loose contaminant falling on the pellicle surface. Contamination of a more serious nature or physical damage to the pellicle will necessitate removing the pellicle and replacing it after cleaning or putting on an entirely new pellicle.

The primary means of minimizing the effects of particles in the op-

Metal-oxide-silicon
field-effect transistor

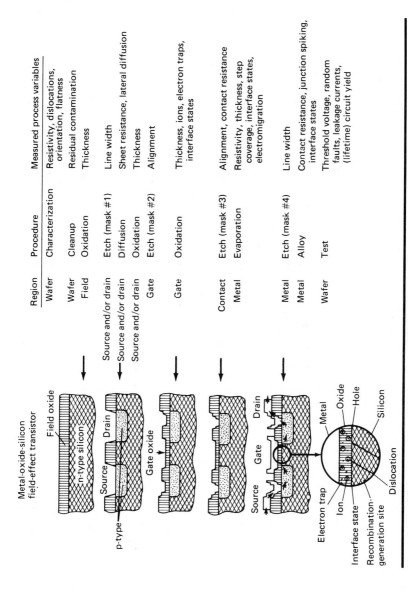

Region	Procedure	Measured process variables
Wafer	Characterization	Resistivity, dislocations, orientation, flatness
Wafer	Cleanup	Residual contamination
Field	Oxidation	Thickness
Source and/or drain	Etch (mask #1)	Line width
Source and/or drain	Diffusion	Sheet resistance, lateral diffusion
Source and/or drain	Oxidation	Thickness
Gate	Etch (mask #2)	Alignment
Gate	Oxidation	Thickness, ions, electron traps, interface states
Contact	Etch (mask #3)	Alignment, contact resistance
Metal	Evaporation	Resistivity, thickness, step coverage, interface states, electromigration
Metal	Etch (mask #4)	Line width
Metal	Alloy	Contact resistance, junction spiking, interface states
Wafer	Test	Threshold voltage, random faults, leakage currents, (lifetime) circuit yield

Figure 14.32 Measurement application in a MOSFET process.[12]

tical path is to provide a distance between the pellicle and the reticle mask sufficient that the particles cannot form images or significantly disrupt the aerial images. The increased distance will, unfortunately, allow more space for particles to find their way to the reticle and mask. A formula for determining a sufficient standoff distance has been proposed.

$$D = \frac{NFP}{280}$$

where D = minimum standoff, mm
$\quad P$ = particle size, μm
$\quad N$ = refractive index (1.0 for air, 1.5 for glass)
$\quad F$ = condenser f number at mask or reticle

Stepping aligners are better suited for pellicles because of their greater depth of focus, allowing particles to "fuzz out" within the optical path and be less of a disturbance to the transmission of microimages.

There are spatial relations among the pellicle, the mask or reticle, and the condenser lens, and they are affected by the presence of particles on the various optical surfaces. This demonstrates the need for working in especially clean environments, even with pellicle protection on both sides of reticles or masks.

REFERENCES

1. N. Yew, "Electron Beam Lithography in the Production Line," IGC Conference, September 1978.
2. G. Zinsmeister, "Hard Surface Mask Technology," Technical Handout from IGC Conference on Microlithography, Amsterdam, Holland, September 1979.
3. A. S. Penfold, "The Coupling of Photoresist to Photomask Blanks," Telic Corporation Publication No. TM-061-75, August 1975.
4. M. Hohga and I. Tanabe, "Fabrication of High Precision Fine Pattern Photomasks and Evaluation of Photoresist Processing," Kodak Interface Conference, 1977.
5. R. F. Leinen, II, "Overview of Advances in Photomask Substrate Flatness Requirements and Flatness Measurement Techniques," Society of Photo-optical Instrumentation Engineers *SPIE*, vol. 100, 1977, p. 74.
6. A. D. Zimon, *Adhesion of Rust and Powder,* trans. by M. Corn, Plenum, New York, 1969.
7. CAD system photo supplied by VIA Systems.
8. J. P. Avenier, "Digitizing, Layout, Rule Checking—The Everyday Tasks of Chip Designers," *IEEE Proc.,* vol. 71, no. 1, 1983.
9. W. Engl, H. Dirks, and B. Meinerzhagen, "Device Modeling," *IEEE Proc.,* vol. 71, no. 1, 1983.
10. F. Uchiya, Hoya Corporation, Japan.
11. H. Abe and K. Nishiuka, "Microfabrication of Photomasks by Gas Plasmas," Kodak Publication G-47, 1976.
12. D. Nyyssonen, "Process Control Metrology for LSI's," *IGC Conf.,* June 1981.

Advanced Imaging

Advanced imaging covers lithography approaches that begin in the submicron resolution area and extend to the fabrication of nanostructures or nanometer-size images. Imaging approaches in this area include excimer laser, electron beam, ion, and x-ray lithography.

Electron beam lithography has been used for many years in IC imaging, primarily as a mask-imaging technique. Low throughput capability and high capital cost relative to optical imaging have prevented e-beam direct writing from becoming a production wafer-imaging method. Electron backscatter from secondary electron emission has caused line geometry control problems associated with proximity effects. High- and low-voltage e-beam imaging, to avoid backscatter, results in other problems, such as damage to the device.

x-Ray lithography, while offering the shortest wavelengths of all the imaging methods, has been limited because of mask, source, and resist technology problems. Sources have not been sufficiently bright, and resists have not been adequately sensitive or process-resistant. The x-ray mask is complex to manufacture and it does not permit resolution consistent with the theoretical limits set by wavelength. Development of solutions to those areas will permit x-ray imaging to emerge.

Ion lithography has been limited by many of the problems that affect x-ray imaging, i.e., inadequate source energy, need for better collimation, and mask-manufacturing problems. Good resists for ion imaging exist, but exposure systems do not.

Excimer laser imaging is rapidly developing as a technology driven by the incentive to extend optical lithography as far as possible and preserve as much of the existing base of learned information as possible. Depending on the gases used in the laser cavity, excimer lasers permit multiple wavelengths. They provide an intense source of highly collimated energy that is ideally suited for nano-level lithography. Although there are challenges in the area of compatible optical materials to deliver excimer laser beams to the wafer, considerable progress is being made. Quartz, calcium fluoride, and related deep-uv transmission materials are being used to produce lens, condenser, and achromat elements.

Where i-line lithography leaves off at about 0.5-μm resolution, excimer lasers begin. Resolution down to the 0.1- to 0.2-μm range will be possible for excimer imaging as a production method in ULSI device fabrication.

Beyond the beam-imaging technologies are more direct fabrication techniques that eliminate one or more steps of the process and some that eliminate lithography altogether. Two examples are laser pantography and direct ion doping. Laser pantography uses a laser beam directed through a gas cell to facilitate laser-assisted etching and deposition. The beam diameter establishes the resolution of the etched or deposited structure, and continuous line movement of a stage results in pattern structure formation without the use of lithographic science as it is traditionally defined. Pantography has the potential for removing much of the complexity and resultant defects attributable to lithography.

A serious limitation of laser pantography is throughput. If the system is limited to a single beam, multiple-beam pantography will be needed to extend the technology to the production floor. Figure 15.1a is the schematic of a pantography, and Fig. 15.1b is a SEM photo of structures produced with the system. Figure 15.1c shows the reduction of process steps made possible with laser pantography.

Advanced Optical Imaging

The implicit objective of optical imaging technology is to extend useful resolution beyond that currently achieved in production of integrated circuits. The major challenge facing optical technology is to continue to supply the imaging needs of the IC fabrication process. The progress made with e-beam technology in the mask market has stimulated optical equipment manufacturers to rise to the challenge of their markets by designing and building imaging equipment to provide increased resolution. The increase in the use of step-and-repeat wafer exposure systems, which allow production imaging of submicron lines, is an example of the response of the optical industry to the continued needs for better resolution. The advent of x-ray exposure and its potential for wafer fabrication likewise poses a challenge to the companies now supplying optical exposure equipment, such as projection aligners and steppers.

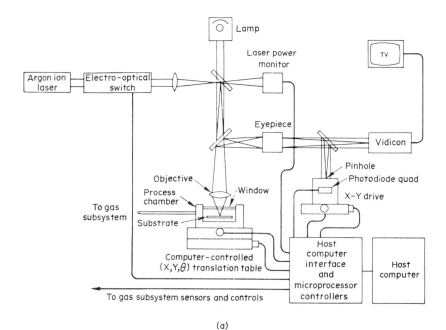

(a)

REVERSIBLE REROUTING OF SEMICUSTOM CIRCUITS by LASER MICROCHEMISTRY

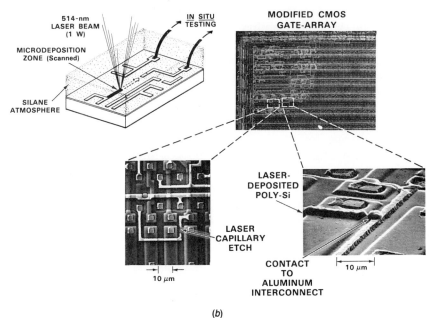

(b)

Figure 15.1 Laser pantography: (a) system; (b) SEM photo of structures made with the system; and (c) process imaging.[12] (*Courtesy of D. Erlich, M.I.T. Lincoln Laboratories, Lexington, Mass., and B. McWilliams, Laurence Livermore Laboratories, Livermore, Calif.*)

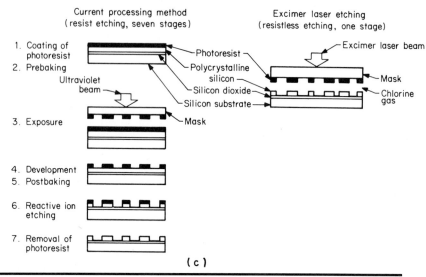

Figure 15.1 *(Continued)*

Thus the overall growth of the IC market, driven by a continuing decrease in cost per bit, has created large equipment and materials markets and intense competition within them. The benefit, beyond increasing the economic viability of semiconductors, is stimulation of research to design "better mousetraps" as these supplier companies vie for their shares of the market. Lower cost per bit is the function that forces more efficient IC production.

As geometries go below the 0.5-μm level, the degree of difficulty in optical imaging increases because that level of resolution begins to approach the theoretical limit of diffraction-limited optics, which is approximately one-half the distance of the imaging wavelength. Since exposing wavelengths are typically 405 to 248 nm, the optical lithography limit for usable resolution would be 185 to 200 nm, or about one-fifth of a micrometer. If the objective is to resolve 0.5-μm structures on wafers after etching and the usual allowable tolerance for line control is 15 to 20% per line side, then 0.5-μm imaging at 248- to 405-nm wavelengths (with line control of 1800 to 2000 Å, or one-half the exposing wavelength) would in fact represent the optical limit. The obvious way to extend the limit is to shorten the exposing wavelength, which is exactly what e-beam and x-ray imaging are all about. The optical industry's response to being outstripped by two orders of magnitude in wavelength is twofold: (1) design process modifications that allow resolutions with existing long wavelengths (405 and 365 nm) that exceed the optical limits and (2) shorter wavelengths. This has materialized in the form of deep-uv imaging. In this chapter we will review micron and submicron imaging techniques with exposing wavelengths of 193- to 436-nm, as well as review the various ways of exploiting shorter-wavelength lithography to obtain submicron resolution in IC fabrication.

Submicron optical imaging

The primary problems in getting submicron resolution with optical imaging equipment are the standing-wave patterns generated in the resist, the change in focus of projected images, and the light scattering and lateral exposure that cause changes in the width of lines. Further, control of submicron images, in the face of these problems, is practically unachievable without some basic changes in process technique. Another formidable problem is covering steps with no change in line width at submicron levels with single-layer resist processes.

Trilevel processing.   One solution proposed by Moran[1] is the "trilevel technique," which involves coating a thick layer of photoresist, about 2.5 µm, onto the surface to be etched or implanted. This is followed by plasma deposition of an oxide about 0.12 µm thick. The result is a surface that is very flat and is ideal for patterning high-resolution images in photoresist or electron or x-ray resists. This third layer patterned on the deposited oxide completes the third level of the trilevel structure. After imaging a pattern in the 4000-Å-thick resist, the oxide is reactive-ion-etched, followed by reactive ion etching of the thick underlying layer of photoresist, in this case a positive optical resist. After the thick layer of resist is completely removed, a submicron pattern in thick resist remains; it covers steps without a change in line width or any perceptible sloping of the resist sidewalls. In short, a three-layer patterning structure is used with conventional imaging equipment to achieve submicron patterning. An outline of the trilevel process is shown in Fig. 15.2. Note that the process is simplified by using the reactive ion etch step to remove both resist and oxide layers. The entire process adds only two extra steps to normal processing, since in the standard process a resist-imaging step would be used. The oxide deposition step and the reactive ion etch are extra.

Figure 15.3 shows an example of the resolution, step coverage, thick resist coating, and vertical resist sidewalls obtained with the trilevel process. Figure 15.4 shows a cross section of the trilevel structure.

Deep-uv imaging

Resolution.   The strategy for extending practical resolution in VLSI and ULSI devices obtainable with optical lithography is to reduce the

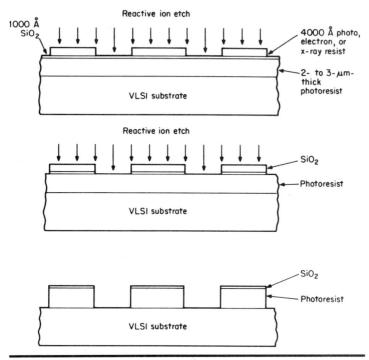

Figure 15.2 Trilevel process sequence.[1]

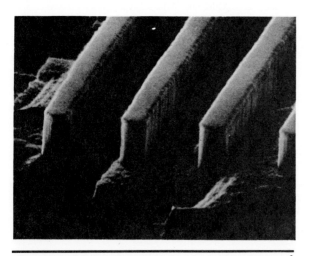

Figure 15.3 Trilevel resist features, 1-μm side, over steps.[1]

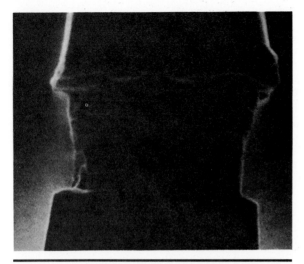

Figure 15.4 Cross section of trilevel structure (resist on top, P-glass in the middle, and SiO_2 on the bottom.)[1]

wavelength of the exposing radiation. This is the relation between wavelength and resolution expressed by

$$R = \frac{K\lambda}{NA}$$

where R is the minimum practical geometry and K (0.8) is a contrast value tied to the resist material. Assuming an exposing wavelength of 193 nm, and an NA of 0.4, resolution in single-layer resists would be calculated as follows:

$$R = \frac{0.8(193 \text{ nm})}{0.4}$$

$$= 0.386 \text{ } \mu\text{m}$$

This resolution can be extended by further reducing wavelength (152-nm fluorine laser), increasing the numerical aperture of the lens, or reducing the value of K. Reducing K to 0.6 by using existing resist chemistry yields

$$R = \frac{0.6(193 \text{ nm})}{0.4}$$

$$= 0.290 \text{ } \mu\text{m}$$

or increasing the lens NA to 0.5 yields

$$R = \frac{0.6(193 \text{ nm})}{0.5}$$

$$= 0.232 \ \mu m$$

Shorter wavelengths, and therefore better resolution, will be possible only with the development of better uv-transmitting materials for use in lens (all refractive or catadioptric) production.

UV lens materials. Fused silica, calcium fluoride, and barium fluoride are the only candidate materials available, and Table 15.1 lists their optical properties. Fused silica is the most thermally stable and is hard enough to polish with relative ease. Calcium fluoride is second to fused silica as a material for achromatizing in a lens system. All deep-uv transmitting materials must also have resistance to color center formation, which leads to crystal damage in the optic followed by reduced uv transmission. Special coatings are used to reduce this problem.

Numerical aperture. Increased NA is a lens design function. The Wynne-Dyson lens design (shown in Fig. 8.45a) offers promise as a production lens for wavelengths down to 152 nm, which approaches the transmission limit of fused silica.

The lens design used in the experiments described in Fig. 15.7 is a 52× modified Schwartzchild reflective objective. An NA of 0.65 is available with this lens. The limited field size and off-axis imaging preclude the use of this design for a production lens but make it ideal for deep-uv resist material and ULSI process research.

Depth of focus. The following Rayleigh criterion is typically used to calculate focus depth for structures at or near the limit of the lens resolution:

$$D_0 F = \frac{\pm \lambda}{2(\text{NA})^2}$$

Substituting NA out of the equation shows that a reduction in resolution of ½ effectively reduces focus depth to ¼, a square-law relation-

TABLE 15.1   Deep-uv Refractive Materials

	Refractive index at 250 nm	Coefficient of thermal expansion (X10-6/ C)	Cutoff wavelength, nm
Fused silica (SIO_2)	1.5076	0.5	147
Calcium fluoride (CaF_2)	1.4676	18	123
Lithium fluoride (LiF)	1.4188	32.3	105
Barium fluoride (BaF_2)	1.519	18.4	134

ship. For practical lithography at 0.5 μm, this leaves only about 0.5-μm focus depth for g-line to i-line imaging but over 0.1-μm focus depth at 193 nm. The total focus budget must be divided, however, between several parameters, the most significant being wafer nonflatness, typically +0.5 μm for very flat wafers. Tip/tilt errors and wafer topography contribute another 0.5 μm, thereby using the entire budget.

The value of K. The last parameter to further reduce resolution is the K value in the formula on page 495. By moving to a "top-layer" imaged resist strategy where only the upper 1500 to 2000 Å of the resist is used for image transfer, a K value of 0.5 or lower is easily achieved. One such process seeds silicon atoms into the resist layer after excimer laser exposure, followed by RIE pattern transfer. Differential uptake of silicon by the resist makes this possible, and several candidate chemistries based on conventional optical resists (positive novolak type) are under investigation.

Excimer laser steppers and technology. Extending optical lithography has been accomplished by using shorter exposing wavelengths and increasing the numerical aperture of lenses. Increases in NA up to about 0.50 are possible by using catadioptric (reflective and refractive lens elements in the same system) optics. However, since depth of focus decreases quadratically as the NA increases, it is undesirable to use an NA much above 0.45. Further, to obtain the large-field sizes needed for VLSI chips, an NA above 0.45 is unlikely, especially in refractive optics. The most likely way to improve resolution then is to reduce the wavelength. Excimer lasers, operating at a variety of deep-uv wavelengths as shown below are likely candidate light sources. The gas lasing medium is shown, along with the wavelength.

Excimer laser wavelength, nm	Gas molecule	Relative uv photon energy
193	Argon fluoride	6
248	Krypton fluoride	4
308	Xenon chloride	5
351	Xenon fluoride	4

The ultraviolet region consists of near-uv (250- to 400-nm), uv (100- to 250-nm) and far- or deep-uv (4- to 100-nm) wavelengths of light. Early sources of energy in the uv and far-uv regions were filtered mercury, mercury-zenon, and deuterium lamps. These sources were used with elliptical mirrors and fly's-eye lenses to maximize the use of all ray bundles for both uniformity and intensity. These sources typically deliver exposure intensities below the photoablation thresholds of organic polymers, and are therefore not suitable for ablation reactions.

Excimer lasers are commonly used uv light sources. They emit in-

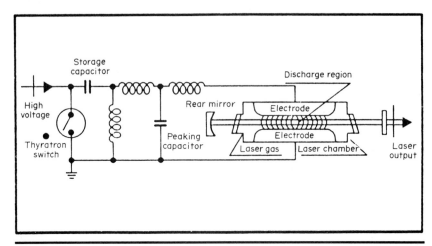

Figure 15.5 Diagram of a typical excimer laser head. (*Courtesy of Lambda Physik.*)

tense ultraviolet radiation at several wavelengths, with varying photon energies, as follows:

Excimer lasers are also efficient sources. They are pulsed, rare gas–halide systems that typically are based on simple, discharge-pumped designs. The schematic of a typical laser head is shown in Fig. 15.5.

Excimer lasers. Excimer lasers were the first uv light sources to deliver high average power, which permitted the expansion of this technology from basic research to production. High average power levels from excimer lasers are achieved by using several techniques, primarily by increasing both the volume of the excited species and the pulse repetition rate. Practical limitations on volume currently limit excimer lasers to exposure areas of several 100 cm^3 and single-pulse energies of 0.1 to 1.0 J at a 20- to 30-kV charging voltage. The maximum repetition rate currently available is in the 500- to 700-Hz region with 100-W lasers. The pulse duration of most excimer lasers is 10 to 20 nsec, with a beam divergence of 10 mrad. Unstable resonator optics can be used to reduce divergence to below 1 mrad.

Improvements in gas circulation rates will permit rates well above 1000 shots/sec in the near future. Increasing both the discharge power density and the pressure will increase the concentration of excited species and thereby permit higher power and higher reprates. A typical beam profile from an excimer laser is shown in Fig. 15.6.

Advantages of excimer lasers. As uv energy sources, excimer lasers offer several key advantages. First, they can provide deep-uv radiation at relatively high average power. Early uv lasers, such as double- or

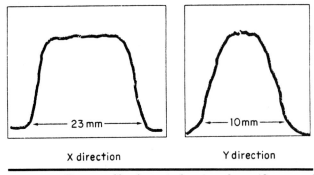

X direction   Y direction

Figure 15.6 Beam profile of a typical excimer laser. (*Courtesy of Lambda Physik.*)

triple-harmonic Nd:YAG and ion lasers, had output power of less than 1 W. Excimers, at similar wavelengths, offer up to 50 W, energy levels that are consistent with the production throughput rates needed in VLSI manufacturing. Excimers have been shown to be essentially speckle-free, unlike earlier lasers operating in the uv spectrum. Speckle, an interference effect caused by high spatial coherence, causes excursions in beam energy profiles and resulting non-uniformities in resist images. Excimer lasers have relatively low spatial coherence and are therefore free of speckle.

Excimers, as sources of deep-uv radiation, are much more efficient than light sources currently used in photolithography. The mercury arc lamp, with strong g-, h-, and i-line peaks, falls off dramatically in the far-uv region, just where the excimer begins. Mercury vapor and mercury halide lamps may have high conversion efficiencies (about 50%), but their high infrared content requires considerable output filtering. Excimer lasers, on the other hand, produce monochromatic radiation without the requirement of filtering.

Perhaps the most significant advantage of the excimer laser as an energy source in VLSI processing is the range of short wavelengths available, all of them shorter than those used today for advanced optical microlithography. This translates directly into better resolution. Argon fluoride, for example, offers a 193-nm wavelength at power levels exceeding those of the mercury lamp at g-, h-, and i-lines. Resist normally exposed in 200 msec with mercury vapor sources can be fully exposed with a 500 mJ/pulse output power excimer laser operating at any principal excimer wavelength in 10 to 20 nsec.

For image resolution below 1 μm, lenses with numerical apertures of 0.4 to 0.5 are needed with the exposure wavelengths currently used. These lenses have very limited depths of focus, and that problem becomes intolerable when 0.5-μm lithography is attempted. With brighter excimer laser light sources, lenses with different numerical apertures, while retaining simple pulse exposure capability and utilization of shorter wavelengths, may become possible. The pathway to better resolution and shorter exposure time appears to be with short excimer laser light source wavelengths.

Finally, the excimer laser, especially at the argon fluoride wavelength, offers the capability of using its highly energetic photons to dissociate organic materials, such as photoresist, by photoablation. These photochemical reactions occur in organic materials without significant thermal effects and are therefore attractive whenever heating effects may alter or damage an underlying layer. The principal reason for the ability to achieve low-temperature surface reactions at high resolution is based upon the high photon energy at far-uv wavelengths.

In summary, excimer lasers offer several key advantages over conventional longer-wavelength lasers, including:

1. Lack of thermal damage to the substrate
2. High depth precision (± 1000 Å) when cutting or ablating
3. Submicrometer imaging or structuring capability
4. An available range of wavelengths to match specific application conditions

The invention of the excimer laser in the late 1970s has led to widespread use of uv laser radiation in all major research disciplines. The number of excimer lasers and excimer beam delivery systems is increasing rapidly as the technology advances. In 1982, the discovery of "ablative photo decomposition" (wherein high-energy photons from an excimer laser beam cause structural breakup in organic solids) accelerated development of the technology. As a result, many useful applications that rely on the phenomenon were identified.

Beam generation. An excimer laser head (Fig. 15.5) typically consists of a metal enclosure containing two nickel-plated electrodes running parallel to the optical axis. Optical resonance is provided by two mirrors placed at either end of the pressure vessel; one mirror transmits 100% of the beam, and the other is coated for partial transmittance to permit beam extraction. Also included in the laser cavity are high-voltage storage capacitors, a thyratron switching circuit, and other electronic circuitry needed to run the system.

The laser beam is produced by introducing a mixture of three gases. For example, an argon fluoride (193-nm) beam is produced by combining fluorine, argon, and a buffer gas like neon or helium at 2500 torr. Lasing action occurs when this mixture, circulated with a fan across the electrodes, is excited by a high-speed, 20,000-V electric discharge.

The gas molecules then transit to the electronically excited upper state to become excited dimers, from which the term "excimer" is derived. The excited dimer is a diatomic molecule of, in this case, argon fluoride (ArF).

Laser radiation at 193 nm is generated when the tightly bound and excited ArF species transits back to the repulsive ground state at which it is weakly bound. While doing so, it gives off energy as photons that produce the excimer beam. Recombination of the atoms then results in the original gas species. The cycle is repeated from 1 up to 500 Hz, depending on the laser system circuitry. The entire process is microprocessor-controlled, and special safety circuits prevent over-pressurization and provide automatic shutdown in the event of a leak. Further, the active species (fluorine) is diluted to a very low concentration, and human detectability is about 10 times the toxicity level for this particular gas. The main by-product of this reaction is heat, which is removed by a water- or air-based heat exchanger. A special carbon filter removes all gas by-products.

Delivery systems. The raw beam from a laser head is generally unsuitable for the processing of applications, primarily because of limitations in beam shape, intensity, and uniformity. Excimer laser beam delivery systems, in which beam characteristics are managed with computer-controlled optical and mechanical subsystems, are finding use in industry.

The optical beam delivery system used in the experiments described in this section is designed to expose surfaces with highly controlled excimer laser light at three wavelengths: 193 nm (argon fluoride), 248 nm (krypton fluoride), and 308 nm (xenon chloride). The imaging resolution of the system is 0.5 μm. Easy exchange of optical modules permits rapid conversion from one wavelength to the other. Figure 15.7 shows the optics schematic of the system. Three optical paths optimized for different wavelengths are involved to permit exposure, pre-exposure beam targeting, and viewing during operation. A separate imaging port is provided for real-time photographic or video recording. Figure 15.8 shows materials imaged with the XLR-100 system.

Excimer lens and imaging systems. The use of a catadioptric optical system is favored for excimer lithography because of the inability to chromatically correct refractive lenses and the very limited choice of optical materials. In a catadioptric system, a mirror surface is the primary element. Chromatic correction is achieved by using a single-element achromat in conjunction with a fused silica lens. The achromat is typically made of lithium fluoride, one of the few crystalline glasses that transmit deep-uv radiation. This lens design, a 1:1 Wynne-Dyson, is shown in Fig. 8.45a. The achromat allows use of the full laser band-

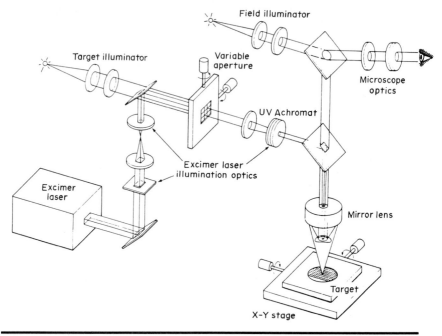

Figure 15.7   XLR-100 excimer laser optical paths. (*Courtesy Image Micro Systems.*)

width, thereby avoiding bandwidth narrowing schemes required in all-refractive excimer imaging. The lens system incorporates oil-filled interfaces between elements to compensate for the differences in the optical thermal expansion properties of silica and the lithium fluoride.

Images in resist produced with this lens are shown in Fig. 8.44*a,b*. The level of resolution expected from this imaging technology is 0.5 μm. The field size of the 1:1 Wynne-Dyson lens is large enough for 40 × 15 or 21 mm^2 die. The NA is 0.32.

In summary, new lenses for microlithography will focus on shorter exposing wavelengths, highest possible numerical aperture, and maximum field size to accommodate both chip dimensions and wafer throughput. The chart in Fig. 15.9 shows a typical projection from a lens manufacturer for the planned evolution of microlithographic lenses.

Excimer-based stepper systems.   Excimer laser steppers for 0.5-μm production optical lithography ($K = 0.8$ μm) have been built by no fewer than nine manufacturers. Most of these systems are chromatic and require line-narrowed lasers. Spectral narrowing, unfortunately, reduces intensity, but it does avoid the problem of fabricating a high-quality achromatic lens of acceptable field size. All of the systems

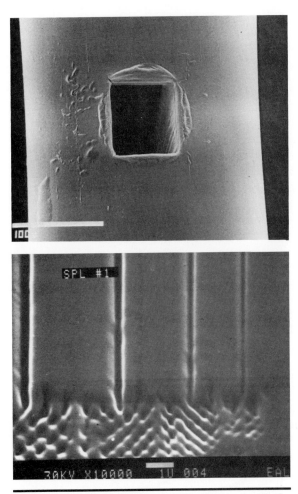

Figure 15.8 SEM photos of materials structured with the XLR-100 excimer laser system. (*a*) Single-pulse ablation of sub-0.25-μm lines in 0.5 μm of Shipley S1400. (*b*) Optical fiber ablated with 193-nm excimer laser system. (*c*) Copper conductor cut at 248 nm with excimer laser delivery system. (*Courtesy of Image Micro Systems.*)

built operate at the 248-nm krypton fluoride wavelength. One example of an achromatic 10× excimer laser stepper is shown in Fig. 15.10.

Resist images produced with this system are shown in Fig. 15.11. Resolution down to 0.325 μm (line-space pairs) was obtained in resists 0.4 μm thick. The resist was of sufficient thickness to produce good dry-etched results as shown in the SEM photos. Despite undercutting caused by 248-nm absorption of the resist, the RIE-etched SiO_2 was

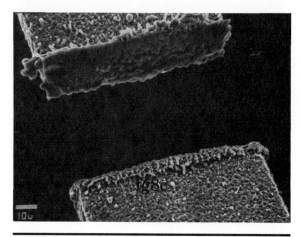

Figure 15.8 *(Continued)*

replicated to exactly match the resist pattern width. Maintaining this degree of geometry control through pattern transfer steps is an essential requirement for high-yield production of ULSI devices in silicon and gallium arsenide. Image reversal of A2-5214, performed by a sim-

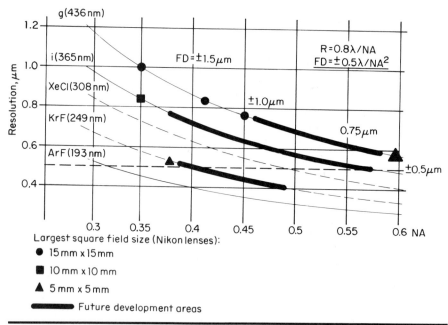

Figure 15.9 Short-wavelength lenses for microlithography: resolution versus numerical aperture (future development). *(Courtesy of Mike Powell, Nikon.)*

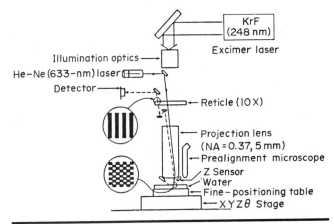

Figure 15.10 Diagram of an excimer laser exposure and TTL alignment system developed for laboratory use. (*Courtesy of Toshiba Corporation.*)

ple bake step followed by flood, was used to produce the patterns shown in the photos.

The excimer laser exposure and TTL alignment system shown in Fig. 15.10 was developed for laboratory use. A KrF (248-nm) excimer laser beam was introduced to the illumination optics by mirrors, which did not employ special equipment to eliminate speckle noise. Then, a reticle pattern was printed to a wafer on an x-y-z-θ stage through a 10:1 reduction achromatic projection lens, of 0.37 NA and 5-by 5-mm field size, fabricated from quartz and CaF_2 materials. The vertical position of the wafer surface was monitored by an oblique incident beam type optical-positioning sensor for focusing.

The TTL alignment scheme was as follows: Checker grating patterns and stripe grating patterns were prepared on a wafer and a reticle, respectively. The grating pattern on the wafer was illuminated by an He-Ne (632.8-nm) laser alignment light, which passed through the reticle window and the projection lens. The distance between the focal plane of the alignment beam and that of the exposure light beam was measured in advance to be 5.24 cm. In contrast to the difficulty with the chromatic lens in the TTL system, described earlier, the achromatic lens which was used provided the short distance.

Only the two first-order diffracted beams from the wafer checker gratings were focused to the stripe gratings on the reticle by employing bending mirrors, which corrected the distance by 5.24 cm. Thus, the interference between the diffracted beams and the reticle gratings generated the so-called "moire pattern," which was used as the alignment signal that changed periodically as a function of the relative po-

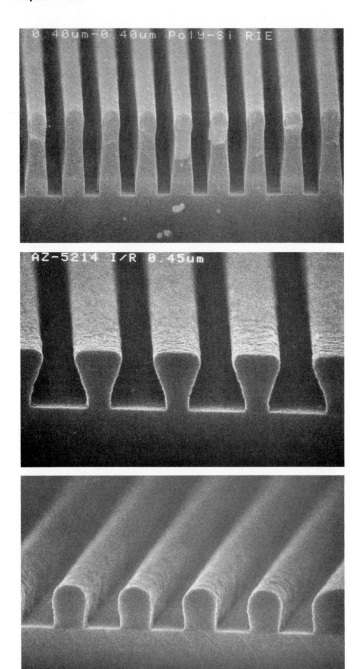

Figure 15.11 Resist images produced with an excimer laser stepper at 1×. (*Courtesy of I. Higashikawa, Toshiba Corporation.*)

sitions of the two gratings on the wafer and the reticle. The special feature of this alignment method is its high S/N ratio sensitivity and its defocus-free property. These have already been confirmed by using the g-line exposure.[10] The xy stage was driven by dc motors in the step-and-repeat mode, and the fine positioning of the TTL alignment was carried out by piezo actuators.

Exposure time in excimer stepper systems is critical, and good wafer throughput is needed to keep productivity high and payback periods as short as possible. One strategy to reduce exposure time is to source a more sensitive resist, one with exposure times below 1 sec. Research indicates that single-pulse (17 nsec) exposures may be enough to expose deep-uv resists.

An alternative is to replace the unnarrowed laser and achromat with an injection-locked system. This is a two-laser system in which the first laser produces low power in a very narrow bandwidth and then "seeks" the second laser. The second laser then generates full output power but retains the spectral narrowing from the master oscillator in the first laser. The bandwidth possible with injection-locking excimer lasers is less than 0.04 Å. Exposure times can be reduced by an order of magnitude.

Advanced Nonoptical Imaging

The need for finer geometries and tighter tolerances in integrated circuit manufacture has led to the development of imaging approaches that extend the useful resolution well into the submicron region. The new technologies that have been directed in pursuit of the submicron region in the nonoptical area are electron beam, x-ray, and ion beam imaging. The initial impetus for nonoptical imaging came from the needs of mask making, in which tolerances are the most stringent, much as positive photoresist first found application in mask making.

Current IC devices are using 1-mm lines routinely, and the trend toward smaller feature sizes continues. This means not only a requirement for masks with 1-μm and smaller element sizes, but much tighter fitting layer-to-layer registration. Added to this requirement is the need for faster turnaround time on both 1× working masks and 4× and 10× reticles, driven mainly by the increasing markets for ICs and the growth of existing markets. New markets also mean different design approaches to circuit layout, and each of these must be evaluated experimentally. Thus another mask-related requirement is rapid turnaround on these new "test" designs.

Beyond the initial need for advancing the state of the art in terms of resolution and line tolerance in mask making, the industry faces the challenge of bringing about the same levels of imaging on the wafers, just as it has done in the past. Therefore, nonoptical imaging must address *wafer*-imaging equipment and processes to carry IC geometries into the submicron region. In this section we will review the use of electron beam, ion beam, and x-ray technologies in solving the future imaging needs of the IC industry. We will cover the equipment and processing currently used with these technologies and discuss their advantages and disadvantages compared with advanced optical-imaging approaches. In addition to technology, economics must be a part of the equation that is used to put together an IC process, and economic comparisons with optical methods will be made.

Electron beam lithography

Electron beam lithography is the latent image exposure of a radiation-sensitive film to a beam of focused electrons in a vacuum. This is followed by development of the resist film and subsequent etching. Electron beam imaging systems offer some major advantages in IC fabrication, notably very high resolution, low defect densities, and high-accuracy pattern overlay capability. The resolution potential of electron beam systems is well below 0.5 µm; it approaches 0.1 µm with special resist processing. An example of electron beam imaging resolution is shown in Fig. 15.12.

The low defect density advantage of e-beam imaging systems derives from the fact that the systems image materials directly in a vacuum and eliminate the use of separate mask substrates that add con-

- 0.3-µm lines
- 5000 Å PMMA resist on GaAs
- 10,000× magnification

Figure 15.12 Images produced with electron beam exposure.

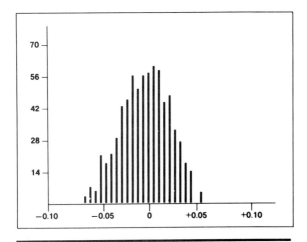

Figure 15.13 Overlay accuracy of electron beam exposure. (*Courtesy of Perkin-Elmer EBT, Hayward, Calif.*)

siderable defects. Pattern data for the lithography are fed direct from the system computer to the wafer stage and beam control systems that direct the exposure steps. The combination of stage movements and beam shaping permits a variety of direct writing strategies.

The other main advantage of e-beam lithography is pattern overlay accuracy. Level-to-level overlay accuracy is well within a 0.1 ± 0.05 μm specification. A high level of machine and software control permits this high overlay accuracy. The overlay accuracy of a typical system is shown in Fig. 15.13. Key controls in the machine are optical column sensing and correction, as well as software controls such as proximity effect correction, design rule changes, and overlap control.

The three key advantages cited above make e-beam lithography ideal for mask and reticle manufacturing, in which high resolution, overlay accuracy, and low defects are major requirements. The IC industry needs more complex circuits with increasingly smaller geometries and more rapid turnaround time, trends that have been satisfied by e-beam lithography. In addition to masks and reticles, e-beam lithography produces a considerable number of specialty devices in the industry, many of them the prototypes needed when design changes must be evaluated quickly before one is finalized. Rapid pattern data changes, followed by direct imaging without the need to make a mask, is a natural advantage for e-beam systems.

Electron beam imaging is performed in one of two main writing strategies: vector and raster scan. In vector scanning, shown in Fig. 15.14, the stage vectors from one write area or address to another. This saves time by avoiding the nonexposed area. While the stage will

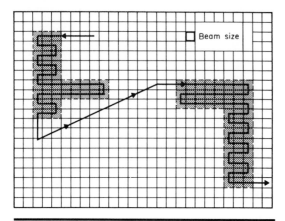

Figure 15.14   Vector scan e-beam exposure method.[8]

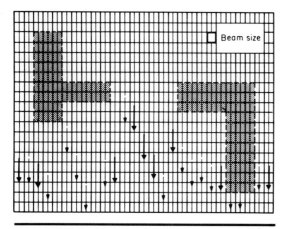

Figure 15.15   Raster scan e-beam exposure method.[8]

vector between address points, a raster scan may still be used to write the feature.

Raster scanning e-beam writing uses a dual raster technique. The xy stage moves constantly under the beam, sweeping over an area orthogonal to the stage direction. The beam is programmed to turn off and on as required to expose resist. Figure 15.15 shows a schematic of the writing strategy, and Fig. 15.16 shows some typical beam profiles.

Beam shaping.   The addition of a variable aperture in a vector scan system to shape the Gaussian current distribution permits square corners to be imaged and various sizes to be patterned faster. A beam line, shown in Fig. 15.17, is produced by using one series of beam-

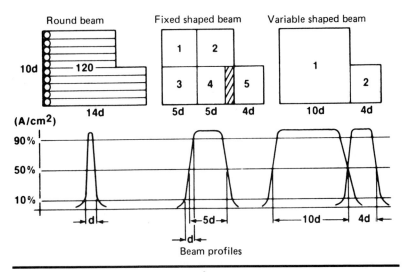

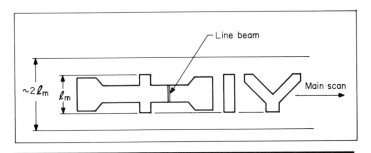

Beam profiles

Figure 15.16   Typical e-beam exposure shapes.

Figure 15.17   Beam width and length.

shaping apertures. In actual exposure, two shaping apertures interact to allow a specific amount of electron energy through. One aperture determines the energy uniformity, and the other is superpositioned to determine the width, length, and orientation. The lenses in the e-beam column then reduce the image magnification and project it onto the resist-coated wafer.

The strategy of writing with a shaped beam is such that highly uniform exposures are made across wide surface areas. Synchronously scanned and blanked, the beam energy exposes various patterns with sharp edge acuity, because the shaping blades distribute energy uniformly. Magnetic and electrostatic deflection of the beam are computer-controlled in concert with stage movement. The careful selection of stage strategy allows for optimization of system throughput changes in subfield size, deflection field, and other key system parameters critical to throughput.

Figure 15.18 shows various lithography pathways used to transfer patterns to wafers.

Electron beam lithography, as compared with optical methods, has several potential advantages, including high resolution, rapid turnaround time for the development of new product designs, increased product yield, mask savings, improved alignment capability, and better control over defects. Along with these potential advantages are some possible disadvantages, including major capital investment, the restraints of working in vacuum, variations in e-beam current as well as variations in the beam itself, and limitations in the process capabilities of e-beam resists and imaging over high steps. E-beam technology is realizing its potential technologically and economically in mask making. The solution to some of the problems mentioned will enable e-beam writing to advance to direct writing of wafers for a sizable number of major IC applications. Figure 15.19 illustrates an e-beam optical column.

Another type of e-beam system using projection rather than scanning is the "demagnifying" projection type proposed by IBM. This sys-

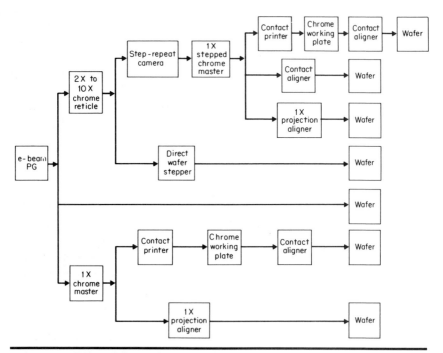

Figure 15.18   E-beam lithography mask pathways.

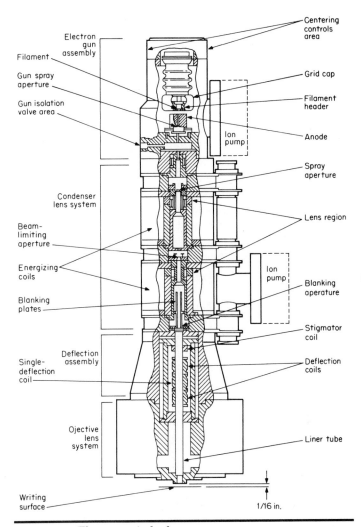

Figure 15.19 Electron optical column

tem uses a shadow mask that incorporates patterns several times the actual desired image size. Although it does not pose the alignment problems of the photocathode approach, it is still in the early stages of development, and too much remains to be discovered to properly assess its potential. In theory, it would be relatively low in cost and could overcome the wafer flatness problem by use in a stepper mode. In addition, because the mask would be several times the actual pattern on the wafer, line tolerances and mask life also would be long, since no degrading effects would be present. The estimated exposure time for this type of system is about 60 sec with resolution of 0.5 μm. This remains an internal project for IBM and is not commercially available.

x-Ray lithography

x-Ray imaging takes advantage of the ultrashort wavelength of an x-ray (approximately 4 to 50 Å) as a means of obtaining very high pattern resolution in x-ray-sensitive resist materials. It is accomplished in much the same way as with all other radiation sources: by placing a source at one end of an imaging system and the resist-coated wafer at the other end. Then the distance between these two fundamental points is filled with a mask that contains the patterns, and if necessary, devices that intensify, collimate, and appropriately filter the exposing radiation.

The use of x-ray technology has been a function of the development of four basic ingredients: good, strong x-ray sources, an effective x-ray mask, sensitive polymeric materials with good IC process compatibility, and a commercial imaging system that accommodates the many needs, mechanical and physical, of advanced IC fabrication.

The short x-ray wavelength is the key to extending the resolution needed in ULSI device fabrication. Optical lithography is limited by diffraction effects, and resolution limits because of optical diffraction are in the 0.05- to 0.1-μm region. The x-ray wavelength does not suffer from this problem.

x-Ray imaging offers high resolution potential (below 0.1 μm), high throughput with small synchrotron sources, and reduced defects, since x-rays are transmitted by low-density particulates like dust and resist flakes.

x-Ray system and mask technology.   The potential for sub-submicron (less than 0.1-μm) lithography, called "nanolithography," is strong in the case of x-ray imaging. The soft x-rays used for production processing are 40 to 100 Å long (1 sine wave). Figure 15.20 shows the key parts of a production x-ray lithography system. A high degree of automated sensing, aligning, mask positioning, wafer movement, and wafer-to-mask gap control allows such a system to be production-viable. The field of exposure is variable, but about 200 mm for full wafer exposure on 200-mm slices is needed. As a resist-coated slice is moved into position on the six-axis stage, x-rays from a point source move through the mask and into the coating. The x-ray tube must be cooled because the focused electrons from the cathode strike the anode and generate considerable heat as x-rays are created, even when the anode is rotated. The soft x-ray photon, upon absorption in the resist,

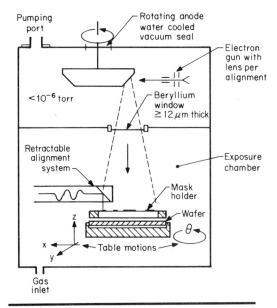

Figure 15.20 x-Ray imaging system components.

releases its energy (or converts it) into low-energy secondary electrons. The absorption of x-rays by resist (or any material) is proportional to electron density. That allows for relatively high absorption rates by low-electron-density organic polymers and thereby provides a wide selection of starting materials for resists. The same principle paves the way for high-density masking materials, such as the gold-based absorber masks, and low-transparency low-atomic boron and silicon serve as the transparent areas of x-ray masks.

Production masks are simple in structure, as shown in Fig. 15.21. The base substrate is solid glass or silicon; it is covered with a pellicle membrane and a top patterned gold layer. In the exposure system, the mask-wafer gap is filled with helium (or evacuated) to reduce x-ray absorption and maintain maximum intensity from source to resist.

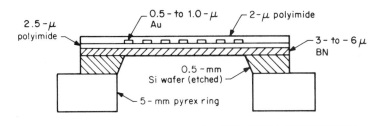

Figure 15.21 x-Ray mask cross section.

Working from source to resist, movement is through the source vacuum, a 25- to 50-μm-thick beryllium window, and the helium envelope and finally into the mask sandwich structure. Past the mask, there is a small proximity gap before the resist surface is reached.

Resolution of any x-ray production system is a function of total system image contrast capability, just as it is in optical lithography. The x-ray source and all elements in the path to the wafer must be optimized for absorption or transmission differentials so as to deliver high signal contrast into the resist layer. Strong sources with relatively long source-to-mask distances are needed in production for good, sharp resolution brought about by the collimation effects. High resolution is provided by a combination of the following:

1. Minimal penumbral shadow

2. Resist contrast

3. x-Ray source spectrum uniformity

4. Magnification effects from the mask

5. Mask transmission properties

6. Accelerating voltage and secondary electron ranges

Mask problems have been one of the most serious challenges in making x-ray imaging a production wafer process technology. Also, masks have had poor contrast properties because of incomplete (90%) absorption of the x-rays. The mask is made of a thin x-ray-absorbing layer of gold supported on a suitable frame. The mask is typically made by depositing LPCVD boron nitride onto a silicon wafer, followed by backside wafer plasma etching of the nitride. A Pyrex ring is used as a support at this point, followed by spin-coating a layer of polymide. The absorber layer is then deposited onto the polymide. A typical absorber film consists of 300 to 5000 Å of tantalum. After sputtering on the absorber, the remaining exposed silicon is etched, which leaves the mask blank ready for resist patterning and etching. Ion milling or reactive ion etching is used to fabricate the mask elements, and resolution limits imposed by etch sidewall angles restrict final mask pattern element size. Resolution is also limited by the thickness of the gold absorber layer, in which minimum thickness is needed for maximum resolution but sufficient thickness for x-ray stopping power is also a factor.

An x-ray mask is complex and difficult to produce if resolution in the 0.25-μm and below region is required. Further, mask-to-wafer distance is critical to pattern fidelity, as is alignment of various layers. Since optical alignment is used, overlay accuracy and the final overlay budget (all contributing elements root-sum-squared) is of the same order as in optical lithography.

Resist energy absorption. The relative productivity of vacuum ultraviolet (VUV) storage ring source is determined largely by actual resist exposure time, a function of effective energy absorbed at the wafer surface. Since most exposure systems lose various amounts of energy from the source to the resist surface, measurements of the spectrum and intensity must be made right at the point at which the energy enters the resist film. The complexity of an x-ray storage ring results in considerable alteration of the energy between source and resist. Various elements used to direct, focus, and intensify the electron beam and x-ray stream remove various amounts and types of energy. Most of the low-energy radiation is absorbed by the beryllium window; energy greater than 1000 eV (12-A) is all that gets by the window. Moving further toward the wafer, the x-rays move through the helium-filled chamber and through the mask, both of which absorb some radiation. Finally, at the resist surface, hard x-rays move right through the film (almost all films are organic resists), leaving only soft x-rays in the 1000- to 2000-eV (7- to 12-A) range to perform the imaging function.

Effective energy at the resist surface is calculated by converting the horizontal milliradian energy from the ring source into square centimeter energy at the wafer and multiplying this by the lowest line. Watts (power) is plotted versus photon energy, in electron volts. For example, at an output level of 750 MeV, the resist sees approximately 3 mJ/(mA • min) across its uniformly energized surface. The aspect ratio of this image is particularly impressive, and experiments with image potential in which various resists and gold absorber films are used indicate that up to 10:1 ratios are possible, depending upon the resist system. The mask-to-wafer gap in these tests was approximately 40 μm, and PMMA was used as a test resist. The resolution is not seriously affected by penumbral blur but is more affected by diffraction effects.

The relative sensitivity or response of resist materials to x-ray radiation may be calculated as follows:

$$I = I_0 e^{-\mu_m P z}$$

where I = x-ray intensity *after* passing through the resist layer
  I_0 = x-ray intensity *before* passing through the resist layer
  μ_m = mass absorption coefficient
  P = resist density
  z = resist thickness

The x-ray photon travels into the resist and scatters as a function of the photoelectron energy generated within the resist during absorption. The photoelectrons in the organic or polymeric matrix bounce around, giving off secondary electrons, and it is the secondary electrons which are responsible for the area-differentiating or latent-image-producing reactions. The range of a primary photoelectron is approximately 150 nm. That dimension in no way restricts the resolution potential of x-ray imaging, since structures below 500 Å (50 nm) across have been produced in laboratory experiments with relatively crude resist-developer combinations.

An example of the high resolution and aspect ratio capability of x-ray lithography is shown in Fig. 15.22. The "posts" of resist are 0.5 μm wide and 4.0 μm high.

x-Ray sources. In the x-ray sources case of x-ray imaging, the energy sources are either high-power synchrotron (storage ring) or electron impact type. The synchrotron puts out a wide spectrum of radiation from infrared to long-wave x-rays. The radiation comes from electron energy loss in orbit and yields highly collimated x-rays that remain close to the orbital plane of source electrons. This poses an access problem, but incentive is high because of the very strong source of energy available.

The impact source of x-rays also delivers a broad spectrum of wavelengths by directing an electron beam at a target material. The x-ray radiation given off by the e-beam impact matches the physical con-

Figure 15.22 Submicron, high-aspect ratio images produced with x-ray exposure. (*Courtesy of Micronix.*)

stants of the target element and is generally below the intensity of the storage ring, or synchrotron sources.

The imaging strategy chosen for x-ray patterning can vary from full-field proximity printing to 1:1 projection to step-and-repeat imaging. As with photo-optical technology, stepping with x-rays takes advantage of the greater uniformity of radiation possible in a smaller image field as well as much greater energy intensity. The full-field x-ray system optimizes for wafer throughput at the expense of overlay accuracy and pattern resolution, similarly to full-field optical printing.

The electron-impact type of x-ray source uses high-energy electrons at 25 keV. These are directed at a metal (palladium) target; the result is a high heat interaction that consumes 95 to 98% of the energy, with a small fraction left as x-rays. The low efficiency is offset by using optimized geometries for the anode and collection cavity, high-velocity cooling systems, and moving anodes. The cost and complexity of the approach makes synchrotron radiation sources attractive.

A synchrotron or storage ring uses radial acceleration of electrons to obtain synchrotron x-rays. Highly collimated x-rays are given off and collected at several points around the ring. Short resist exposure times are possible because of the relatively high (20 mW/cm^2) energy flux density at the target or wafer plane.

The storage ring concept truly solves the major problem of energy intensity needed to provide economical exposure times, assuming that a much smaller ring than the one discussed is finally produced for volume manufacturing. The areas needing further development beyond streamlining the ring are the mask and wafer-to-mask alignment mechanism. The alignment technology to provide a highly precise automatic system exists, and mask material research and actual x-ray mask fabrication have developed rapidly. The actual power or energy available for resist exposure can be improved, since only a very small amount of flux is available in the region (7 to 12 A) in which image formation takes place. A smaller ring would provide a higher radius of curvature, which would reduce the amount of useful energy. Increasing the power of the beam is probably the solution, along with the possibility of breaking the ring into a greater number of segments to reduce the radius where energy leaves the beam. However, major redesign would be needed for that approach. Power increases to 1000 MeV with the current system would improve the energy output by a factor of 5, bringing the exposure time down to only a few seconds with positive novolak-type and most other resist candidates. Many other design options for storage rings are under consideration, including 1-m rings with superconducting magnets to reach the higher energy fields needed to get the beam of electrons through the smaller radius.

Development of special resists for x-ray imaging will solve the problem of matching the characteristic sawtooth x-ray absorption profiles of most resists. Matching the source output (or x-ray energy spectrum at the wafer) with the resist absorption profile will naturally maximize resolution and exposure throughput.

Development of new masks to optimize energy transfer is expected. The gold used in the mask to absorb x-rays must be thick enough to prevent any leakage of energy into the resist. Gold is expensive and may be replaced by a more cost-efficient material. The gap between mask and wafer is helium-fitted because oxygen and nitrogen, major components of ambient air, absorb considerable x-radiation. Even helium absorbs some useful energy, and a material may be substituted that has even less effect on the contrast of the signal. The contrast of the mask itself is an area that needs improvement. Mask contrast is measured by the ratio of energy passed through the mask to the resist to the energy absorbed by the resist through the mask absorber or intended opaque area. This ratio is called the residual absorbed dose (RAD). Because greater contrast means higher edge definition, low RAD numbers and inherently high-gamma resist systems are desirable.

Many of the improvements cited above will culminate in field production use of a second-generation x-ray lithography system for submicron geometries on ULSI devices. The ultimate resolution of the x-rays is really limited by the secondary photoelectron emission in the resist, mask contrast (mask-to-absorber pattern thickness), and the penumbral blur caused by a finite source spot size.

Unlike photolithography, standing waves are not a factor, and unlike electron beam imaging, proximity effects are not a problem. The other real benefit beyond wavelength is relative immunity to defects and dust that cause such problems in optical lithography. A dust or organic or even silicon particle must be about 25 to 30 times thicker than the gold absorber layer before it will absorb an equivalent amount of 7-Å x-radiation. In other words, x-rays will penetrate through resist flakes, dust colonies, cotton and synthetic fibers, stains, and the many contaminant species that infiltrate the clean room and end up on imaging surfaces. Almost all of these defects are completely transparent to x-rays.

Principle of storage ring energy. Synchrotron radiation studies evolved in the early 1950s with the understanding of accelerating charges and the relativistic treatment of radiation from such charges. In a vacuum ultraviolet storage ring, high-energy electrons are bent as a beam by dipole magnets. The radius of curvature for a nine-dipole magnet system is 1.9 m. The accelerating beam of electrons gives off radiation in a continuous spectrum from infrared to soft x-rays. This radiation comes off tangent to the electron beam as it accelerates between the bending magnets; it is focused into a straight line by quadrapole magnets. The diameter of the storage ring varies with the need to produce a minimum size. Early rings had a 15-m diameter; 1- to 3-m rings would be more practical, and it is possible to obtain multiple substations for x-ray exposure at various points on the ring. Thus, although the initial cost and complexity of a synchrotron source are high, so is the potential productivity.

The uniformity of energy from the beam line as it exits from the storage ring is not ideal, even though the effective source size is approximately 1 mm^2. The small apparent source size is possible because the arc reduces the penumbral blurring. However, the exposure energy assumes a Gaussian-like distribution pattern in the vertical plane, and optical elements of the beam line are used to correct this problem. Horizontal radiation is, fortunately, quite good. The intensity of the energy from a storage ring x-ray source varies with the size of the ring (radius of curvature), electron count in the beam, and final beam energy. The National Synchrotron Light Source (NSLS) vacuum ultraviolet ring, shown in Fig. 15.23, was designed to have 1-Å maximum circulating current, and beam life, as a function of vacuum quality, typically runs about 2 hr.

The line drawing of the NSLS beam shows the mirror box for increasing energy uniformity and a differential pumping station to remove possible hydrocarbon and helium contamination from the system and keep the vacuum quality high enough at the point at which the vacuum joins the storage ring. The 18-μm-thick beryllium window helps protect against rupture from the helium, and added valves and gauges further ensure against overpressure in case of a system rupture. The helium absorbs heat, which is considerable because the x-ray radiation, coming directly off the storage ring, is very intense. Heat on the mask must be prevented to maintain close alignment with the wafer. The exposure chamber in which the mask and wafer reside is therefore surrounded by its own vacuum, a factor that helps make alignment and mask changing much easier. A more advanced version of this system would replace the manual alignment system with autoalignment, which is possible with a step-and-repeat system.

The mirror is a special gold-coated, high-reflectivity cylinder that scans the beam up and down on the wafer in the shape of an arc. It improves both the collimation and the intensity of the beam. Actual

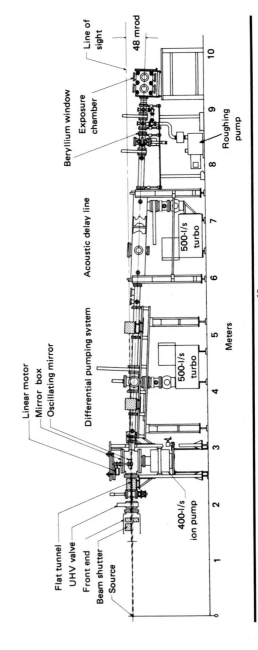

Figure 15.23 Beam line of the National Synchrotron light source.[18]

beam intensity is sensed by monitors at various points in the beam line; other sensors check beam alignment, water pressure, temperature, vacuum, and other critical parameters, all of which are displayed on an instrument separate from the system computer.

Laser-generated or "z-pinched" plasmas offer a high spectral brightness x-ray source at a 0.8- to 2.5-μm wavelength. The source is not as collimated as x-rays coming off a synchrotron, but it has other advantages. Synchrotrons produce hard (0.8-nm) x-rays in a collimated beam and have a long working distance between the source and the wafer. A collimated x-ray beam results in slightly better resolution, compared to that from a point x-ray source, because of the reduced penumbral blur. In addition, superposition errors from circuit topography and nonflat wafers are less severe when a synchrotron-based x-ray stepper is used. However, with a nearly collimated x-ray beam, the image field size cannot be altered readily to compensate for process-induced wafer run-out. (The ability to correct for process-induced errors will likely be crucial to achieve the required overlay for deep submicron lithography.)

A laser-based x-ray plasma source is less expensive than a synchrotron source and therefore, intuitively, more cost-effective. Conversion efficiency from IR photons to x-rays typically is 20% for a well-engineered system. Coupling a laser-generated x-ray plasma source with a stepper, compared to synchrotron-based x-ray steppers, is similar to what the semiconductor industry is accustomed to. More specifically, a laser-generated plasma source (e.g., using a solid-state infrared high-intensity laser to generate the plasma[7]) is more compact than a z-pinched source of synchrotron and may result in a more compatible clean-room footprint. Images produced in a laser-based x-ray plasma source stepper are shown in Fig. 15.24.

A number of devices have been made with x-ray imaging, including magnetic bubbles, CMOS/SOS, MOS LSI, and EPROMs. As the mask and exposure throughput problems are solved, x-ray imaging could emerge as a major IC fabrication technology. Its potential is significant considering that its resolution potential exceeds that of the e-beam; its cost (resolution/throughput product) is potentially low; it allows high aspect ratios in resist; and it is insensitive to dust and similar contaminants. One area to be refined is the alignment overlay, which must be improved so it becomes consistent with the resolution capability provided by the rest of the system.

Ion beam imaging

The use of ion beams to expose submicron structures in resists and etch those structures in wafer surfaces has been reported on for a

Figure 15.24   SEM of 0.5-μm resist images produced with a laser-based x-ray plasma system. (*Courtesy of Hamshire Instruments.*)

number of years. The techniques are focused ion beam lithography (FIBL) and masked ion beam lithography (MIBL), both using a stream of ions to expose a resist layer that is subsequently developed and etched. FIBL is a direct-write technique, and MIBL is a flood or full wafer area exposure technique.

Ion beam lithography.   Focused ion beams used as writing tools for submicron patterning offer several distinct advantages over other beam-imaging methods. Ion beam lithography involves the penetration of ion species into resist and semiconductor layers. The atomic mass of impinging ions is very much like the mass of material it penetrates, which eliminates most of the energy scattering that occurs with most all other beam lithography approaches. For example, the direction of an electron beam onto a resist-coated wafer results in the commonly observed phenomenon called proximity effect, or the scattering of relatively light electrons in the resist film. Further, ions can be focused and collimated to increase resolution. Since almost all ICs have small and large patterns in close proximity to each other, the effect is one of changing pattern size disproportionately. Compensating for the proximity effects caused by electron beams and other high-intensity beam-based exposure systems requires rather complex and expensive software routines.

In ion beam exposure, the only real scattering of energy occurs when there is a nuclear collision between one of the ions in the beam and the atom in the substrate. Collisions, fortunately, are quite rare, and travel distance is only ~0.01 µm. In addition to greatly reduced scattering of energy, ion beam energy transfer from beam to resist is extremely efficient because of the similarity of atomic masses of ions and resist and substrate. Energy efficiency is further enhanced by the elevated energy inherent in a focused beam of ions.

Many ion beam systems provide energy levels up to 150 keV. This very high energy level solves another major problem typically encountered in other beam exposure systems: exposure throughput. The energy available to the resist removes a key barrier, common to most beam exposure technologies, by making available nearly any resist of choice for use with ion exposure. Electron beam and laser exposure technologies, for example, require resists chemistries that are very wavelength-specific in order to provide economic wafer throughput.

The combination of very high energy levels, delivering a wide choice of production proven resist systems, very low energy scatter (no proximity effects), and high-resolution beams, gives ion beam exposure high potential for submicron VLSI lithography. The electron beam diameter permits high-resolution imaging, but secondary electron scattering and long write times prevent this technology from being used in wafer fabrication.

Ion beam imaging is based upon the use of highly collimated beams of protons, or hydrogen ions. The principal exposure arrangement is shown in Fig. 15.25. Proton or ion beam imaging is used with the well-established positive optical resist chemistries. These resists actually have greater sensitivity to ions than to the photons or electrons with which they have been used for years. As the protons enter the resist, they lack the high-energy scatterable electrons and concentrate all of their mass into producing a latent image in the resist layer.

Scattering of ion beam energy does occur in the mask, which is often a thin (0.4-µm) layer of single silicon crystal. Despite some scatter in the mask membrane, very high resolution is possible. Part of the high resolution is attributable to a reduced sensitivity to critical dimension change as a function of resist thickness and is much less than with optical lithography. This means that the need for inherent high contrast in the resist is much less than for competitive exposure technologies. Giving up contrast in the resist will allow the improvement of another functional parameter, thermal stability. The use of multilayer metal structures, with their incumbent surface reflectivity, makes the benefit of lower sensitivity to resist thickness a characteristic of ion beams.

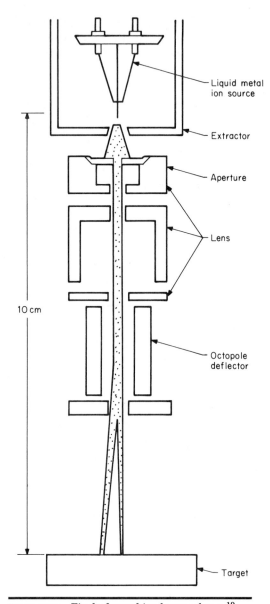

Figure 15.25   Finely focused ion beam column.[19]

Focused ion beam imaging.   Focused ion beam lithography uses a writing scheme similar to e-beam writing, except that a focused ion beam is deflected across the resist-coated wafer. The basic outline of an ion beam imaging system is shown in Fig. 15.26. The ions are emitted from a source and formed by the lens system; then they are positioned and scanned on the desired area after passing through an electrostatic deflector. Enlargement of the beam at the field edge permits only about a 1-mm^2 area to be scanned at once, and these "areas" are combined to constitute a complete wafer exposure. Layer-to-layer alignment is accomplished by using registration marks that are keyed by particle detectors that sense secondary electron signals, and the gain, offsets, and rotation of the electrostatic deflection field are adjusted to permit finding other address locations. The stage is monitored by a laser interferometer and can be moved to address other fields. The ion source is similar to the lanthanum hexaboride source used in e-beam imaging, and the same types of resists (PMMA and COP) are used. Resolution to 400 Å was obtained by coating a 600-Å layer of PMMA over a 400-Å gold layer, imaging the resist, and sputter etching the gold in an argon plasma. By using a similar imaging scheme, 2200-Å images were formed in COP resist and etched. As in x-ray imaging, a mask that absorbs the exposing energy is used to cast a shadow on the wafer, but the penumbral distortion of point-source x-ray imaging is not encountered in ion beam imaging because of the electrostatic deflection and focusing. However, submicron images probably will require a step-and-repeat exposure in order to meet a 0.02- to 0.05-μm maximum registration error between mask layers. As a direct-write technique, FIBL is a low-throughput technology suitable for production of high-resolution masks. A high-speed writing technique could be used with FIBL for applications-specific IC (ASIC) or custom device lithography.

Masked ion beam imaging.   In masked ion beam lithography (MIBL), a collimated ion beam is used to flood the mask, which is placed in close proximity to the resist surface. The exposure takes place as ions pass

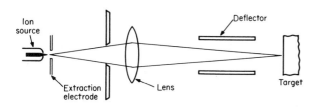

Figure 15.26   Ion beam exposure principle.

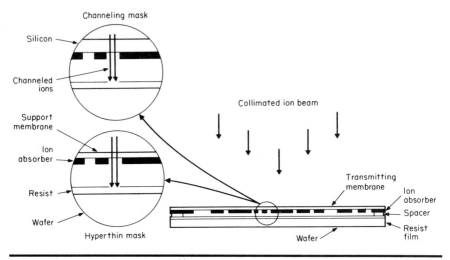

Figure 15.27 Ion beam mask types.[11]

through the openings in the ion-absorbing film. The unfocused power source and mask type are similar to those of x-ray imaging systems. Estimated exposure time for a 6- to 8-in wafer is 2 min, using 0.5-µm images. Figure 15.27 shows two types of masks used with this method: a channeling type that uses a 1-µm layer of single-crystal silicon oriented to allow ions to pass through the crystal lattice with little scattering and an amorphous layer of aluminum oxide that is formed by anodizing aluminum and etching it back to form a hyper-thin film or membrane about 700 to 2000 Å thick. Both materials are used as support structures for the ion-absorbing layer of gold and are chosen because of their ability to minimize ion scattering. The mechanical spacer shown in the figure is 20 µm thick, and reducing the thickness will permit better resolution. The imaging system using the masks produced resolution of about 0.5 µm, with only about 0.15 µm of widening in the PMMA image space width and nearly vertical resist walls. The masked ion beam system appears to have met the necessary requirements for submicron IC fabrication. Continued refinement of the mask-and-resist technology will be required before MIBL is reduced to a production technology. When the mask scattering and distortion problems are solved, MIBL may prove to be a nonoptical alternative to obtaining submicron images without moving to e-beam or x-ray imaging.

Direct ion device fabrication. One unique aspect of ion beam technology that has the potential to displace lithography is the placement of both matter and energy in the wafer substrate. The ability to place ions with great precision without the use of resists or masks will have a major impact on VLSI device fabrication. Direct, maskless, and resistless ion implantation, which is possible with computer-driven beams, offers the following key benefits:

1. Elimination of over 50% of all process steps
2. Elimination of registration errors
3. Increase in available surface area (to increase density) by elimination of lateral etch distances
4. Removal of all resist-related process problems, such as resist flakes and residues
5. Great cost savings from elimination of wet processing
6. Significant increase in manufacturing yield
7. Complete software control of IC pattern delineation
8. Greatly increased production throughput

Direct ion implantation may also prove to deliver lower annealing temperatures, a benefit for high-density devices that are much more sensitive to substrate dimensional changes. The custom software needed to run prototype chips will be simple and fast, since a reticle or mask is not needed. Designers will be able to take CAD programs and quickly "get them in silicon," via the computer-driven beam to test a new chip layout. The configuration of a direct doping source is shown in Fig. 15.28. The ion source operating principle is to provide a liquid metal reservoir which also serves as a heater. The ion delivery system provides energy in the 30- to 150-keV range. The probe is submicron in size and can deliver several different ion species from the reservoir. The ion probe itself is computer-programmed to deliver precise

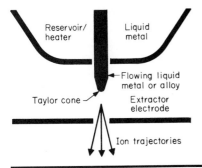

Figure 15.28 Direct ion doping source and principle.[20]

amounts of ions to coordinate with an accuracy of approximately 0.1 μm.

Direct ion doping has tremendous potential to not only provide the quality of pattern delineation needed in submicron design rule devices but also to generate high throughput with zero distortion. Even after subtracting all the yield loss from resist process and associated defects, ion beam doping will deliver still more good die per wafer by avoiding the etch step and its negative yield impact. Finally, the doping process is free from pattern drift, mask dimension changes, and all of the overlay errors that often cause a die to be rejected. Overall yield impact should be substantial.

The ion column consists of the liquid metal ion source (top of the diagram in Fig. 15.28). This part of the system is a microscopic needle, often a sharpened piece of tungsten. The tip of the tungsten is wetted with the liquid ion source materials, such as gold or gallium. After wetting, a positive potential of 3 to 6 kV is applied in the area above the extractor and below the needle tip. The energy field causes the molten metal source on the needle tip to ionize when the potential is strong enough to overcome the energy or surface tension stress holding the liquid metal on the needle. The positive potential effectively creates a strong electrostatic force that tends to energize molecules from the molten metal and convert them to ions. Just before ionization occurs, a cone-shaped formation (Taylor cone) generates on the end of the needle tip; it is created by the stress forces of the applied high-energy field. The cone diameter is typically about 500 Å. The cone is surrounded by an electric field of its own equaling approximately 1 V/A, and this small field in turn converts the liquid metal dopant into a vapor.

It is from the dopant evaporation that ion currents which pass through the collimating ion optical elements onto the target below are created. The size of the ion stream reaching the target, and hence the resolution potential of this source as a lithographic tool, is a function of the efficiency of the ion optical elements and the source size. Assuming a zero-distortion ion column, the target beam resolution would equal the source resolution, or be about one-twentieth of a micrometer (500 Å).

The most common distortion encountered with these ion optical elements is chromatic aberration. Since there are tradeoffs between optimizing lens design to eliminate chromatic aberration and increasing lens resolution with small acceptance angles, beam intensity (and hence throughput) is partially compromised. One answer is to pursue alternative ways to energize the liquid metal dopant to raise current density, such as working in the area of high-voltage discharge or more carefully controlled flow of liquid metal to the needle tip. Some researchers are investigating hydrogen ion sources that have the potential of raising ion beam current density 100 times existing levels, to about 100 Å/cm^2. The area of liquid metal and other source technology is critical to the evolution of ion beam writing at high speeds with good resolution. Indium, gallium, and gold are used as dopant sources that generate current densities of about ½ Å/cm^2 at 20 keV. The beam diameters of these systems range from 0.1 to 1 μm.

Below the needle tip, ionized dopant moves through the aperture of the electrostatic deflector lenses, past the octopole deflector, and onto the target. One of the advantages of this system is the ability to change the beam energy from approximately 1 to as high as 30 keV. Further, the ion beam can be manipulated to be parallel with the ion optical axis and electrostatically scanned across the target. Electrostatic scanning is used to reduce astigmatism. Experiments with beam manipulation have included deflection of 30-keV beams across 860-μm fields with 0.1-μm resolution, which illustrates the lithographic potential for software-driven ion source writing. Beams have also been raster-scanned to generate a scanning ion image.

Resist imaging with ions.   The generation of both primary and secondary ion images allows the application of focused ion beam lithography to mask making (primary) and wafer fabrication with ions passing through a mask (secondary). In either case, all of the previously cited benefits of ion beams, compared to laser and electron beams, are realized. Ion lithography can be adapted to proximity printing, in which channeled ions are used, or to ion optical imaging, which is analogous to step-and-repeat optical reduction printing. In focused ion-beam lithography, there is still some lateral and reflective scattering of energy in the resist and off the substrate, but the energy of these scattered secondary electrons is so low as to be insignificant to the patterning process.

Many resist types can be used for ion beam lithography. They include positive and negative resists from e-beam and optical technologies. The reason for good exposure throughput with such a wide range of resist chemistries lies in the efficient movement of an ion energy through polymeric structures. For example, positive resists are rendered developer-soluble by a process of chain scission. As the ion passes through the resist molecule, the larger atomic number and reduced velocity of the molecule, compared to those of an electron, allow fewer ions to produce an equivalent amount of chain scission. In short, for an equivalent amount of energy density at any given point in the resist layer, fewer ions than electrons or photons are required.

Increased resist sensitivity is not without its costs, however, and ion beam exposures of resist layers have resulted in the same type of pattern geometry control problem as is encountered with very high sensitivity negative optical resists. Even with highly focused ion beams, there are inevitable dose fluctuations, uniformity variations, and even beam position shifts. Since the resist is so much more responsive to the ion energy, it is adversely affected by these inconsistencies. Thus, random statistical variations occur in focused ion beam lithography, and tests indicate that a dose of approximately 800 ions/pixel is necessary to maintain pattern dimension control of 10%, 3σ. For a fixed dose, even variations in ion spatial arrangement and ion scatter could degrade the edge acuity of a resist pattern element. Figure 15.29 shows the effects of granularity on feature resolution obtained by making Monte Carlo simulations of ion trajectories of hydrogen and

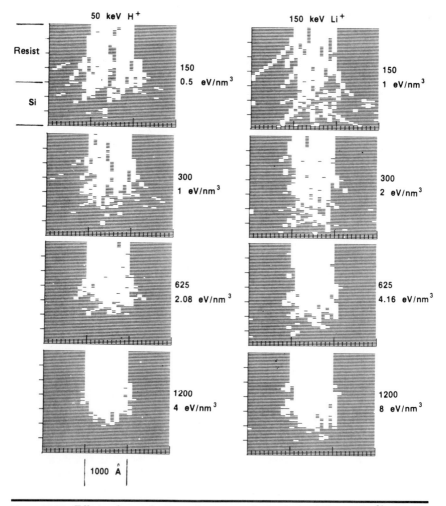

Figure 15.29   Effects of granularity on feature resolution in ion lithography.[21]

lithium ions. Ion incidence was random in the area about 0.1 μm^2, and the level of exposure was adjusted to four separate positions. The exposure was adjusted by varying the ion dose (increase) and resist sensitivity (decrease) so the apparent sensitivity of energy response would be the same. The variations in exposure that resulted occurred inside the 0.1-μm pixel as unexposed regions and outside the pixel as overexposures. Note the gradual improvement in resolution with a reduction in resist sensitivity and increase in ion dose. That is analogous to optical resists, since the highest-sensitivity materials typically exhibit the lowest resolution. Although the resolution inside the highest dose example above was excellent, the resulting wafer throughput would not be competitive with e-beam writing systems unless current density was increased. That is an area of active research and will no doubt be incorporated into the next generation of focused ion beam lithography systems.

Step-and-repeat ion beam printing. Step-and-repeat imaging has already proved itself in optical lithography as a moderately high throughput tool with special resolution benefits related to the small image field size, which relaxes much of the pressure for high-quality mask and large-field optics with expected distortions. Typical production is approximately 1000 chips per hour, 8 by 8 mm in size. Resolution is to well below 0.5 μm, with alignment accuracy in the $\pm$ 0.1-μm area or better. The applications for such a system are perhaps more varied than those for channel or full-field beam writing. They include the following:

1. Metal silicide formation for gate and interconnection layers by ion mixing
2. Oxidation of local areas by nitrogen ion bombardment
3. Electrical property alteration in gallium arsenide
4. Resist image reversal by using negative resists
5. Ion-induced change in integrated optical components
6. Device formation redirect doping on gallium arsenide, silicon, nitride, and other dielectric materials
7. Ion etching of molybdenum, nickel, and other metallic layers

Flashing the beam through the mask results in very short resist exposure times (milliseconds). The step-and-repeat alignment and exposure will overcome wafer nonflatness and mask overlay problems. The ion beam originates from a rare gas ion source, passes through a condenser, and then through the patterned mask or reticle. The ions then move through the optical projection lens system, which demagnifies the mask's ion image. Finally, the image is focused onto the resist-coated wafer, which is moved stepwise by the xy positioning stage. A long-focal-length, small-aperture imaging system helps overcome dimensional control problems more common in 1:1 and 5:1 optical and projection in all types of proximity printing systems. Another benefit of $10\times$ ion stepping is the increase in current density at the wafer plane, a limitation of focused ion beam imaging at $1\times$ magnification. The power density figures measured at the wafer surface can be approximately 1000 times higher than at the mask; there are approximately 10^{16} ions/cm^2, and the ion source has an energy level that delivers the very short millisecond resist exposure times. The ion energy at the mask is 4 to 10 keV, accelerated up to 100 keV at the wafer.

Ion beam stepping provides 6-in wafer exposures to very high resolution and alignment tolerances with relatively good exposure throughput. The step-and-repeat strategy appears to be giving the same benefits in ion as in optical lithography but at the lower resolution CD tolerance levels needed in VLSI and ULSI device fabrication processes.

Superconductors

The discovery of an yttrium-barium-copper-oxygen ceramic that becomes superconducting at 35 K ($-238°$C) resulted in a 1987 Nobel prize to J. Georg Bednorz and K. Alex Mueller of IBM. Research to discover an ambient temperature superconductor is intense. Many materials, mostly ceramics, exhibit superconducting properties, some at a temperature as high as that of liquid nitrogen (77.2 K, or $-198.8°$C).

Superconducting films operating at room temperature, or even at liquid nitrogen temperature, could have tremendous potential in improving the performance of integrated circuits. High-speed computers would use superconductors to greatly improve operating speeds. One possible application in lithography would be in making a very compact synchrotron for x-ray imaging. Magnetic levitation and electric power generation and transmission are other areas that would benefit greatly from superconductors.

Sputtering targets for high-temperature superconducting materials are available; they make possible the application of thin-film technology and the incorporation of superconductors in integrated circuits. Interconnections on a chip and between chips (in lead frames) will make substantial improvements in computer systems and many other solid-state electronic systems. Delay times between chips on a master slice or printed-circuit board have become limiting parameters in speeding up computational machines. Superconductors will change many of the methods now used in IC fabrication and provide overall performance improvements.

Limits of Lithography

As production geometries on integrated circuits approach the nanometer level, questions about the practical resolution limits for these devices arise. Several factors combine to create what may be regarded as the practical resolution barrier:

1. Operation temperature of the devices
2. Effects from high-energy etching environments
3. Parasitic effects inside the chip
4. Doping fluctuations
5. Electron mobility
6. Dielectric insulation limits
7. Material solubility problems
8. Device "drift" characteristics over time

V. Leo Rideout, of IBM, Yorktown Heights, categorized these limits as follows:

Physical limits	0.01–0.02 μm
Technical limits	0.1–0.2 μm
Process complexity limits	0.2–0.4 μm

Many forces will determine exactly when the resolution of ICs will begin to slow down noticeably, including the success or relative failure of many of the frontier technologies now under investigation. However, despite the complete success of new image formation techniques, the basic physical and technical limits remain, and it appears unlikely that it will be economically feasible to manufacture ICs with minimum geometries below 0.1 μm. The most likely evolution is for ICs to become more three-dimensional, with an increasingly larger number of individual layers.

REFERENCES

1. J. Moran, "Processing Sequence for the Tri-Level Resist," Bell Laboratories, Murray Hill, N.J., 1979.
2. B. J. Lin, "A Double Exposure Technique to Macroscopically Control Submicrometer Linewidths in Positive Resist Images," *IEEE Trans.*, vol. 25, no. 4, 1978, p. 419.

3. Canon Technical Data Sheet on the PLA-520 F/FA Proximity Mask Aligner with Deep UV Illuminator, Canon, Newport Beach, Calif., 1980.
4. B. J. Lin, "AZ-2400 as a Deep-UV Photoresist," IBM Thomas J. Watson Research Center, Yorktown Heights, N.Y., 1978.
5. B. J. Lin, "AZ-1350J as a Deep-UV Mask Material," IBM Thomas J. Watson Research Center, Yorktown Heights, N.Y., 1978.
6. Y. Nakane, T. Tsumori, and T. Mifune, "Deep-UV Photolithography," Kodak Interface Seminar, 1978.
7. N. Yew, "Electron Beam Lithography in the Production Line," IGC Conference Technical Handout, Chatham, Mass., September 1978.
8. J. A. Reynolds, "An Overview of E-Beam Mask Making," *Solid State Tech.*, August, 1979, p. 87.
9. B. P. Piwczyk, "Electron Beam and X-Ray Lithography Update," *Electronic Packaging and Production,* May 1978, p. 37.
10. G. P. Hughes, "X-Ray Lithography for IC Processing," *Solid State Tech.*, May 1977, p. 39.
11. R. L. Seliger and P. A. Sullivan, "Ion Beams Promise Practical Systems for Submicrometer Wafer Lithography," *Electronics,* March 27, 1980.
12. Toshiba Corporation, Tokyo, Japan.
13. D. Markle, "Deep UV Lithography: Problems and Potential," *Ultratech Stepper,* April 1988.
14. I. Higashikawa et al., "Recent Progress in Excimer Laser Lithography," Toshiba Corp./VLSI Research Center, Kawasaki, Japan. (MRS Meeting, Boston, 1987).
15. R. D. Moore, "EL Systems," *Solid State Tech.*, September 1983, p. 127.
16. W. Bottoms, Varian/Lithography Products Division, Gloucester, Mass.
17. P. Burggraaf, "E-Beam and x-Ray Lithography for the 1980's," *Semiconductor Int.*, May 1980, p. 39.
18. J. Silverman et al., "x-Ray Lithography Exposures Using Synchrotron Radiation," *SPIE,* vol. 393, March 1983.
19. A. Wagner, "Applications of Focused Ion Beams to Microlithography," *Solid State Tech.,* May 1983, p. 97.
20. Technical Literature, Ion Beam Technologies, Beverley, Mass. 1983.
21. Ref. 19.

INDEX

About the Author

David J. Elliott is Operations Manager, Eastern Region, Cymer Laser Technologies, a manufacturer of line-narrowed excimer lasers for microlithography. He was formerly involved in manufacturing and marketing excimer laser systems at Leitz-IMS and wafer steppers at GCA Corporation. He also worked at Shipley Company, in Newton, Massachusetts, for 16 years in technical and marketing capacities. His work involved developing, testing, and marketing photoresists for IC manufacturing processes.

Mr. Elliott has conducted technical symposia in Russia, China, eastern and western Europe, and the United States in the areas of IC device lithography and process technology. He has written numerous technical articles and also three books on IC fabrication for McGraw-Hill (*IC Fabrication Technology, IC Mask Technology,* and *Microlithography*).

Mr. Elliott holds patents in the field of IC device processing. He received his MBA degree from Boston College, his BS degree from Northeastern University, and an AA degree from Allan Hancock College. He served in the U.S. Air Force from 1962 to 1966.